ALLE ZEIT WACH 1842

Jeff R. Wright Lyna L. Wiggins
Ravinder K. Jain T. John Kim (Eds.)

Expert Systems in Environmental Planning

With 62 Figures

Springer-Verlag
Berlin Heidelberg New York
London Paris Tokyo
Hong Kong Barcelona
Budapest

Prof. Jeff R. Wright
Purdue University
School of Civil Engineering
1284 Civil Eng. Building
West Lafayette, IN 47907-1284 USA

Prof. Lyna L. Wiggins
Department of Urban Studies and Planning
Massachusetts Institute of Technology
Cambridge, MA 02139 USA

Dr. Ravinder K. Jain
Army Environmental Policy Institute
Installations, Logistics, and Environment
U.S. Army Corps of Engineers
Champaign, IL 61826-6569 USA

Prof. T. John Kim
Expert Planning & Information
Systems Laboratory
University of Illinois
907 $^1/_2$ W. Nevada Street
Champaign, IL 61820, USA

ISBN-13: 978-3-642-77872-8 e-ISBN-13: 978-3-642-77870-4
DOI: 10.1007/978-3-642-77870-4

Library of Congress Cataloging-in-Publication Data. Expert systems in environmental planning / J.R. Wright ... [et al.] (eds.). p. cm. Includes bibliographical references and index.

1. City planning – Environmental aspects – Data processing – Congresses. 2. Expert systems (Computer science) – Congresses. I. Wright, J.R. (Jeff R.) HT166.E96 1993 363.7'004'633–dc20 92-37632 CIP

Softcover reprint of the hardcover 1st edition 1993

32/3145-5 4 3 2 1 0 – Printed on acid-free paper

Preface

The connections between economics, planning, and the environment are receiving increased attention among scholars and policy makers in many countries. The common denominator among these three variables is the earth's life support systems, the ecosystems on which the world depends. When we describe our physical surroundings as a collection of possible uses, we are establishing linkages between economics, planning, and the environment. Because possible alternative uses compete with each other, and conflicts arise over scarce land resources, the varying environmental impacts of alternative uses are major concerns for the current as well as the next generation. How to achieve sustainable development is the pressing question for today's environmental professionals. Environmental planners and engineers help us study the implications of our choices, and new technologies and techniques that improve the practice of environmental planning should enhance our ability to protect our future.

The depletion of the earth's natural resources and loss of biodiversity, the degradation of air, land, and water quality, the accumulation of greenhouse gases leading to changes in our climate, and the depletion of the ozone layer comprise only a partial list of environmental issues that concern our policy makers. To support their decisions, environmental planning must be a multidimensional and multidisciplinary activity that incorporates social, economic, political, geographical, and technical factors. Solutions for problems in these areas frequently require not only numerical analyses but also heuristic analyses, which in turn depend on the intuitive judgements of planners and engineers.

With the advancement of expert systems technology, it becomes possible to represent experts' knowledge about particular problems in a computer system designed specifically to inform decision-makers about environmental issues. Characteristics of expert systems include: the representation and use of expert knowledge, symbolic and heuristic reasoning, inclusion of domain-specific complex rules, effective manipulation of large knowledge databases, and facilities for explanation of reasoning and conclusions. Many mental activities carried out by experts, such as thinking, feeling, understanding, and consciousness, cannot easily be replicated by an algorithm or structured system. Human experts' common-sense knowledge and other human qualities are not replicated by current expert systems. In developing an expert system, one has to make sure that the system does not require common-sense knowledge and that the task is well understood. As Waterman (1986) has indicated, when developing an expert system the task is to require only cognitive skills. Genuine experts must exist and agree on solutions, be able to articulate their methods, and their knowledge must lend itself to incorporation into a manageable fashion in order to develop an expert system.

Waterman (1986) has suggested that expert systems development stages include: a demonstration prototype, a research prototype, a field prototype, and a production model, followed by a commercial system. Many systems in the late 1980s were still at early stages of development (demonstration and research prototypes). These systems are sometimes referred to as "toy systems." Few expert systems have reached the point of a production model or widespread commercial usage. Several questions then naturally arise: What can be expected from expert systems in the future? How can these systems be tested and validated? What benefits can be derived from the development of expert systems when most of them are not really operational?

The evolution of expert systems to address complex environmental planning and management problems is analogous to the evolution of computational methods adapted to problem solving in these same areas. Early on computer technology was used for solving only very specific problems, and then, only fairly routine computational aspects of those problems. Later, as our confidence in algorithms grew, analytical models were developed using computer technology. Eventually complex optimization problems were attempted using increasingly sophisticated-computational systems. During these early stages, progress was slow and the utility of the computer seemed more potential than real. Over the years, this has changed dramatically and now the utility, cost effectiveness, and robustness of computer technology to address complex computational and optimization problems is taken for granted. In a similar manner, developing expert systems may take some time before it is clear that expert systems can, in fact, perform difficult tasks at an expert level of performance in a specific domain.

Early expert systems may be best used to augment other well-accepted computer aided system utilities. We expect that secondary benefits will accrue from our early attempts at developing expert systems. Since many environmental planning and engineering problems arise in a complex economic, social, and technical context, solutions then require participation of individuals representing diverse disciplines. It seems that developing expert systems may provide a vehicle to integrate the knowledge of diverse disciplines to address complex environmental problems.

Since developing an expert system requires a detailed description of the knowledge needed to carry out an important task, codifying this knowledge is of value in itself. Systematic description of knowledge will sometimes uncover alarming gaps and inconsistencies, and this description may codify and distill a vast amount of expertise in an area. Knowledge can then be preserved in spite of personnel turnover and can be shared within an organization and systematically extended.

With a goal of documenting and sharing these possibilities, an international conference on Expert Systems in Environmental Planning and Engineering was held in Cambridge, Massachusetts, on September 21-22, 1989. The editors of this volume were the conference organizing committee. The conference was funded by the U.S. Army Construction Engineering Research Laboratory and was hosted by the Lincoln Institute of Land Policy and the Computer Resources Laboratories, School of Architecture and Planning, Massachusetts Institute of Technology. The objective of the conference was to provide an intellectual forum for active researchers of expert systems in environmental planning and engineering.

This book is a collection of papers presented at the Cambridge conference in combination with related work, and is intended to inform readers of the current state-of-the-art in expert systems in environmental planning. The book is a sequel to *Expert Systems: Applications to Urban Planning*, edited by T.J. Kim, L.L. Wiggins, and J.R. Wright, published by Springer-Verlag in 1989. The earlier book contains several introductory chapters on expert system technology, and we refer readers who are not familiar with expert systems to this work.

This book is organized into four parts: (1) Land Use and Resource Management, (2) GIS and Spatial Modeling, (3) Environmental Planning and Management, and (4) Knowledge Acquisition, Evaluation, and Validation. The first three parts focus on specific domains within environmental planning, and the fourth emphasizes the process of design, implementation, and evaluation of expert systems.

This volume would not have been possible without the help and oversight of Ms. Susan E. Cockerill. We are indebted to Susan for preparing the manuscript for publication including copy editing, typesetting, and redrafting of all graphics materials. Her management skills and her attention to detail are without equal, and are greatly appreciated.

Permission to quote and reproduce selected figures, tables, and/or other material has been kindly granted by the *Journal of the American Planning Association, Computers, Environment and Urban Systems, Environmental Impact Assessment Review*, the American Chemical Society, and Envo Publishing. We also wish to thank the editors of the several journals who graciously permitted us to reprint the included articles.

Finally, we wish to again acknowledge the continuing support of the U.S. Army Construction Engineering Research Laboratory and the Army Environmental Policy Institute for work in this area.

Jeff R. Wright
Lyna L. Wiggins
Ravinder Jain
Tschango John Kim

Contents

Editors

Ravinder K. Jain is the founding Director, Army Environmental Policy Institute. An environmental engineer and research & development manager for 25 years, Dr. Jain received degrees in Civil Engineering (B.S. and M.S.C.E) from California State University. He earned a master's degree in Public Administration from Harvard University and a Ph.D. at Texas Tech University. A research affiliate of the Massachusetts Institute of Technology, Dr. Jain is also an adjunct professor at the University of Illinois at Urbana-Champaign. He has served as chairman of the Environmental Engineering Research Council, ASCE, and is a member of the American Academy of Environmental Engineers, and a fellow of ASCE. Dr. Jain's many awards and honors include the Army's highest research award—the Army Research & Development Achievement Award. He is a recipient of the Founder's Gold Medal and was named "Federal Engineer of the Year" for 1989. He has co-authored or edited seven books.

Tschangho John Kim is a Professor in the Department of Urban and Regional Planning at the University of Illinois at Urbana-Champaign. For fourteen years, he has been teaching computer applications to urban and regional planning, transportation planning, land use planning, and economic analysis of public plans and policies. Dr. Kim has also been actively involved in research in solving large scale planning problems using supercomputers. Current research projects include building expert systems for the coal trade and facility siting, and developing an efficient solution algorithm for non-linear programming models. Dr. Kim holds a bachelor's degree in Architectural Engineering and both a MCP and Ph.D. degree in Urban and Regional Planning from Princeton University. His publications include the books *Integrated Urban Systems Modeling: Theory and Applications*, *Advanced Transport and Spatial Systems Models: Application to Korea*, and *Expert Systems: Applications to Urban Planning* and numerous articles and reports in urban, regional, and national planning.

Lyna L. Wiggins is Assistant Professor of Urban Studies and Planning at the Massachusetts Institute of Technology where she teaches courses in quantitative methods, computer applications in planning and utility systems planning. Her research has included development of demand forecasting models and facility siting methods for energy facilities. She has designed decision support systems for environmental planning and infrastructure planning and management. Recent research has included an examination of the adoption of microcomputer technology in urban and regional planning agencies, and an analysis of organizational factors on system implementation and effectiveness. Dr. Wiggins holds bachelor's degrees in Mathematics and City and Regional Planning from California Polytechnic State University, San Luis Obispo, a master's degree in Statistics from Stanford University, and a Ph.D. in Urban and Regional Planning from the University of California at Berkeley.

Jeff R. Wright is a Professor of Civil Engineering at Purdue University where he teaches and conducts an active research program focused on the application of computer technology to the solution of engineering management problems. Dr. Wright has designed and implemented decision support systems for a wide variety of public-sector engineering problems, including facilities location, scheduling and sequencing, route design and management, spatial design and analysis, and systems operations. His present research interests include the integration of analytical engineering management models (including knowledge-based systems) with geographical information systems. Dr. Wright holds bachelor's degrees in Social Psychology and Civil Engineering and a master's degree in Civil Engineering from the University of Washington, and a doctorate degree from Whiting School of Engineering at The Johns Hopkins University through their program in Systems Analysis and Economics for Public Decision Making.

Contributors

Ilker Adiguzel is the Program Manager of the Engineering and Materials Division at the U.S. Army Corps of Engineers Construction Engineering Research Laboratory in Champaign, IL. He holds B.S. and M.S. degrees in Civil Engineering from the Middle East Technical University in Ankara, Turkey and a Ph.D. degree in Civil Engineering from the University of Illinois. His current research interests include topics related to information management, applied operations research, theory of decision-making, distributed database management, knowledge-based expert systems, and microcomputer-aided engineering applications.

Marc P. Armstrong is an Associate Professor at the University of Iowa where he holds a joint appointment in Geography and Computer Science and teaches courses in computer cartography, geographic information systems, and computer graphics. Before taking the position at Iowa in 1984, he was a research programmer in the Department of Landscape Architecture at the University of Illinois. Dr. Armstrong holds a B.A. in Geography from the State University of new York, an M.A. in Geography from the University of North Carolina, and a Ph.D. in Geography from the University of Illinois. He has published articles in the *International Journal of Geographical Information Systems, Environment and Planning B: Planning and Design, Environment and Planning C: Government and Policy, URISA Journal, and Computers, Environment, and Urban Systems.* Dr. Armstrong also serves on the Editorial Board of Cartography and Geographic Information Systems. His current research activities involve parallel processing and cartographic visualization methods for spatial decision support systems.

Edwin Blewett is a private consultant living in Vancouver, Canada. He works primarily in the natural resource sector, principally fisheries, and in urban planning. Dr. Blewett has conducted consulting assignments in strategic planning and decision making; financial and economic analysis and modelling; market research; management information systems; performance measurement and evaluation; government policy; and forecasting and evaluation of forecasts. He has been a member of an Industrial Inquiry Commission and has acted as an intermediary between natives and third parties in native land claims. He has consulted with private businesses from small and medium-sized firms to major corporations and all levels of government and has provided expert advice to law firms. Dr. Blewett holds a Ph.D. in Economics from the University of British Columbia and has recently been working on the further development of the expert system described in his chapter of this book.

Kurt A. Buehler is currently employed as a Civil Engineer at the U.S. Army Construction Engineering Research Laboratory, Environmental Division, Spatial Analysis and Systems Team, in Champaign, IL. He is working on the development of graphical user interfaces for the Geographical Resources Analysis Support System (GRASS) geographic information system (GIS). He is also involved in researching future directions for GIS software development at USA-CERL. Mr. Buehler holds a B.S. and M.S. in Civil Engineering from Purdue University, where he specialized in spatial decision support systems.

Philip H. Byer received his degrees in Civil Engineering from the Massachusetts Institute of Technology and is an Associate Professor in the Department of Civil Engineering and Institute for Environmental Studies at the University of Toronto. For the past five years, Byer has been Chair of the Advisory Committee to the Ontario Minister of the Environment. His current research focuses on multi-objective evaluation of public projects and risk management.

Qian Chen is a Visiting Scientist in the CSIRO Division of Building, Construction, and Engineering in Melbourne, Australia. Dr. Chen is working on applications of artificial intelligence in environmental engineering. He studied Theoretical Physics from 1959 to 1965 in Beijing University, China. From 1968 to 1978, he was an Asst. Professor in the Dept. of Engineering Mechanics, Tsinghua University in China. Dr. Chen later joined the Beijing Municipal Inst. of Labour Protection to lead a unit of 15 scientists and engineers working on noise and vibration control. He is also the Deputy Chief of Technology Division, Beijing Research Center for Environmental Physics. Dr. Chen joined CSIRO in 1988. He has published more than 80 books and papers on noise and vibration control.

Keechoo Choi received his B.S. and M.S. degrees in Urban Engineering at the Seoul National University, focusing on transportation planning and engineering. He currently is a Ph.D. student in the Department of Urban and Regional Planning at the University of Illinois at Urbana-Champaign. His research involves applying computer tools and quantitative techniques to transportation planning and urban information systems, and combining transportation paradigms and information technology for improved transportation decision support systems. Mr. Choi is a member of the Institute of Transportation Engineers (ITE) and the Urban and Regional Information Systems Association (URISA). He is the author of the chapter on "Transportation, Spatial Interaction, Telecommunication, and Information Systems: A Research Agenda," in the book *Regional Science*, edited by Boyce, Nijkamp, and Shefer.

Victor B. Ciesielski received his B.Sc. and M.Sc. degrees from the University of Melbourne (Australia) and his Ph.D. from Rutgers University. Currently, he is a Senior Lecturer in the Department of Computer Science at the Royal Melbourne Institute of Technology. Dr. Ciesielski's research interests include applications of artificial intelligence, neural networks, and the integration of symbolic and neural computation. He has developed knowledge-based systems in a number of diverse areas, including insurance underwriting, geomechanics, medicine, law, and environmental conservation.

James Richard Davis is a Principal Research Scientist with the Division of Water Resources, Commonwealth Scientific & Industrial Research Organization (CSIRO), where he is also leader of the Resource Management Systems Group. His broad interests lie in developing improved methods for planning and managing natural resource use; in recent years, he has concentrated on developing management advice programs based on artificial intelligence techniques. These programs have been applied to problems of land use planning, wildfire management, pest control, shrub regrowth, and Army land management. His recent publications include chapters in *Microcomputers for Local Government Planning and Management* (Hargreen, Melbourne), *GIS in the Defense of Australia* (Pergammon Press, Melbourne), and *Introductory Readings in Geographic Information Systems* (Taylor and Francis, Philadelphia). He has published papers in the *Journal of Environmental Management, ASCE Journal of Water Resources Planning and Management, and ASCE Journal of Irrigation and Drainage.* He is on the editorial board of *AI Applications in Natural Resource Management.*

Glenn Elliott received his B.Sc. in Computer Science from the Royal Melbourne Institute of Technology, Australia, in 1988. He is currently employed by the State Electricity Commission of Victoria (SECV) as a Management Representative with the Power Computing Solutions Division which specializes in software development for the SECV's thermal power stations. Glenn has consulted on a number of expert system projects for the SECV since 1988.

Thomas Evans is a Ph.D. candidate in the Environmental and Water Resources program of the Department of Civil Engineering at the University of Texas at Austin. He holds B.S. and M.S. degrees in Mechanical Engineering from Rice University and the University of Texas at Austin, respectively. Before his conversion to Civil Engineering, Mr. Evans worked principally in experimental studies of heat transfer and fluid mechanics. His present research efforts involve exploring the relationships between engineering and public policy in environmental and water resources issues.

Hsai-Yang Fang is Professor and Director in the Geotechnical Engineering Division of Lehigh University's Fritz Engineering Laboratory. Dr. Fang is considered one of the foremost authorities in foundations engineering with 30 years of experience in this area, including a prominent role as a consultant to the Federal Highway Administration, the Pennsylvania Department of Transportation, U.S. Steel, and Bethlehem Steel, as well as to the governments of Columbia and El Salvador in the wake of devastating earthquakes in those countries. Dr. Fang has published more than 100 scholarly papers in ASCE, ASTM, Transportation Research Board Publications, and other international journals and proceedings in such areas as geotechnical engineering, earthquake and soil-pollution interaction, and construction materials. He is the co-developer of the GEOTOX knowledge-based expert system for hazardous site identification and has been awarded U.S. patents for inventions related to his field. Dr. Fang is the recipient of many honors and awards from professional societies, universities, organization, and governments throughout the world.

John J. Fittipaldi received his MCP degree in City and Regional Planning from the Georgia Institute of Technology. As a principal investigator in the Environmental Division of the Army Construction Engineering Research Laboratory, he engaged in environmental planning, policy research, and environmental planning applications of knowledge-based systems. Presently, as a senior fellow with the Army Environmental Policy Institute, he serves as an environmental policy analyst. Mr. Fittipaldi is a Certified Environmental Professional and holds an Adjunct Faculty appointment at the University of Illinois, Department of Urban and Regional Planning.

Sang-Yun Han is an Assistant Professor in the Department of Public Administration at Kyung Hee University in Seoul, Korea. He holds a B.S. degree in Political Science from the University of Iowa, and an MUP and Ph.D. in Regional Planning from the University of Illinois. He wrote his Ph.D. dissertation on expert urban information systems and has co-authored a number of articles in this area. He is currently developing an expert system for land pricing in Korea.

Eric J. Heikkila is Assistant Professor of Urban and Regional Planning at the University of Southern California where he teaches courses on quantitative methods and policy analysis. His primary research focus in on the interplay between urban, economic, geographic, and demographic structures, and he has authored many scholarly articles in these areas. Prior to his current position, Dr. Heikkila worked for five years as an urban planner and financial analyst for the City of Vancouver. He received his doctorate in Economics from the University of British Columbia.

Michael E. Leary earned a B.Sc. in Human Ecology and an M.Sc. in Urban Planning and is now a Chartered Town Planner working in the United Kingdom public sector and an Associate Lecturer in the School of Town Planning at Oxford Polytechnic. His research interests include expert systems for urban design and the use of GIS in local government. While working at Oxford Polytechnic, he attracted government funding for a project to evaluate the potential of expert systems in planning. Mr. Leary is a regular contributor to journals and conferences in the fields of computing and urban planning.

David R. Maidment is an Associate Professor of Civil Engineering at the University of Texas at Austin. His received his M.S. and Ph.D. degrees in Civil Engineering from the University of Illinois and his bachelor's degree from the University of Canterbury in Christchurch, New Zealand. His research interests lie in the application of expert GIS to civil engineering planning problems, in particular those concerned with environmental and water resources issues.

Geoffrey T. McDonald is Director of the Institute of Applied Environmental Research within the Division of Environmental Sciences, Griffith University, Brisbane, Australia. He is a member of the Royal Australian Planning Institute and his research field is environmental planning, especially land use planning theory and practice. Professor McDonald's training was in economics, economic geography, and GIS at the University of Toronto.

George K. Mikroudis, a civil engineer and author of several papers on expert systems, hazardous waste, geotechnics, structures, and computer-aided engineering, is an Assistant Professor in the Engineering Department of Southeastern University in Athens, Greece. Dr. Mikroudis is also president of Avans, Inc., also in Athens. His research projects have included the GEOTOX expert system, expert systems for geotechnical applications, environmental assessment, and structural design, as well as the development of a system architecture for cooperative multi-domain systems. He received a Ph.D. and M.Sc. in Civil Engineering from Lehigh University and a diploma in Civil Engineering from the National Technical University in Greece. He is a member of ASCE, Sigma Xi, and the Technical Chamber of Greece.

Dundee Navinchandra is currently a Research Scientist at the Robotics Institute, Carnegie Mellon University. He is also an Adjunct Faculty member in the Department of Civil Engineering at CMU. Dr. Navinchandra received his Ph.D. in Engineering Systems (1987) and his M.S. (1985) from the Massachusetts Institute of Technology (MIT) and his B.Tech. from the Illinois Institute of Technology (IIT) in 1982. His research interests include knowledge representation, design theories, project management, and the application of artificial intelligence to engineering problems. He has published a book and several papers in these areas. Dr. Navinchandra is originator of the concept of "green engineering"; the study of product design for environmental friendliness without compromising product quality. Dr. Navinchandra co-founded the Intelligent Engineering Systems Laboratory at MIT in 1985. He is Publications Secretary of the AAAI special interest group on manufacturing and has co-chaired the annual AAAI workshops on Design and Concurrent Engineering since 1987.

Leonard Ortolano teaches courses in environmental policy implementation and water resources development in the Department of Civil Engineering at Stanford University. Dr. Ortolano's current research focuses on environmental policy implementation in developing countries. He is also participating in a study to develop a "critiquing expert system" to aid operators of a water supply network serving the San Francisco Bay area. Dr. Ortolano is the author of *Environmental Planning and Decision Making*, published by John Wiley & Sons in 1984.

Sibel Pamukcu received his B.S. degree in Civil Engineering at Bogazici University, Istanbul, Turkey in 1978; he then worked at the National Highway

Department of Turkey as a project engineer for one year. Prof. Pamukcu received his M.S. degree in Civil (Geotechnical) Engineering in 1981, and his Ph.D. in the same area in 1986, both from the Louisiana State University. Currently, he is an Assistant Professor of Civil Engineering at Lehigh University, and his research interests include environmental geotechnology and marine geotechnology. Prof. Pamukcu has authored and co-authored more than 30 papers in refereed journals and conference proceedings in geotechnical engineering and environmental geotechnology, and has co-authored two chapters ("Soil Stabilization and Grouting" and "Advances on Earthquake Effects on Soil Foundation Systems") in the second edition of the *Foundation Engineering Handbook* published by Van Nostrand Reinhold, Co. Prof. Pamukcu is an active member in TRB (Transportation Research Board) Committees A2L02, ASTM, and ASCE and is on the editorial board of *Journal of Marine Geotechnology*.

Catherine Perman currently works as a partner for Albathion Software in San Francisco. Albathion Software provides consulting services and software for expert systems and multi-media applications. Dr. Perman received her doctorate in Civil Engineering from Stanford University. Her research focused on developing and validating an expert system that diagnoses operating problems at wastewater treatment plants. Prior to studying at Stanford, Dr. Perman worked as an environmental engineer and systems analyst in the petroleum industry.

Stefano Schibuola received his degrees in Civil Engineering from the University of Toronto. He is currently a transportation planner with the IBI Group in Toronto, Ontario, Canada and is active in the areas of urban transportation planning, environmental assessment, knowledge-based systems development, and freeway traffic management.

Ronald Sharpe is currently a Senior Principal Research Scientist and leader of the Knowledge Based Systems Project at the CSIRO Division of Building Research in Melbourne, Australia. Since receiving his Ph.D. in Civil Engineering in 1968, Dr. Sharpe has developed and applied optimization and artificial intelligence (AI) techniques to a wide range of building, urban and regional planning, and industrial projects. A number of these projects have led to full commercial implementations with government and industry groups. Dr. Sharpe has published over 150 papers and books.

Marian Tracz graduated from the Cracow University of Technology in Poland with an M.Sc. degree in Civil Engineering in 1967. He spent several years gaining practical experience in the Municipal Traffic Engineering office in Cracow. He obtained his Ph.D. in 1974. Since 1980 he has been the Director of the Institute of Roads, Railways, and Bridges in the Civil Engineering Department, Cracow University of Technology. Dr. Tracz was promoted to Professor in 1987. He is author or co-author of more than 80 articles and conference papers and 8 books. His research interests include intersection design and performance evaluation, capacity of intersections, traffic modelling and applications of computer simulation in traffic studies, environmental effects of roads and road traffic including traffic safety and noise prediction.

Bronislaw M. Wawrzynkiewicz is currently an Assistant Professor at the Department of Civil Engineering at Kielce University of Technology in Poland. He received his M.Sc. degree in Civil Engineering from the Warsaw University of Technology in 1976 and his Ph.D. degree from Cracow University of Technology in 1989. His teaching responsibilities include traffic engineering and computer programs and techniques for improving traffic flow, as well as models for transportation systems analysis. His research interests are in the areas of expert systems, decision support systems, artificial intelligence, and fuzzy sets theory application, interacting with real world transportation problems. Professor Wawrzynkiewicz is a member of the Polish Operational Research Society, and has published several articles in various journals and conference proceedings.

Sarah U. Wisseman is the Assistant Director of the University of Illinois' Program in Ancient Technologies and Archaeological Materials (ATAM). She received her B.A. in Anthropology from Harvard University in 1973, and a Ph.D. in Classical and Near Eastern Archaeology from Bryn Mawr College in Pennsylvania in 1981. Her research focuses on the applications of instrumental methods to archaeological and art historical problems, particularly those dealing with pottery, metalwork, and mummies from the ancient Mediterranean.

PART ONE

Land use and resource management

Introductory remarks by Ravinder Jain and Lyna Wiggins

Leary presents, in Chapter 1, some initial research findings regarding the application of expert systems to land use planning in Great Britain. A prototype system called the Development Control Advisor is introduced. The system is based on a checklist approach to assessing the merits of a planning application for residential development. The procedure runs through a range of criteria derived from a series of interviews with development control experts. The checklist approach leads the user through a series of questions about a development proposal in order to reach a conclusion about its acceptability.

Another rule-based decision support system for local government planning, an expert system called ADAPT, is presented in Chapter 2 by Davis and McDonald. The system is designed to assist planners in drawing up zoning schemes and it incorporates knowledge-based information and heuristics. The system seems to overcome some of the shortcomings of procedural planning programs. The distinctive features of the system include: (1) use of both quantitative and qualitative information, (2) recognition that the planner is better at resolving complex issues than is an algorithm, and (3) ability to remember and explain reasons for decisions. ADAPT was applied to the production of a local government zoning scheme in Queensland, Australia.

In Chapter 3, Heikkila and Blewett describe an expert system designed to check compliance with municipal building codes. This system helps satisfy a tedious and complex primary function of local government—regulation and enforcement of building codes. This system can help overcome the inexperience of persons working in local government who may not fully understand the complex and numerous building code requirements. The authors describe how expert systems can be used to provide consistency and comprehensiveness in the enforcement of certain state and local regulatory requirements for land use planning.

Kim, Han, and Adiguzel present in Chapter 4 a decision support system that combines a machine learning program, an expert system, and an integer programming model. The feasibility of coupling expert systems with other existing computer techniques is introduced in detail. The prototype system is applied to facility management and planning as an aid to decision makers. The authors convincingly demonstrate that thc intelligent interface provided by expert systems for specifying the modeling tasks in decision support systems may stimulate planners to employ mathematical models more frequently and easily in their problem-solving processes.

All of the expert systems described in Part One, Land Use and Resource Management, are prototypes. With the experience gained from these early developments and with continuing availability of resources, we believe that such systems will evolve into field prototypes and production systems.

1

Expertise and Expert Systems in British Land Use Planning

Michael E. Leary

Introduction

Why is it important to research the potential of expert systems in land use planning, when even a cursory acquaintance with planning systems often leads to an immediate assumption? Planning decisions, especially as the system operates in the UK, are based on a set of policies and plans that are full of technical criteria. These policies and plans require a good measure of interpretation and professional judgement, despite (or perhaps because of) the array of quantitative techniques that planners have borrowed from other disciplines over the last fifty years.

Expert systems we are told in most introductory text books (e.g., Jackson, 1986; Waterman, 1986), are computer programs that behave like real experts and use heuristics rather than algorithms to solve complex problems in various fields of expertise. This makes expert systems different from conventional programs and the differences arise mainly from the incorporation of concepts and techniques from the much broader field of artificial intelligence (AI) [Boden, 1987; Winston, 1984; Rich, 1984]. This field of scientific endeavor is not without critics (e.g., Roszak, 1987; Dreyfus, 1984). The most successful examples quoted are in the fields of chemistry, medicine, geology, and computing. What then is the assumption? It is that because of the putative characteristics of land use planning and expert systems the potential is obvious and can be taken as the starting point for applied research.

This chapter is based on research which seeks to challenge this assumption by examining two areas: (1) what are the characteristics of planning expertise and expert systems that set limits to the extent of application, and (2) what can usefully be achieved before these limits are reached. The chapter first addresses some theoretical issues in expert system development then goes on to present some initial research findings on both the domain and the application of expert systems to a specific area of British land use planning. Although uncertainty and explanation are important aspects of expert systems, the problems they raise are outside the scope of this paper and have been discussed elsewhere (e.g., Leary, 1987; Johnston and Hopkins, 1987).

Knowledge acquisition: methodological issues

Conventional wisdom in the expert systems world holds that such systems can only be developed by a process known as 'rapid prototyping' (Davis, 1982)—an iterative process of trial and error during which the programmer or knowledge engineer shuttles back and forth from expert to computer gradually refining an initial bug-ridden prototype. Two of the many problems with this approach to system development concern the type of knowledge acquired and the type of system produced. Expert systems have been developed by what may be called the 'pure consultancy model' (Figure 1.1).

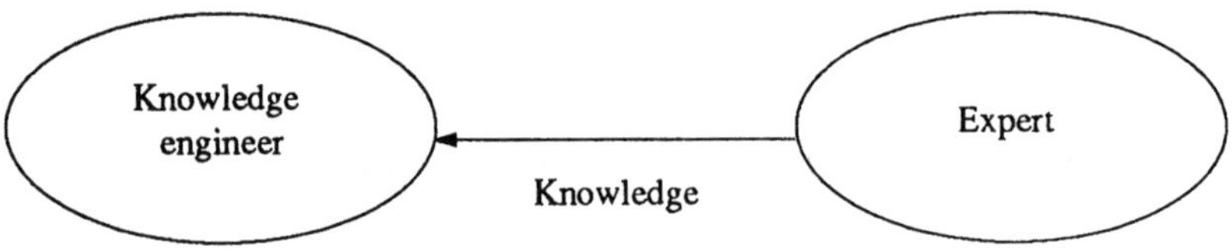

Figure 1.1. The pure consultancy model.

The process of knowledge engineering described above seems to be based on a rather peculiar self-image that the knowledge engineer has of his initial relationship with the expert; he knows nothing about the domain, while the expert knows and can articulate everything. However, it becomes apparent that the expert passes on a simplified version of knowledge, what is known as 'shallow knowledge' (Steels, 1987). Shallow knowledge consists of "rules of thumb" about *how* to do certain things, but not *why* (which is often called 'deep knowledge'). It is clear that the pure consultancy model needs refining to reflect this filtering of expertise as suggested by Figure 1.2. Much of the research in this area of expert systems characterizes this dichotomy as deep knowledge offering the causal explanations which underlie shallow knowledge (Chandrasekaran and Mittal, 1984).

The nature of expertise

Discussions of shallow knowledge seem to imply a derogatory view of this type of expertise which is probably unjustified. There seems to be no reason to believe that there is a single absolute level of causal knowledge that is able to explain all the rules of thumb an expert may use. There are ontological implications for the

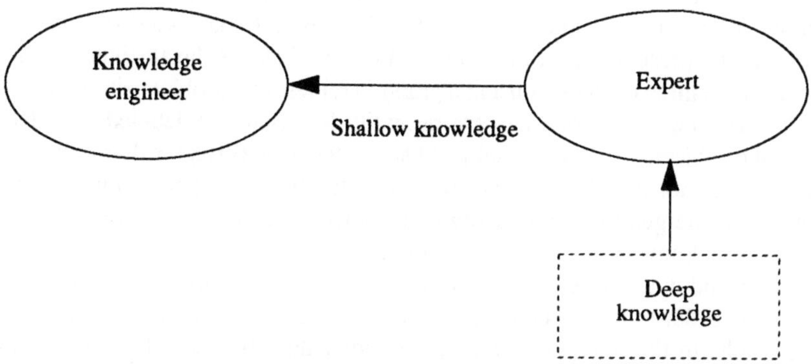

Figure 1.2. The filtering of real expertise.

explanation facilities of any expert system since deep knowledge can be seen as a relative concept which differs depending on the particular level of enquiry and the depth of explanation sufficient to satisfy the user.

In land use planning, ideas about the nature of expertise did not seem to readily to fit a deep/shallow dichotomy. This question is crucial for an assessment of the potential of expert systems since it goes to the heart of the extent of usefulness. To gain an insight into the nature of expertise, we conducted an interview survey and asked a group of planning experts (academics and practicing professionals) about the nature of expertise and the attributes that define experts. Academics must constantly rationalize and verbalize their knowledge in order to transmit it, and therefore they provide a good source of insight into these issues. After eliminating repetitions and overlaps in the survey responses, these attributes of expertise were determined (Leary and Rodriguez-Bachiller, 1988):

1. Judging significance
2. Understanding what is really going on
3. Awareness of causal connections
4. Having lots of detailed 'factual' knowledge
5. Being able to assess a situation and develop a broad strategy
6. Patterns of response developed from experience
7. Resolving conflicts
8. Ability to see a structure in a situation
9. Knowing the limits of knowledge
10. Innovative thinking and creativity
11. Critical ability.

Some of these attributes have already been discussed by others, such as the role of experience (Kolodner, 1984) and strategic knowledge (Clancey, 1983). Other writers have suggested that some of the attributes do not constitute intelligence

(Dreyfus, 1984). The list here does not claim to be exhaustive, but despite the jumble of attributes produced, it is possible to detect some kind of consensus about what represents a core of expertise. Figure 1.3 shows the attributes grouped according to three categories using a graph to represent a problem situation.

The first diagram emphasizes the necessity for wide and detailed knowledge about the subject. On its own, subject knowledge as expertise is doubtful, but in practice it is not possible to separate knowledge from its application to a problem. What emerged from this survey was that while being a necessary component of expertise, knowledge alone is not sufficient.

The second diagram refers to the processes of simplification that an expert carries out in order to handle complexity, variety, and sheer volume of information. This results in the rules of thumb that a knowledge engineer identifies without much thought. Simplification attributes include being able to judge what is important in a situation, being able to abstract patterns from a mass of unrelated detail and seeing short-cuts to a solution.

An unfortunate metaphor (borrowed from computer science) has emerged in the cognitive psychology literature that describes this process as 'compiling' knowledge (Anderson, 1980). Given that a compiler translates source code into binary object code, this is perhaps an overly crude characterization of how the brain deals with colossal amounts of conceptual and sensory information received from the environment, and about which we know very little.

In the third diagram, the group of attributes seems to refer to more explanatory process. Instead of merely simplifying reality for operational efficiency, experts look deeper into a problem, digging beneath the surface, and are able to recognize a greater number of steps in a line of reasoning (not apparent to the non-expert) which allow novel solutions to emerge. It is this kind of expertise that enables experts to detect causal relationships implicit in a situation, and to look beyond the frontiers of conventional knowledge.

It is interesting to note that in trying to express these ideas about expertise many experts are unable to be precise about the nature of the knowledge they are using to solve problems or give a professional judgement. Most put it down to experience, inspiration or intuition. In this respect, expertise displays some characteristics of the 'tacit knowledge' that is a familiar concept in expert systems literature (Berry, 1987) borrowed from Polyani (1967). Other writers have also classified this kind of expertise (or intelligence) as being on a different plane altogether from the application of formal knowledge and methods to problems. Ryle (1946) called it *knowing how* as opposed to *knowing what* and Dreyfus (1984) referred to 'nonformal intelligence'. Simon (1973) drew a distinction between structured and ill-structured problems.

1. The three components of expertise identified in the survey (description/detail, simplification/abstraction, and explanation/exploration) lead to a more convincing picture of how expertise works than is suggested by the deep/shallow dichotomy. Figure 1.4 is an attempt to illustrate the hypothesis that the three elements of expertise can be seen as operating in a dynamic way depending on the difficulty and nature of a particular problem.

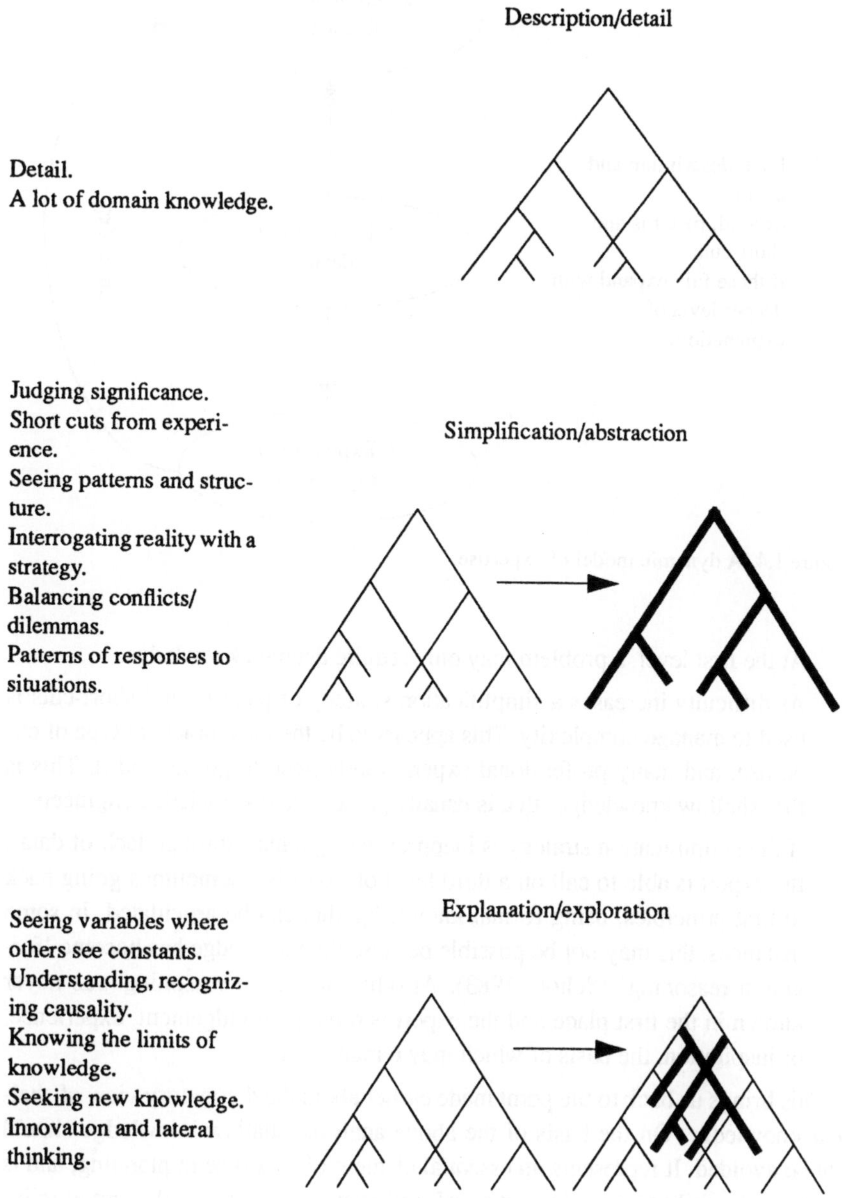

Figure 1.3. Patterns of expertise.

1. First, description and detail.
2. Second, patterns and short-cuts.
3. If these fail, expand with deeper level of explanation.

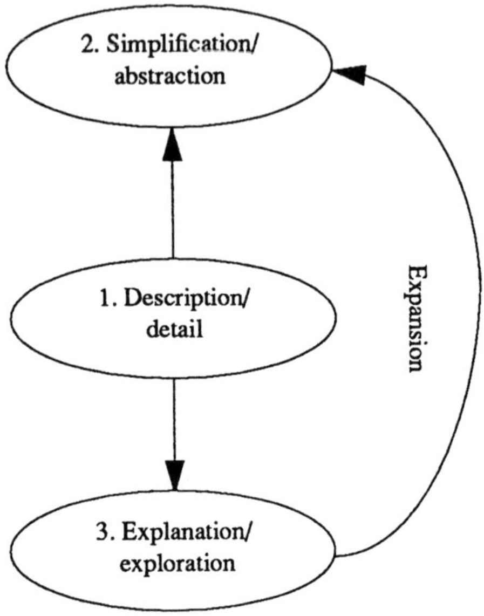

Figure 1.4. A dynamic model of expertise.

2. At the first level, a problem may only require detailed knowledge.
3. As difficulty increases a simplification strategy of patterns and short-cuts is used to manage complexity. This appears to be the most practical type of expertise, and many professional experts rarely need to go beyond it. This is the 'shallow knowledge' that is usually passed on to knowledge engineers.
4. If the simplification strategy is inappropriate (perhaps through lack of data), the expert is able to call on a third level of expertise, sometimes going back to first principles, using formal knowledge that can be articulated. In some instances, this may not be possible because the knowledge has become 'forgotten reasoning' (Schon, 1983). At other times, the reasoning was never known in the first place and the expert is relying on judgement, experience, or inspiration, the basis of which may remain tacit.

This brings us back to the point made earlier about the derogatory view of shallow knowledge. On the basis of the above analysis, shallow knowledge should not be avoided. It represents an essential element of expertise in planning, and is condensed, easily accessible, and developed over many years by the expert to increase practical efficiency. Shallow and deep knowledge do not represent corrupt and pure knowledge; rather, they are simply two kinds of expertise suited to routine and non-routine problem-solving. Using the idea of a dynamic model of expertise in a practical context, we turn to land use planning and examine some of the ideas expressed above and the available shell technology.

Expert systems in planning: a look at the literature

At the time this research was started in 1985, there was virtually no published literature on expert systems in British land use planning. There was some passing reference to knowledge-based systems in Batty (1984) and Barrett and Leather (1984). Some early articles on the potential of expert systems in planning raised interesting issues, but served mainly to promote the new technology (e.g., Cullen, 1986; Leary, 1986). Work of a more theoretical nature was reported by Sharpe and Marksjo (1985), Johnston and Hopkins (1987), and Leary (1988). In the United States, Ortolano and Perman (1987) provided a useful introduction to the subject for planners, while Kennedy (1987) described how planners might begin to make a start in understanding Prolog. J.R. Davis, et al. (1986) provided information on a series of research initiatives on the application of expert systems in the Australian planning context (e.g., Davies, et al., 1986), and Wyatt (1987) suggested how expert systems may be applied in making decisions on the relative importance to be attached to land uses.

Introduction to the domain

The British land use planning system can be divided into two parts: policy making and development control (or the implementation of policy). It should be noted that in Britain the system of planning control and building regulations are completely separate. Planning functions are delegated from central government to local government. Central government is only involved in issuing broad guidelines that are not legally binding and in adjudicating cases where there exists a grievance with the decision of a local planning authority (LPA).

All development of land requires planning permission and, in considering applications from potential developers, the LPA must take account of its own plans for the area and other 'material considerations'. Development control (DC) is a major function of local government in Britain, and is carried out by 400 district planning authorities (a district typically has a population of about 100,000 and each county has about six districts). About 400,000 planning applications are submitted each year, and most of these are straightforward domestic household applications. Where development proposals do not conflict with planning policies, it is often simply a matter of checking application details and ensuring that a routine consultation is carried out.

However, because planning policies are not legally binding (even on the LPA), a large amount of administrative discretion and professional judgement can be exercised in deciding whether to grant or withhold planning permission (Jowell, 1973).

The definition of development is deliberately broad:

"The carrying out of building, engineering, mining, or other operations in, on, over or under land, or the making of any material change in the use of

any buildings or other land."
(Town and Country Planning Act 1971, Section 22)

It is a formal definition that operates by creating a wide definition in principle, extending and limiting the definition by further statute law, and relaxing the requirement for express planning permission in certain circumstances. In seeking to relax the criteria of development a piece of subordinate legislation, the General Development Order (GDO), automatically grants planning permission for a range of minor developments and changes of use. The GDO is of particular interest because it is virtually unique in British planning legislation, because it consists of a detailed and precise set of regulations with little scope for discretion or interpretation. The criteria in the GDO apply throughout Britain, and include details such as the size of porches and garden fences that are regarded as permitted development.

From this brief introduction to the domain and its position in the context of the British planning system, it can be seen that decision-making in DC draws on a combination of precise regulations coupled with administrative discretion and professional judgement. This range of inputs pointed the research effort in the direction of development control as a suitable area of planning for an evaluation of expert system techniques.

An *a priori* model of development control

Because DC represents one half of the planning system and uses a vast range of data and techniques in the assessment of development proposals, it is necessary to focus on specific areas of expertise within the whole domain and establish a working overview to guide knowledge acquisition and selection of hardware and software. Considering the view of development control presented thus far, it may appear as if decisions are purely technical in essence, and if the proposal conforms with the regulations, policies, and well-established standards, then permission for land development is granted. However, the process, in reality, is more complex and raises three complications:

1. Policies and plans are not legal documents, and standards can be relaxed depending on the circumstances.
2. Some material considerations, such as aesthetics, do not have any standards, regulations, or technical criteria.
3. The final decision is always arrived at in the political arena.

This adds layers of increasingly informal and non-technical knowledge to what at first sight may appear a rigid, rule-based system of implementation. Figure 1.5 shows a schematic illustration of this idea. As one moves away from the technical/legal kinds of expertise, the knowledge used becomes less formal and determinate and based more on interpretation, judgement, discretion, and political values. This view of development control is supported by Harrison (1972); other researchers have made similar discoveries about particular facets of DC. For ex-

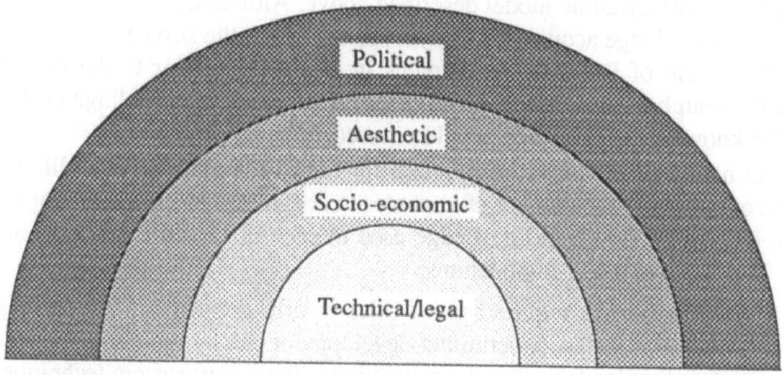

Figure 1.5. Facets of development control.

ample, Underwood (1981) identified the routine nature of development control and talks of DC officers as 'fault-finder clerks'; McLoughlin (1973) found that the advice of colleagues was an important input into decision making; and H.W. E. Davies, et al. (1986) studied the relationship between policy and the practice of development control.

An empirical model of development control

With the conceptual framework outlined above in mind, the task of knowledge acquisition proceeded in stages. First, a glossary of key concepts and terms in DC was established, and was followed by an attempt to understand the domain as it operates in practice. Documentary sources for the first stage included primary legislation (the 1971 Town and Country Planning Act). Secondary legislation (the GDO), reports on DC (RTPI, 1979), planning law text books (e.g., Grant, 1982; Heaps, 1988), a manual of DC for practitioners (Mason, 1986), a popular guide to the need for planning permission (James, 1977), and a guide to submitting planning applications (Salt, 1987) was also used for documentation.

The second stage of knowledge acquisition consisted of a series of structured interviews with planning officers from a nearby district council, and DC experts in the School of Planning at Oxford Polytechnic. The survey was meant to establish how practitioners actually do DC, the kinds of knowledge they use, the levels of expertise employed, and how they perceive their own problem-solving requirements and abilities.

From the survey, the model of DC that emerged was a procedural one that identified stages in a process rather than types of knowledge or expertise. It consisted of several almost self-contained modular tasks (Figure 1.6) that varied in the extent to which they required the different types of expertise of the kinds referred to in the dynamic model described above. Although no universal methodology of knowledge acquisition has yet been devised, the survey did break down whole domain of DC into sub-domains in a manner similar to the method by which researchers such as Grover (1983) and Breukner and Wielinga (1983) advocate knowledge acquisition should proceed.

The nature of the expertise required for each of the sub-tasks is different in terms of structure, the number of variables to consider, and the amount of discretion, judgement, and tacit knowledge used in each of the different problems areas. This has two major implications:

1. Certain sub-tasks require sets of rules clearly defined by law or by good practice. For instance, permitted development can be represented as a decision tree that lends itself to formalization using expert system techniques because the basis for decisions is contained in a set of detailed central government regulations applied fairly consistently throughout Britain.

   In a similar way, other sub-tasks can be identified in the overall development control process (completing the planning application, checking it, calculating fees payable, and identifying the range of people and organizations to be consulted) which, though complex, follow a relatively straightforward self-contained logic which could be treated in a similar way.

2. Other areas of the land use planning domain, such as interrogating the policy environment (derived from volumes of statutory and non-statutory policy, local precedent, and case law), environmental impact analysis (with its value-laden models, design-based elements, and socio-economic complexity) and negotiation with developers, do not seem to show the same degree of explicit structure in the expertise they require. Expertise for these tasks seems to draw more on the tacit knowledge and judgement of the expert, and a much broader range of factors, which vary depending on the context and characteristics of each particular case, must be taken into account.

In Figure 1.7, the levels of expertise identified earlier have been fitted (at the risk of some oversimplification) to the modules of development control. What this suggests is a sequence that might be followed in the development of expert systems for this and other domains. The software requirement for level one expertise could be adequately satisfied by conventional databases which are becoming increasingly common in many LPAs. Expert systems may even have some role to play here in the creation of friendly interfaces for new users. A large (and useful) proportion of level two expertise could well be modeled using available expert system shells with or without uncertainty facilities.

Some elements would be difficult to reproduce in an expert system, including interpretation of poorly defined concepts in the GDO, subjective weightings given to policies, and aspects of on-site assessment such as aesthetic impacts and the relative importance of physical features in the landscape or developed environment. In terms of expert system applications, these aspects of level two and all of

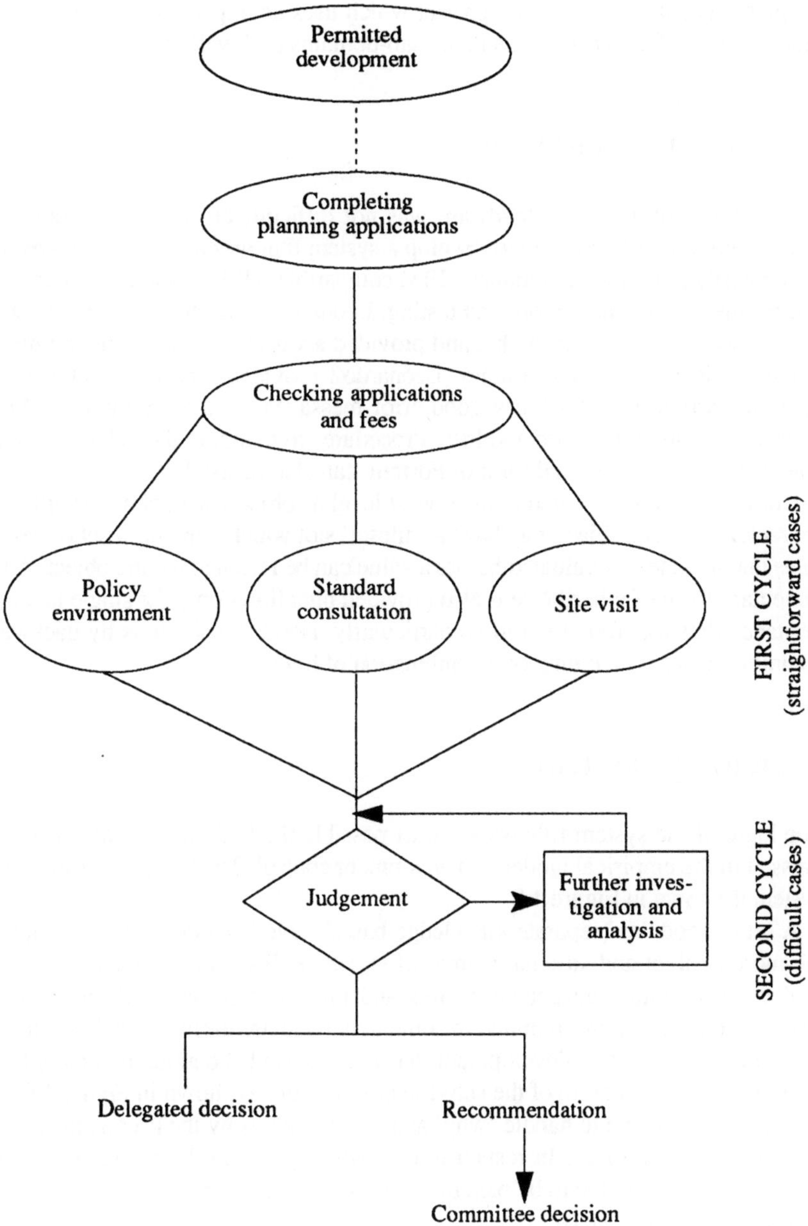

Figure 1.6. Development control sub-domains.

level three expertise must await the outcome of further research. The rest of this chapter concentrates on a pilot project which uses an expert system shell to explore the role of expert systems in the sub-domains of development control.

Software and hardware

Selection of software and hardware was not difficult, given the constraints of project funding and the need to develop a system that was accessible to most local authority planning departments. IBM compatible AT 286 machines were chosen for the development work and testing. Leonardo2 was the software selected, mainly because it was affordable and provided a variety of knowledge representations, objects, rules, and frames. Leonardo2 provides a built-in editor and a compiler which provides fairly good error messages for the programmer. Variables can be alpha-numeric and lists. Procedures to compute the value of a variable, written in a simplified form of Fortran, can also be used.

Knowledge is represented at the lowest level as objects with slots for different characteristics. An object may have a 'ruleset' slot which contains a set of nested rules that have to be evaluated before a value can be returned for the object. A rudimentary form of inheritance is also provided but allows only for single level inheritance that was found not to be particularly useful. Inference is by backward chaining, forward chaining, or a combination of both.

A prototype system

The logic of the system follows as far as possible the logic of the domain as expressed in the empirical model of development control. The basic structure of the system is shown in Figure 1.8.

A set of modules (separate knowledge bases) have been developed to capture the more explicit and structured areas of expertise. Each sub-domain is, in turn, broken down into manageable chunks and programmed as a rulebase. These branches have been programmed in some detail and are discussed below. One of the branches (permitted development) has been expanded diagrammatically to illustrate the logical nature of the sub-domain structure as shown in Figure 1.9.

The system is able to handle 'why' and 'how' queries by the user as the examples below demonstrate. In response to a 'why' query on the ruleset that deals with the appropriate fee to be paid the system would respond with:

```
I am using the rule
  IF info_required is fees
  AND fee_type is 'outline_permission'
  AND outline_type is 'private dwellings'
  AND dwelling_number is <= 2
  THEN fee = dwelling_number * 33
to establish the value of advice given
```

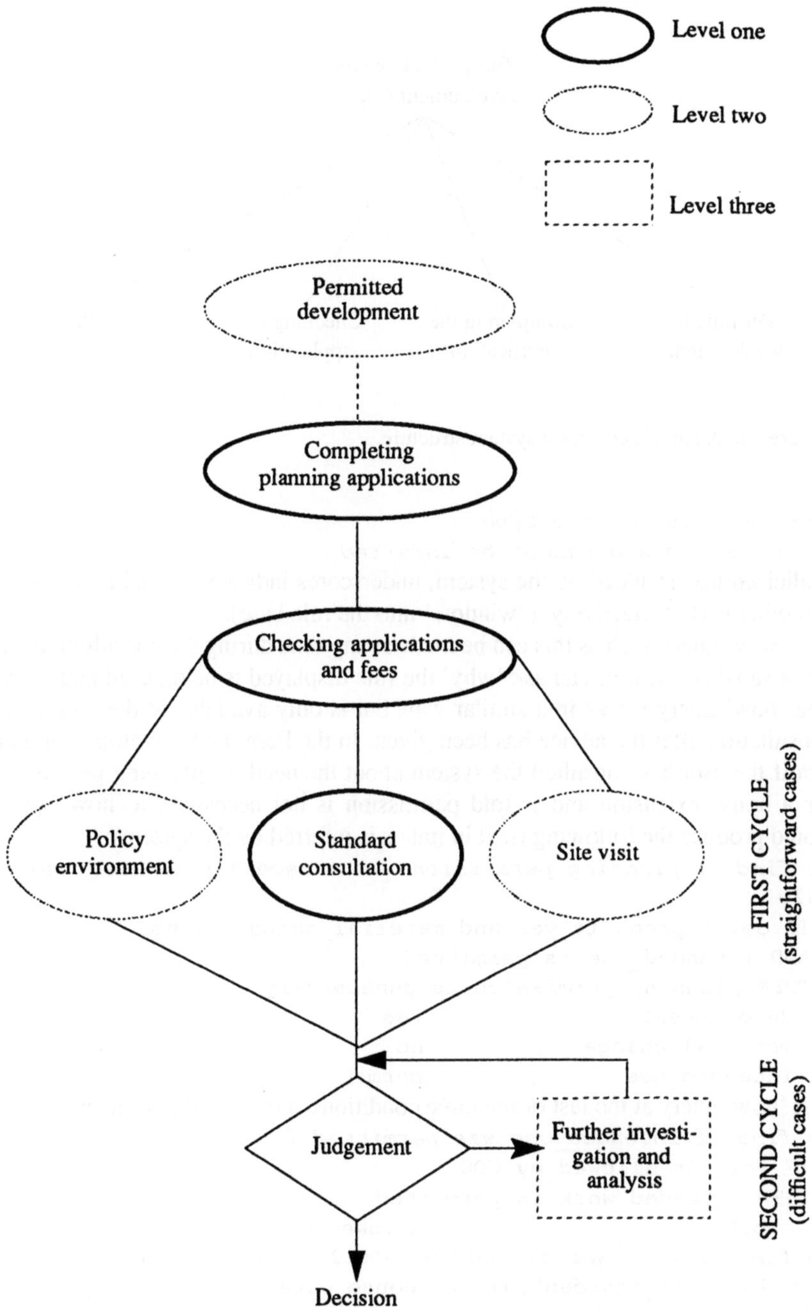

Figure 1.7. Levels of development control expertise.

Figure 1.8. A simplified expert system structure.

and the value of fee type
is not known and cannot be inferred
(italicized text is added by the system, underscores indicate a variable or object, and other text is effectively a 'window' into the rule base).

A 'why' query such as this can be made at any time during the consultation and if not satisfied the user can ask 'why' the rule displayed is being used and so on. The 'how' query works in a similar way, but is only available at the end of the consultation after the advice has been given. In the Permitted Development System if the user has consulted the system about the need for planning permission for a house extension and is told permission is not necessary, a 'how' query would produce the following (text in italics is inferred by the system):

```
To find if planning permission was unnecessary I used the
rule
  IF development is yes and material_change is no
  AND intended_use is permitted
  THEN planning_permission is unnecessary
   development                 yes
   material_change             no
   intended_use                permitted
```

A 'how' query at the last of the three conditions above would produce:

```
To find if intended_use was permitted I used the rule
  IF work is allowed_by_GDO
  THEN intended work is permitted
    work                       allowed_by_GDO
To find if work was allowed_by_GDO I used the rule
  IF intended_residential is 'house extension'
  AND extension_ok is yes
  THEN work is allowed_by_GDO
   intended_residential        house_extension
   extension_ok                yes
```

and so on until the system arrives at details such as size, height, and position for

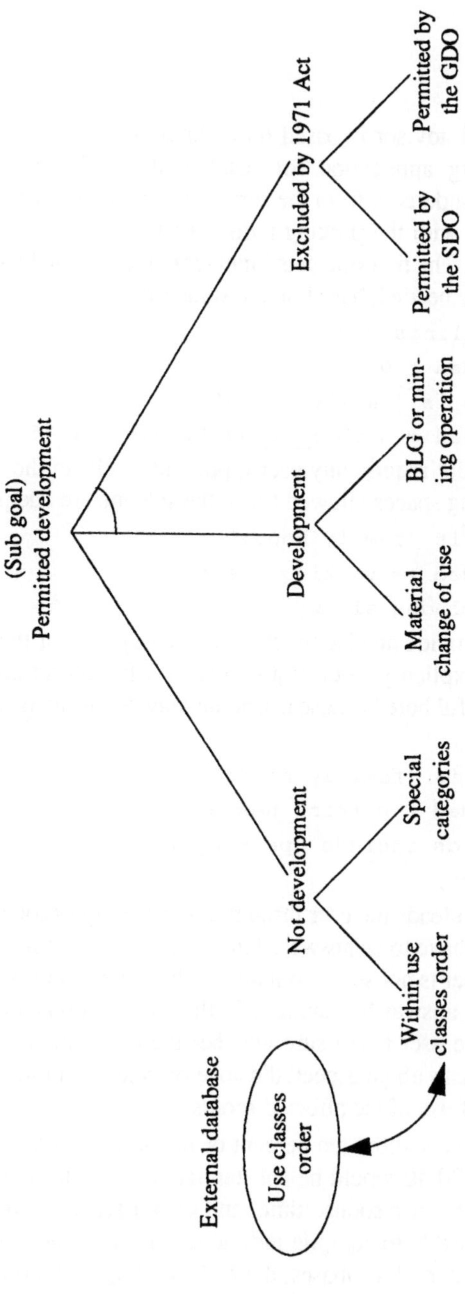

Figure 1.9. Decision tree for permitted development.

the extension that allow the system to determine if the building proposed is permitted development.

The DC advisor

The development control advisor is based on a checklist approach to assessing the merits of a planning application for residential development, and runs through a range of criteria derived from the series of interviews with DC experts. A pair of rules that deal with the parking requirements for a proposed housing scheme are shown below. The first rule gets input from the user and works out the number of parking spaces needed, based on the standard:

```
IF number_of_dwellings > 0
AND spaces_provided > 0
AND dwelling_type is 'family house'
THEN parking_needed = number_of_dwellings * 1.75
```

The second rule does not require any user input and works in the background to determine if the parking spaces allowed for in the scheme are adequate:

```
IF dwelling_type is 'family house'
AND spaces_provided >= parking_needed
THEN decision includes allow
```

Leonardo2 does not provide an else function so the opposite of the above rule has to be programmed explicitly. 'Reject-grounds' is a list object that can store more than one value, useful here because a scheme may be unsatisfactory on several criteria:

```
IF dwelling_type is 'familiy house'
AND spaces_provided < parking_needed
THEN reject_grounds include 'parking';
decision is reject;
```

The checklist approach leads the user through a series of questions about a development proposal that have to be answered in order to reach a conclusion about its acceptability. If the user is not sure about any of the criteria, the system tries to help the user to infer an answer by asking a further set of supplementary questions. For example, if the user is not sure whether there will be a problem with overlooking the system asks about aspect, distance between windows, position of windows, and the type of use of the affected rooms.

The prototype only covers about ten percent of the domain and has about 500 objects, of which about 30-40 would be substantiated in a consultation session. There have been problems in response time, and compilation can take up to 30 minutes, though it is possible to compile individual rulesets. Although there are facilities for linking to external databases, the collaborating LPA holds all its DC records on a mainframe and this link-up was not possible. The prototype is at present being evaluated by the collaborating LPAs.

Some initial conclusions

In programming the prototype, it became increasingly obvious that the gulf between AI and shell technology is a wide one. Searching the knowledge base is not heuristic, but a blind, depth-first exhaustive search. Problems of combinatorial explosion can be expected even in the relatively straightforward sub-domain of permitted development, because all the possible combinations have to be explicitly programmed in a series of similar rules. The only sense in which heuristics play a part in the system is within rules rather than between them. It is the rules themselves which embody the experiential knowledge of the expert. In deciding which rule to apply, the system simply tries them all until it reaches the end of the knowledge base. What the shell provides is a useful way of organizing a logical decision tree. Although the rulesets offer a degree of structure, there are no other built-in devices by which the programmer can impose order on what can quickly become an unwieldy set of rules.

Findings so far indicate that development control has several features that make it suitable for the application of expert systems. Contrary to views commonly expressed in expert-systems literature, development controllers (the domain experts) are able to articulate the basis of much of their knowledge and did not feel threatened by the new technology. They see expert systems having a place in giving advice to other planners with little development control experience, trainees, developers and the general public, although a different system would be required for each user group.

The prototype systems were less well received by senior planning managers and information technology managers. Planning managers recognized that on its own and without access to existing DC databases, the effectiveness of the system would be drastically reduced. They were also skeptical that much of the judgemental knowledge could be formalized. Information technology managers were reluctant to accept that expert systems had any potential. They thought existing systems were adequate and could be improved, and were extremely cool toward the idea of systems into which planners had direct programming input.

Of modules tested and demonstrated, the permitted development system was seen as being most useful by DC experts. The GDO is a mass of intertwined conditions and exclusions, and taking the user through the relevant choices visually on a screen was an effective way of offering advice. Most of the planners who have already tested the system said they actually learned something, remembered some piece of the GDO they had forgotten, or learned something they never knew before. It quickly became clear during the testing sessions that experience breeds degrees of belief and suspicion that provide the expert with a level of 'meta-rules' of the type 'IF the criteria is at the margin, THEN probe a bit deeper'. This is an interesting finding that did not emerge during the interviews and demonstrates that this kind of prototype does have a role in knowledge acquisition.

Acknowledgments

This paper is mainly based on a research project funded by a grant from the Economic and Social Research Council, reference D 00 23 2254.

Thanks are due to the following Planning Departments for their assistance: South Oxfordshire District Council, Barnet London Borough, Leicester City Council, Surrey County Council, and Westminster City London Borough.

References

Anderson, J.R., 1980. *Cognitive Psychology and its Implications.* W.H. Freeman, San Francisco.

Barrett, S. and C. Fudge, 1981. "Examining the policy-action relationship," in S. Barrett and C. Fudge, *Policy and Action*, Methuen, London.

Barrett, S. and P. Heather, 1984. "Information technology in planning practice," Economic and Social Research Council, London.

Batty, M., 1984. "Formal reasoning in urban planning," Papers in Planning Research No. 76, Dept. of Town Planning, University of Wales, Institute of Science and Technology.

Berry, D.C., 1987. "The problem of implicit knowledge," *Expert Systems 4(3)*: 144-150.

Breukner, J.A. and B.J. Wielinga, 1983. "Analysis techniques for knowledge based systems. Part 2: Methods for knowledge acquisition," ESPRIT Project 12, Report 1.2, University of Amsterdam.

Chandrasekaran, B. and S. Mittal, 1984. "Deep versus compiled knowledge approaches to diagnostic problem-solving," in M.J. Coombs, ed., *Developments in Expert Systems*, Academic Press, Inc., London.

Clancey, W.J., 1983. "The epistemology of a rule-based expert system—A framework for explanation," *Artificial Intelligence 20*:215-51.

Cullen, I., 1986. "Expert systems in planning analysis," *Town Planning Review 57*(3):239-251.

Davies, H.W.E., D. Edwards, and A.R. Rowney, 1986. "The relationship between development plans," Development Plans and Appeals, working papers in *Land Management and Development*, University of Reading.

Davis, J.R., J.R.L. Hoare, and P.M. Nanninga, 1986. "Developing a fire management expert system for Kakadu National Park in Australia," *Journal of Environmental Management* 22:215-227

Davis, R., 1982. "Expert systems: where are we and where are we going?", *AI Magazine 3*:3-22.

Dreyfus, H., 1984. *What Computers Can't Do*, Harper and Row, New York.

Grant, 1982. *Urban Planning Law*, Sweet and Maxwell, London.

Grover, M.D., 1983. "A pragmatic knowledge acquisition methodology," paper presented at the 8th International Joint Conference on Artificial Intelligence, 436-438.

Harrison, M.L., 1972. "Development control: The influence of political, legal, and ideological factors," *Town Planning Review 43*(3):254-274.

Heaps, D., 1987. *An Outline of Planning Law*, 9th edition, Sweet and Maxwell, London.

Jackson, P., 1986. *Introduction to Expert Systems*, Addison-Wesley, Workingham, England.

James, D.E.H., 1977. *Notes on the Need for Planning Permission*, 2nd ed., Oyez, London.

Johnston, D.M. and Hopkins, L.D., 1987. "Expert systems in planning analysis: the logic of uncertainty," *Town Planning Review 58*(3):342-346.

Jowell, J., 1973, "The legal control of administrative discretion," Public Law, 178.

Kennedy, M., 1987. "Logic programming and computers: Prolog," *Journal of Planning Education and Research 7*(1):35-45.

Kolodner, J., 1984. "Towards an Understanding of the Role of Experience in the Evolution from Novice to Expert," in M.J. Coombs, ed., *Developments in Expert Systems*, Academic Press, Inc., London.

Leary, M., 1986. "Expert systems: What potential for planning?", *The Planner, 72*(12):28-29.

Leary, M., 1987. "Development Control: The Role of Expert Systems," *Town Planning Review 58*(3):331-342.

Leary, M., 1988. "Knowledge and reasoning in development control and urban design: An expert systems approach," *Environment and Planning B(15)*:383-398.

Leary, M. and A. Rodriguez-Bachiller, 1988. "The Potential of Expert Systems for Development Control in British Town Planning," in D.S. Moralee, ed., *Research and Development in Expert Systems IV*, Cambridge University Press, pp. 198-210.

McLoughlin, B.J., 1973. *Control and Urban Planning*, Faber and Faber, London.

Mason, R., 1986. "Development control: A manual," *Law and Procedure*, Unit 2B, Royal Town Planning Institute and School of Planning, Oxford Polytechnic.

Ortolano, L. and C.D. Perman, 1987. "A planner's introduction to expert systems," *American Planning Association Journal*, Winter, 98-103.

Polyani, M., 1967. *The Tacit Dimension*, Anchor Books, New York.

Rich, E., 1983. *Artificial Intelligence*, McGraw-Hill, New York.

Royal Town Planning Institute, 1979. "Development control in the 1980s. Final report of the development control working party."

Ryle, G., 1946. *The Concept of Mind*, Pelican, London.

Salt, A., 1987. *Planning Applications: The RMJM Guide*, BSP Professional Books, London.

Schon, D., 1983. *The Reflective Practitioner*, Basic Books, New York.

Sharpe, R. and B.S. Marksjo, 1985. "An artificial intelligence approach to dynamically locating urban facilities," paper presented to the Regional Science Conference, Manchester University, England, September 4-6.

Simon, H., 1973. "The structure of ill-structured problems," *Artificial Intelligence 4*(3).

Steels, L., 1987. "Second generation expert systems," in M.A. Bramer, ed., *Research and Development in Expert Systems III*, Cambridge University Press.

Underwood, J., 1981. "Development control: A case study of discretion in action," in S. Barrett and C. Fudge, eds., *Policy and Action*, Methuen, London.

Waterman, D.A., 1986. *A Guide to Expert Systems*, Addison-Wesley.

Winston, P.H., 1984. *Artificial Intelligence*, Addison-Wesley.

Wyatt, R., 1987. "An expert advisory system for strategic planning," in P. Newton, R. Sharpe, and M.A.P. Taylor, *Desk Top Planning: Advanced Microcomputing for Social and Physical Infrastructure Planning*, Hargreen Publishing, Australia.

2

Applying a Rule-Based Decision Support System to Local Government Planning

James Richard Davis and Geoffrey McDonald

Introduction

The enthusiasm during the 1960s to develop a rational, procedural approach to land use planning arose from the success of developments outside planning. These included operations research and the development of efficient optimizing algorithms during the Second World War, the potential to handle massively repetitive calculations using electronic computers, and the advent of systems analysis for structuring complex problems. This interest in what came to be called 'procedural rationality' was not confined to planning but was shared by most research fields concerned with decision-making.

There were continuing attempts during the 1960s, 1970s, and early 1980s to develop *models of decision processes* involved in selecting the best plan. Some of the early models were based on mathematical optimization methods (e.g., Schlager, 1965) because of a belief, rooted in technical rationality, that selecting the best land use plan could be modeled in the same way as selecting the most cost-effective production schedule for a factory. However, the majority of decision-making models have been based on non-optimizing techniques, principally rating and weighting (RAW), because of the more modest data and computational demands and the greater transparency of these techniques. Numerous, essentially similar RAW models have been developed over the years in diverse disciplines concerned with decision-making (e.g., Hill, 1968; Edwards, 1977; Anderson, 1984; Cantor, et al., 1986) together with some non-quantitative methods (e.g., Holmes, 1972; Blair, 1978).

Land use planners have used two other types of models which differ fundamentally from decision models in that they predict rather than optimize or satisfy. The first type is *models of human activities*. The large urban land use and transportation models which dominated much of the early effort to apply computers to land use planning constitute good examples of this type. The second type is *models of natural processes*, such as fire prediction (Rothermel, 1972), soil movement (Wischmeier and Smith, 1978), and flood estimation (Mein, et al., 1974). We distinguish these two types because the latter are potentially more deterministic since they do not require the prediction of human activities and they have been adopted more recently than the first type by planners, as the need for land use planning has spread to semi-urban and rural areas. The role for these two types of models is to provide predictions of some of the consequences of alternative plans. Although obtaining reliable predictions from both types of models has proven to be difficult (via Lee's [1973] well-known criticisms of human activity modeling), a better understanding of the processes involved, more powerful computers and modeling techniques, and more modest modeling aims have led to a steadily increasing use of human activity and natural process models by planning agencies (Newton, et al., 1988).

The application of decision models was quickly confronted by major practical problems (Batey and Breheny, 1978), many of which were related to the limitations of rational problem solving discussed much earlier by Simon (1947). Many of the practical problems revolved around the use of mathematics for decision modeling. Land use planning must resolve the often conflicting objectives of diverse interest groups. These objectives are usually imprecise, difficult to ascertain, and subject to change. They are often qualitative in nature and, even when quantitative, often based on different measures. In addition, the planning environment meant that basing decisions on the predictions of consequences, either using predictive models or educated guesses, required decision models to take uncertainty into account. AIDA (Friend and Hickling, 1987) was one of the few decision models which could incorporate uncertainty (Faludi and Mastop, 1982). Even if these problems were overcome, it would still be literally impossible to choose the optimal plan because of the vast number of choices to be considered (Harris, 1989). In short, mathematical algorithmic decision models, especially optimizing ones, are not appropriate for the unstructured decision environment of land use planning.

Curiously, these difficulties led not to a re-examination of the decision-making process from which new decision models could emerge, but to a hiatus in the development of new decision models for land use planners. Currently, land use plans are almost always drawn up without assistance from formal decision models, although optimization has continued to be applied in planning solutions to essentially engineering problems (e.g., some water resource problems), and McHarg's map overlay method continues to enjoy popularity probably because of its transparency and, more recently, because of its ease of application using geographic information systems (GIS).

Decision support and expert systems

Similar lessons have been learned by researchers in other fields concerned with decision-making. For example, researchers attempting to install management information systems (MIS) in business organizations in the 1960s and early 1970s found that databases were useful for routine, structured problems but were of limited use to senior decision makers facing unstructured problems. Studies into the needs of these decision makers in the mid-1970s eventually led to a new type of computer program, termed a decision support system (DSS) by Bonczek, et al., 1981. The number of DSS definitions indicates a diversity of views on the important features of computer programs for assisting senior managers (Ginzberg and Stohr, 1982). One widely accepted definition is "an *interactive* computer-based system that *helps* decision makers utilize *data* and *models* to solve *unstructured* problems" (Sprague and Carlson, 1982). The italicized words emphasize the features identified in these studies as being important in the design of DSS.

An unstructured problem is not susceptible to algorithmic solution "because it (the problem) hasn't arisen before, or because its precise nature and structure are elusive or complex, or because it is so important that it deserves custom-tailored attention" (Simon, 1960). Instead, decision makers typically need to research possible solutions, evaluating and modifying the possibilities until an acceptable solution is obtained. Central to the development of DSS is the belief that the role of the DSS is not to undertake the search, but to assist the decision maker in undertaking it (hence the word 'support'). DSS developers have sidestepped the need to explicitly state an algorithm for choosing between alternatives by leaving this component of the decision-making process to humans. Much of the information needed to recognize an acceptable solution is so qualitative, uncertain, and incomplete as to be impossible to capture in a computer program, and it is often futile to encode a choice algorithm in decision models when the responsibility for the decision ultimately must reside with the decision maker.

In the context of land use planning, this latter point recognizes the need for the planner to feel a sense of ownership of the resulting plan(s) because of his/her need to explain and justify these plans to politicians and the general public. Consequently, decision-making using decision support systems must be a cooperative effort between human and computer, and this shifts the focus of decision models from the technical preoccupations noted by Batey and Breheny (1978) to issues of human-computer interaction. One important result of this shift is the need to involve the decision maker in the design of the program at an early stage, usually through a series of prototypes.

Even with this change in focus, DSS have still not been widely adopted in commercial organizations, and the majority of applications still occur in relatively well-structured decision situations (Landry, et al., 1985). This is because DSS are still not based on a well-grounded theory of multiperson decision-making in a multiobjective, corporate environment (Landry, et al., 1985), there is little understanding of designing effective human-machine interfaces, and DSS are difficult and expensive to develop. In spite of the similarities between decision-making in a corporate environment and in planning agencies, DSS research has had little impact on land use planning to date.

Technically, DSS consist of a database, a model base, and a human-computer interface. The database contains factual information about the problem area (e.g., current land uses), the model base contains predictive models (e.g., models to predict soil erosion), and the interface not only interprets a wide range of requests from the decision maker but can also justify results from the data and model bases. Because of their interest in both handling qualitative information and in a natural language dialog between human and computer, DSS researchers were quick to recognize the relevance of expert systems (e.g., Bonczek, 1976; Reitman, 1982). Expert systems were probably the most successful product to emerge from artificial intelligence (AI) research in the late 1970s. Because expert systems are well described in a number of text books (e.g., Hayes-Roth, et al., 1983), we shall provide only a brief introduction and discuss a few of their significant features here.

Early attempts by computer scientists to develop computer models of cognitive behavior, by applying traditional procedural programming, did not result in recognizably 'intelligent' behavior (McCorduck, 1979). After much controversy, the AI community concluded that human intelligence results not from applying clever algorithms to problems but from rapid access to a large amount of information gained from experience (Winograd, 1975). Armed with this conclusion, AI researchers were still unable to replicate the behavior of humans in solving everyday problems because of the vast amount of information that we typically access in making even simple decisions. Ironically, researchers had greater success in modeling expert behavior because of the smaller amount of highly structured information that constitutes expertise in a specialist problem domain.

Modern expert systems consist of at least two major components. The specialist domain knowledge is maintained in a knowledge base, generally in a highly modular form, and is applied to problems using a knowledge management system termed the inference engine. Separation of the knowledge base from the inference engine provides a number of advantages including ease of modification of the specialist knowledge and the ability to keep a trace of the knowledge that is applied to a particular occurrence of the problem.

Most specialized expertise is qualitative rather than quantitative in nature. Production rules (Hayes-Roth, 1985) are widely used methods that have been devised for representing this information (Barr and Feigenbaum, 1981). Production rules are condition-action (IF-THEN) statements, each of which expresses a separate module of knowledge. For example:

```
IF      wind strength is high
        season is summer
        temperature is hot
THEN    fire danger is high
```

is a simple rule taken from a fire modeling expert system (Davis, et. al., 1986).

Specialized knowledge about a problem is typically obtained by either interviewing experts or by inducing rules from case examples. Most expert systems allow the source of the information (such as the name of the expert), the degree of certainty associated with that module of information, and other descriptive textual information to be attached to each rule.

The inference engine will typically obtain information from the user (and from sensors, if part of a real-time system) about a particular occurrence of the problem and then use its knowledge base to infer logically valid conclusions about that occurrence. In contrast to conventional computer programs, expert systems infer rather than calculate solutions. One particularly powerful feature of these programs is their ability to logically 'justify' conclusions, using the trace of the knowledge used in obtaining the solution.

Expert systems have been developed for a variety of problem types, including design, classification, and evaluation problems; Hayes-Roth, et al. (1983) provide a classification of these types and Stefik, et al. (1983) have classified system types. They have been successfully applied to a variety of problem areas, notably medical diagnosis, mineral prospecting, computer design, and recently agricultural management (Waterman, 1985). These applications demonstrate how this technology can provide planners with both human activity and natural process models when sufficient specialist expertise is available. Others have shown expert systems can be used to model decision-making (e.g., Bohanec, et al., 1983).

Barath and Futo (1984), DeMers (1985), and Tanic (1986) provide some of the few examples of the use of this technology for land use planning. Other authors, e.g., Smith (1984), Langendorf (1985), Batty (1985), Cullen (1986), Ortolano and Perman (1987), and Sharpe, et al. (1987) have referred to the possible value of expert systems; however, in general, land use planners have been slow to adapt this technology to either their predictive or their decision modeling requirements.

Local land use control

The responsibility for developing and administering local land use planning tools throughout Australia has been delegated to local government, although the legal and administrative arrangements differ from state to state. In Queensland, the Local Government Act (1936-1986) requires all local governments to prepare strategic plans and zoning schemes. Strategic plans indicate preferred land uses and infrastructure development, and contain both a set of goals and objectives for land use in the local government area and the data on which the plan is based.

Zoning schemes are legal statements of existing use rights and so indicate to potential developers the likely government response to development proposals. Each scheme consists of a map delineating distinct areas within a local government area, a table listing the controls to be exercised over proposed land use activities in each of these areas, and a set of conditions attached to those land uses that may occur in each area. The three typical controls found in zoning schemes are 'permitted', 'permitted with the consent of the local government' and 'prohibited'. Permitted land uses are those considered to be 'as-of-right' in a particular area (often predominant existing uses). Permitted with consent uses are those that may be acceptable in that area depending on the details of the proposal. Prohibited uses are those that will always be unacceptable in that area (e.g., noxious industries in a residential area).

Zoning schemes are normally drawn up in an ad hoc rather than a planned way. Zoning is one of the major instruments available for achieving the longer term land use objectives shown in the strategic plan, and so the land use objectives in the strategic plan are used to guide revisions of the zoning scheme. However, efforts to revise zoning schemes in a rational, consistent, and defensible way using a conventional RAW decision model (Davis and Grant, 1987; Laut and Davis, 1988) were unsuccessful. The reasons were similar to those noted by other authors commenting on the application of mathematical decision models to strategic planning.

1. The program was unable to either represent adequately or work with the mass of qualitative information on which zoning schemes are based. RAW methods require all information to be not only quantitative but cardinal, so that the operations of addition and multiplication can be performed. The result is that qualitative, imprecise information is forced into quantitative form to suit the needs of the technique. This is particularly problematic when the importance of an objective has to be reduced to a single number, stated in advance of its implications being known (e.g., Kettle and Whitbread, 1973).
2. A related point was the difficulty planners had in accepting the rating and weighting formula and the associated simplistic, global decision rule. These computerized manipulations did not correspond to the familiar political, incremental method of drawing up a zoning scheme.
3. Even when the information is intrinsically quantitative, the power and efficiency of mathematics is gained at the expense of transparency. In principle, a trace of the mathematical operations performed by a RAW program could be kept, but such a trace would be too symbolic to constitute a useful justification of the resulting plan. Planners need to examine how the results were obtained and what were the implications of these results. Without this transparency, they feel no 'ownership' of the plan.
4. Few mathematical methods take explicit account of the diverse objectives of interest groups, even though interest groups are central to land use planning in a pluralist society. For example, local government planners need to know the extent to which these groups are affected by different proposed zonings.
5. Most of the data and information used in planning are imprecise but few mathematical decision models take this imprecision into account. Planners need to know the extent to which decisions are reliable.
6. It is common for RAW and other decision models to treat each location independently in order to reduce the computational requirements. However, this simplification of the problem ignores important interactions between deci-

sions taken at locations. Nevertheless, the NP-completeness of the problem when interactions are included implies that some simplification, such as an incremental consideration of interactions, is necessary.

7. Perhaps the greatest difficulty with most mathematical decision models lies in the neglect of the role of the planner. These models effectively relegate the planner to entering data and inspecting resulting plans simply because the decision-making is carried out by an algorithm. Thoughtless neglect of the professional competence of the planner is largely a result of the over-emphasis by program developers on technical aspects noted by Batey and Breheny (1978). Not only is it important to acknowledge the role of the planner for reasons of acceptance when designing computer programs for planners, but the inability of these programs to handle complex decisions using uncertain, incomplete, qualitative information implies that this role may also be essential for reasons of effectiveness.

The ADAPT program

ADAPT's design was influenced principally by the need to redress these difficulties with mathematical decision models and not by technological advances in decision modeling (Davis and Grant, 1987). However, in retrospect we can see that it shares many features with expert systems and decision support systems.

There are five steps in setting up ADAPT to assist in the production of a zoning scheme.

i. The local government area is divided into a number of relatively homogeneous regions (termed planning units).

ii. The land use activities to be regulated in the zoning scheme are defined.

iii. The planner decides on the control categories to be used in the study.

iv. Objectives held by different interest groups about appropriate controls to be exercised over land use activities are entered into the program as rules of the form:

```
Assign <land use> to <control category> if <condition>
```

The condition is a set of <data-item relation data-value> triplets ANDed together. The interest group(s) proposing the objective and unlimited explanatory text can be attached to the rule. Thus a typical rule could be:

```
Assign intensive poultry sheds to prohibited
if distance-to-town lt 20, and
town-water-supply is affected.
Group: urban residents
Explanation: the residents have expressed concern about
the smell and noise problems associated with intensive
poultry sheds and the potential of this activity to pol-
lute the town water supply.
```

The condition in this example contains a mixture of quantitative and qualitative data.

v. ADAPT contains a geographic database which is automatically primed with the data-items mentioned in the rules. Other data-items can be added. The fifth step in setting up the program is to load the database with the data-values for as many of these data-items in as many units as possible. It is not necessary that the database be complete. These data can be entered either from the keyboard or, if they are available from an existing geographic database, through ADAPT's import capabilities.

The use of the program once set up is most easily explained by describing a recent application in Gatton Shire, Queensland.

Gatton Shire

Gatton Shire is a small local government area (157,000 ha) 75 kms to the west of Brisbane, the capital of Queensland. Land use in the Shire is predominantly rural in nature, being a mixture of grazing, forestry, national park, and horticulture on the rich alluvial flats of the Lockyer River. However, there is an increasing pressure for the development of small rural allotments (1.5 to 6 ha) for hobby or rural residential purposes. The main urban area in the Shire is Gatton town although there are four villages and several small service centres.

At the time of the application described here, planning consultants had just completed a strategic plan for the Shire. The Council supported the approach from the present authors to align the Shire's Zoning scheme with the strategic plan using ADAPT. However, this was not an official Council planning study.

The Shire was divided into 397 planning units by overlaying maps of those biophysical and social data-items judged to be important to the study. Planning units are thus homogeneous in the most important data-items.

Nine land use activities were selected from a range of uses nominated by the Shire Council (Table 2.1). A tenth, extractive industry, was added during the study. Only the three conventional control categories (permitted, permitted with consent, and prohibited) were used in this study.

Seventy four land use policies were obtained from the objectives in the strategic plan, a questionnaire circulated to Shire Councillors, inspection of the Council's published planning rules, and from discussions with the Lockyer Watershed Authority and relevant State Government departments. One hundred and seventy

Table 2.1. The Land Use Activities Use in Drafting the Gatton Zoning Scheme

Activity	Definition
Intensive agriculture	Including horticulture
Mixed agriculture	
Extensive agriculture	Predominantly grazing and forestry
Small rural allotments	Low density, semi-urban, reticulated water, 0.6 ha
Hobby farms	Semi-rural, 1.5 - 6.0 ha
Urban	Residential and associated industry
Village	Small settlements
Open space	Conservation and recreation
Special purpose	Major reserves of significance and committee public land
Extractive industry	Sand, gravel, and rock extraction

rules were generated from these policies; multiple rules were generated from policies that referred to more than one land use or more than one data item.

The rules referred to 29 data-items (Table 2.2). All but three data-items had been compiled during the development of the strategic plan and were available from the Shire Council or State government departments. These 26 data-items were imported into ADAPT's database; the remaining three were calculated separately and entered into the database manually.

Once these entities had been entered, ADAPT was used to produce a zoning scheme. Take the first land use activity, intensive agriculture, and the first planning unit. ADAPT scanned the set of rules and selected those that referred to intensive agriculture and whose conditions were true on that unit. The latter selection involved a simple test of the condition part of each rule with the corresponding values of the data-items of the first planning unit. This subset of selected rules constitutes the objectives of possibly different interest groups concerning the appropriate controls to be exercised over intensive agriculture on that unit.

This subset can be in correspondence, in conflict, or empty. We define a simple decision situation to be one where the selected rules are in correspondence (i.e., they all propose that intensive agriculture be assigned to the same control category). This is trivially the case if only one rule is selected. The planner is not asked to intervene in this situation; instead ADAPT assigns the control to the activity on that planning unit together with the identifier of the rule(s) that led to the assignment. If the rules are in conflict (i.e., there is disagreement, possibly from

Table 2.2. Data Items Used in Drafting the Gatton Zoning Scheme

Data item	Comment	Source
Land suitability	12 classes	QDPI
Land tenure	8 classes	Shire
Land use	10 classes	Shire
Extractive industry	No. of sites	Mines
Uncleared forest	Percent	Shire
Salinity hazard	Occurrence	QDPI
Sheet/rill erosion	Percent	QDPI
Gully erosion	Percent	QDPI
Land slip	Occurrence	QDPI
Erosive flooding	Percent	QDPI
Subdivision size	8 classes	DOGI
Geology	8 classes	Mines
Rock and minerals	Potential deposits	Mines
Electricity	Existing supply	SEQEB
Zoning	10 classes	Shire
Flooding	1974 floods	Shire/QWRC
Pollution sources	4 activities	Shire
Services	Within 10 kms	Shire
Access	To existing roads	Shire/DMR
Vegetation	12 classes	QNPWS
Vegetation change	4 classes	QNPWS
Altitude	10 classes	DOGI
Aspect	8 classes	DOGI
Groundwater quality	mg/l chloride	QWRC
Water reticulation	Existing/potential	Shire
Munition dump		Mines
Ridgelines	Prominent ridges	Topo. maps
Bushfire risk		Calculated
APL pipeline	Presence/absence	CSR (APL)

Note:
- DOGI: Dept. of Geographic Information
- Mines: Queensland Dept. of Mines
- QDPI: Queensland Dept. of Primary Industry
- QNPWS: Queensland Nat'l. Parks & Wildlife Service
- QWRC: Queensland Water Resources Commission
- SEQEB: Southeast Queensland Electricity Board
- Shire: Gatton Shire Council

different interest groups, over the appropriate control to be exercised over intensive agriculture on this unit) ADAPT appeals to the planner. Although it is conceptually possible for ADAPT to possess decision-making rules to resolve these more complex situations, the program was designed without them in accordance with points 2 and 7 in the local land use control section.

If there are no rules indicating the control to be applied to intensive agriculture on that unit, ADAPT checks whether a default control category has been set up for this activity. These defaults, which represent the common planning practice of assigning unremarkable activities to either prohibited or permitted with consent, can be optionally provided when the activities are set up. The decision-making situation is regarded as complex if no default control is available, and ADAPT appeals to the planner to assign a control category.

Planners can do three things when asked to make a complex decision. First, they can make a decision immediately if sufficient information is available. Second, they can ask ADAPT for information about the situation and, if sufficient information is available, make a decision. The information can be attributes of the spatial unit obtained from the geographic database, or attributes of earlier decisions (usually concerning the same land use activity). Third, they can put the decision off and ADAPT will record that no control has yet been assigned to this activity on this unit. If a decision is made, ADAPT will record it in the usual way and request the planner to supply a reason for the decision. The program cycles through land use activities on all planning units until decisions have been made about controls over all activities, i.e. a draft zoning scheme has been developed.

At any time, the planner can ask ADAPT such questions as:

- Where has an activity been prohibited?
- To what extent have the rules advocated by the Council been met?
- How many hectares have been allocated to a particular activity with consent?
- What reason was given for prohibiting a land use activity on a unit?

Appendix A shows the output from the assignment of controls to all nine activities on planning unit 6.

ADAPT can produce a file of either nominated data-items or decisions about a nominated activity, suitable for mapping using the ATLAS mapping package (see Figure 2.1). The planner would normally revise the proposed scheme in light of the map and answers to questions such as those on the previous page (Appendix B) until a suitable zoning scheme emerged.

Details of the production of the zoning scheme for Gatton Shire are described in McDonald and Davis (1989).

Evaluation

Reviewing ADAPT's performance in Gatton Shire is, at this stage, impressionistic, although a more rigorous evaluation is underway involving planners and elected officials from the Shire and more experienced planners from elsewhere in the State. Some criteria used in evaluation include: Is it acceptable to political, legal, and professional interests? Can it be effectively explained to other professionals and to the community, especially when allocative decisions are involved? Is it authoritative? Is is compatible with the form and content requirements of current planning? Does it improve communications between planners and clients? How much training of personnel would be required?

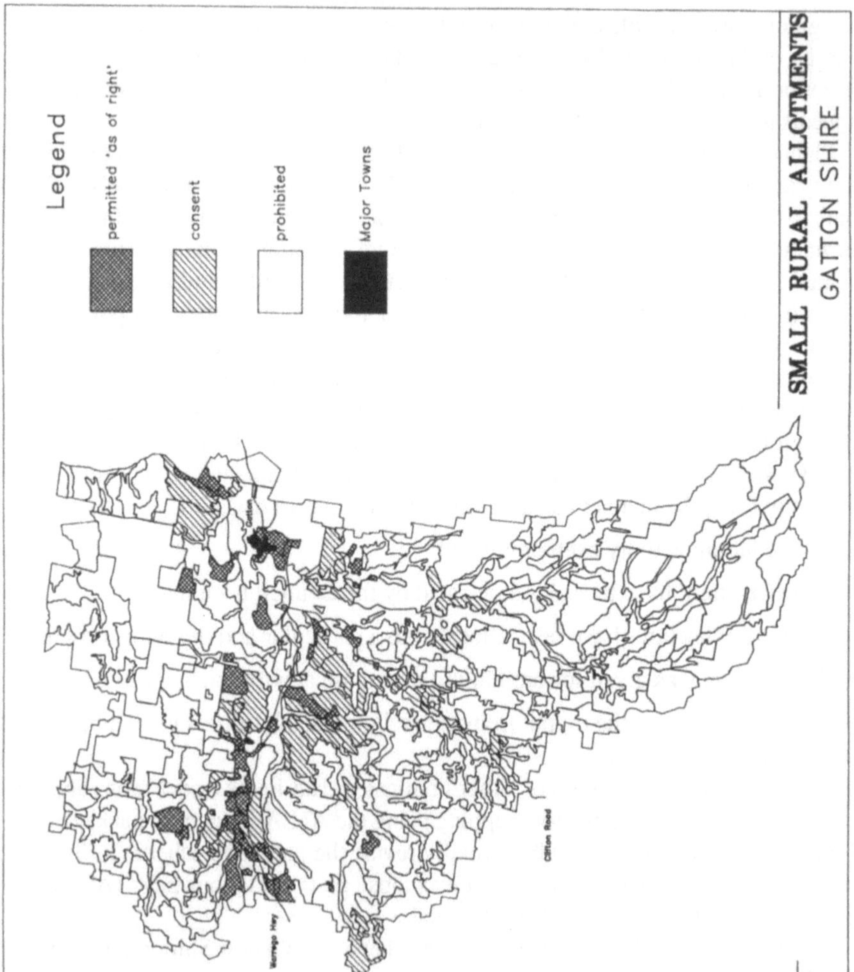

Figure 2.1. This map shows controls assigned to the small rural allotment activity after applying ADAPT a number of times. The reasons for assigning a control at a particular location can be obtained by asking the program. The map was drawn by exporting the table of assignments to the ATLAS mapping package.

Evaluations to date indicate that planners are very impressed with ADAPT. By far the greatest interest is expressed in the promise of codifying the rule set that planners must work with. Planners are predominantly administrators rather than designers, and their day-to-day work is heavily involved in retrieving and interpreting sets of policies and standards. For Gatton planners this problem is not serious, but for a metropolitan area the rule set is normally very large and complex, and only experienced planners are able to recall and interpret it quickly and accurately. ADAPT shows the possibility of accurate retrieval of all rules pertaining to a given site and allows the planner to consider the judgemental aspects of the task, rather than being involved in the endless searching of files to ensure that rules have been retrieved correctly. The results for the community are more speedy responses and satisfactory explanations about reasons for planning decisions or indications as to what standards have to be met in development proposals.

Planners are impressed with the ability of ADAPT to provide a nexus between continually changing sets of policies and standards with plan review and amendment. The fact that ADAPT provides a transparent view of this process is essential. The greatest difficulty in this case study, and for DSS in planning, is in turning nebulous policies into operational statements for the purposes of rule specification. Many planning policies are vague statements with ambiguous value judgements embedded in them.

It is common practice for planners to collect data as required when development proposals are submitted. In operational planning complete and current data sets are extremely rare. In cases where there are incomplete and missing data, ADAPT provides guidance on the collection of data to meet the requirements of rules operating in the area for the proposed activity. It also provides criteria for judging new proposals before data are even collected.

The application in Gatton Shire has indicated ways in which the program's design can be improved. The program is designed to aid the planner by acting as a filing cabinet, recording decisions and their underlying reasons for later review, and improving the planner's efficiency by automatically taking the simple decisions and focusing attention on the more complex decisions.

The reasons, either advanced by the planner when the program is being run or comments attached to rules, are unstructured strings of text. They cannot be searched and so the program cannot respond to common questions such as: How many times has intensive agriculture been prohibited because it was too close to town? The reasons behind decisions can themselves be pictured as higher level decisions and so should be treated in a similar manner to the decisions on appropriate control categories.

Although the definition of a simple decision (based on one or more non-conflicting rules) was acceptable to the Shire planners, it was not strong enough. Many other decisions regarded by ADAPT as complex were regarded by these planners as being simple. In particular, conflicts over existing activities dominated the decisions passed by ADAPT to the planner. Although the original design of ADAPT eschewed the automation of decision-making, we now believe that the program should have provision for conflict resolution rules. However, these rules must be sensitive to the context in which the decision is made and should only refer to large classes of similar decision situations.

ADAPT does not attempt to rectify two of the previously identified shortcomings of traditional decision models. First, the program does not provide the planner with an indication of the reliability of decisions except through comments attached to rules. This is not a major deficiency at present since these comments are available both when the planner is asked to make a decision and when examining past decisions. However, if human activity and natural resource models are included in later versions of the program (see below), then one of the standard methods used in expert systems for recording and transmitting uncertainty (e.g., Bonissone, 1983) may need to be added to the program.

Second, ADAPT treats each planning unit independently and so spatial interdependencies (such as interactions between decisions on adjacent planning units) cannot be included in the assignment of controls to land use activities when the decisions are simple. However, spatial and other interdependencies may well be included by the planners as part of their decision-making about complex situations. There are no special interrogatives available to help the planner explore these interdependencies at present, although we intend to provide a mapping interrogative as part of a current research project. This will allow the planner to see spatial patterns prior to making a complex decision.

Conclusions

This chapter describes the development and an application of a novel computer program that assists planners developing local government zoning schemes. Producing these schemes is one example of a class of planning problem characterized by the attempt to assign an entity (here, a land use activity) to one and only one of a number of categories (here, the control categories) on each of a number of planning units using a set of assignment rules. The rules may be inconsistent and incomplete. ADAPT can be applied to other problems of this type.

For example, strategic planning is the attempt to assign an entity (preference) to one and only one land use activity on each of a number of planning units using assignment rules that reflect the objectives of interest groups. Consequently, ADAPT's assignment rules for strategic planning would be of the form:

```
Assign <preferred> to <activity> if <condition>
```

The definition of simple decisions would need to be extended for strategic planning and conflict resolution rules would need to be introduced.

Although ADAPT's design arose directly from the perceived needs of planners rather than from technological developments, we can see in retrospect that the program is a decision support system. For example, it was only designed after a study of the institutional and professional requirements of planners drawing up zoning schemes. This design is based on two principles.

The first is that the planner should be free from making simple decisions that arise when a zoning scheme is being developed, so that s/he can focus on the complex ones. At present, we define simple decisions to be those where there is no conflict between the recorded objectives of different interest groups.This is not a trivial definition since these objectives are invariably incomplete and interactions, including spatial interactions, between decisions are ignored. Simple decisions are made by the program using a global decision rule. In this sense ADAPT is similar to quantitative decision models, such as the RAW models, which use a global 'largest score' decision rule. However in contrast to quantitative methods, ADAPT leaves complex decisions to the planner to resolve, not only because this creates a sense of 'ownership' in the plan, but also because the factors needing to be considered in these situations are too complex to be encoded as a set of rules let alone a global rule. It is apparent, in light of the application described here, that the definition of simple decisions needs to be expanded and that context-sensitive decision rules (i.e., not a single global decision rule) should be introduced.

The second principle is the need for the program to assist the planner in making complex decisions. At present, the program can answer a range of factual questions posed by the planner, not just about the attributes of the planning units, but also about previous decisions. All entities in the program (i.e., rules, data-items, interest groups, decisions, etc.) can have descriptive comments attached to them and these comments, especially those attached to earlier decisions, can also assist the planner in making a complex decision. We are presently adding an internal mapping capability to the program in order to provide information on spatial relationships.

Although ADAPT is like an expert system in that it stores and manipulates qualitative information in the form of rules, it is not strictly an expert system in either performance or technique. The program neither "exhibits the high level of performance that a human expert achieves" (Brachman, et al., 1983), nor does it infer further facts relevant to the program using conventional expert system techniques such as forward or backward chaining. In fact, expert systems are an inappropriate technology for decision modeling in planning since it is axiomatic with DSS that this must remain a user-centered and not an expert-centered task.

However, ADAPT could still benefit from expert system concepts. For example, the values of data-items, such as erosion hazard, could be inferred from a set of rules instead of, as at present, being recorded in the database. The assignments could then be justified in the usual expert system way by backtracking through these rules. Uncertainty in results from these predictive models could be transmitted to the planner as part of the support offered when making a complex decision.

Earlier decision models for land use planners, based on quantitative methods, were successful for only a narrow range of problems and the development of new approaches for the more common multiparty, multiobjective, qualitative planning problem lapsed during the 1970s and 1980s. However, the development of DSS has demonstrated the benefits of focusing more on the need to support planners when they are making decisions and less on the technicalities of developing mathematically sophisticated decision-making algorithms. Expert systems have the potential to provide some of these requirements (e.g., qualitative information handling, explanation, and uncertainty handling). To date, there have been few applications of expert systems to planning and, as the enthusiasm for quantitative methods has taught us, the true worth of any technology can only be assessed after such experience.

References

Anderson, B., 1984. "Multiobjective, multipublic method of alternative evaluation," Bureau of Reclamation, Washington D.C.

Barath, E. and I. Futo, 1984. "A regional planning system based on artificial intelligence concepts," in *Proceedings, 23rd European Congress of the Regional Science Association*, pages 135-154.

Barr, A. and E.A. Feigenbaum, 1981. *The Handbook of Artificial Intelligence*, Volume 1, Heuristech Press, Stanford, CA.

Batey, P.W.J. and J.J. Breheny, 1978. "Methods in strategic planning, Part II: A prescriptive review," *Town Planning Review 49*:502-518.

Batty, M., 1985. "Formal reasoning in urban planning: Computers, complexity and mathematical modeling," in M. Breheny and A. Hooper, eds., *Rationality in Planning: Critical Essay on the Role of Rationality in Urban and Regional Planning*, pages 98-119, Pion, London.

Blair, P.D., 1978. *Multiobjective Regional Energy Planning*, Martinus Nijhoff, Amsterdam.

Bohanec, M., I. Bratko, and V. Rajkovic, 1983. "An expert system for decision making," in H.G. Sol, ed., *Processes and Tools for Decision Support*, pages 235-248, North-Holland, Amsterdam.

Bonczek, R.H., 1976. "Observations on a generalized intelligent query processor for decision support," paper, Krannert School of Graduate Management, Purdue University, W. Lafayette, IN, NTIS Pub 267 472.

Bonczek, R.H., C.W. Holsapple, and A.B. Whinston, 1981. *Foundations of Decision Support Systems*, Academic Press, New York.

Bonissone, P.P., 1983. "Coping with uncertainty in expert systems: A comparative study," in *Proceedings, American Control Conference*, San Francisco, IEEE, New York, June.

Brachman, R.J., S. Amarel, C. Engelman, R.S. Englemore, E.A. Feigenbaum, and D.E. Wilkins, 1983. "What are expert systems?", in F. Hayes-Roth, D.A. Waterman, and D.B. Lenat, eds., *Building Expert Systems*, pages 31-58, Addison-Wesley, Reading, MA.

Breheny, M.J., 1987. "The context for methods: the constraints of the policy process on the use of quantitative methods," *Environment and Planning A19*:1449-1462.

Cantor, L.W., S.F. Atkinson, and F.L. Leistritz, 1986. *Impact of Growth: A Guide for Socio-Economic Impact Assessment and Planning*, Lewis Press, Chelsea, MI.

Cullen, I., 1986. "Expert systems in planning analysis," *Town Planning Review 57*(3):239-251.

Davis, J.R., P.M. Nanninga, and J.R.L. Hoare, 1986. "Developing a fire management expert system for Kakadu National Park, Australia," *Journal of Environmental Management 22*:215-227.

Davis, J.R. and I.W. Grant, 1987. "ADAPT: a knowledge-based decision support system for producing zoning schemes," *Environment and Planning B14*:53-66.

DeMers, M.N., 1985. "The formulation of a rule-based GIS framework for county land-use planning," in *Modeling and Simulation (16), Part 1*, pages 93-97, *Proceedings 16th Annual Pittsburgh Conference on Modeling and Simulation*, Pittsburgh, PA, April.

Edwards, 1977. "Evaluate alternatives using the SMART system," *IEEE Trans. on Systems, Man, and Cybernetics 7*:326-340.

Faludi, A. and J.M. Mastop, 1982. "The 'IOR School', The development of a planning methodology," *Environment and Planning B 9(3)*:241-256.

Ginzberg, M.J. and E.A. Stohr, 1982, "Decision Support Systems: Issues and Perspectives," in M.J. Ginzberg, W. Reitman, and E.A. Stohr, eds., *Decision Support Systems*, pages 9-31, North-Holland, Amsterdam.

Harris, B., 1989. "Beyond geographic information systems: Computers and the planning professional," *American Planning Association Journal 55*(1):85-90.

Hayes-Roth, F., 1985. "Rule-based systems," *Comm. Assocation of Computing Machinery 28*:921-932.

Hayes-Roth, F., D.A. Waterman, and D.B. Lenat, "An overview of expert systems," in F. Hayes-Roth, D.A. Waterman, and D.B. Lenat, eds., *Building Expert Systems*, pages 3-29, Addison-Wesley, Reading, MA.

Hill, M.A., 1968. "A goals-achievement matrix for evaluating alternative plans," *Journal American Institute of Planners 34*:19-29.

Holmes, J.C., 1972. "An ordinal method of evaluation," *Urban Studies 9*(2):179-191.

Kettle, P. and M. Whitbread, 1973. "An ordinal method of evaluation: a comment," *Urban Studies 10*:95-99.

Landry, M., D. Pascot, and D. Briolat, 1985. "Can DSS evolve without changing our view of the concept of 'problem'?", *Decision Support Systems 1*(1):25-36.

Langendorf, R., 1985. "Computers and decision making," *American Planning Association Journal 51*(4):422-433.

Laut, P.-R., and J.R. Davis, 1988. "An assessment of a rational approach to drawing up local government zoning schemes," *Landscape and Urban Planning 15*: 303-314.

Lee, D.B., 1973. "Requiem for large scale models," *Journal of the American Institute of Planners 39*(3):163-178.

McCorduck, P., 1979. *Machines Who Think: A Personal Inquiry Into the History and Prospects of Artificial Intelligence*, W.H. Freeman, San Francisco.

McDonald, G.T. and J.R. Davis, 1989. "ADAPT: Application of a knowledge based planning system to Gatton Shire," Discussion paper, Division of Australian Environmental Studies, Griffith University, Nathan, Queensland.

Mein, R.G., E.M. Laurenson, and T.A. McMahon, 1974. "Simple nonlinear model for flood estimation," American Society of Civil Engineers, *Journal of Hydraulics Division 100*(HY11):1507-1518.

Newton, P.A., M.A.P. Taylor, and R. Sharpe, 1988. *Desktop Planning, Microcomputer Applications for Infrastructure and Services Planning and Management*, Hargreen, Melbourne, Australia.

Ortolano, L. and C.D. Perman, 1987. "A planner's introduction to expert systems," *Journal of the American Planning Association 53*(1):98-103.

Rothermel, R.C., 1972. "A mathematical model for predicting fire spread in wildland fuels," USDA Forest Service Research Paper, INT 115.

Reitman, W., 1982. "Applying Artificial Intelligence to Decision Support: Where do Good Alternatives come from?" in M.J. Ginzberg, W. Reitman, and E.A. Stohr, eds., *Decision Support Systems*, pages 155-174, North-Holland, Amsterdam.

Schlager, K.J., 1965. "A land use plan design model," *Journal of the American Institute of Planners 31*:103-111.

Sharpe, R., B.S. Marksjo, and J.V. Thomson, eds., 1987. "Special issue on expert systems in planning and design," *Environment and Planning B14*(3):237-358.

Simon, H.A., 1947. *Administrative Behaviour: A Study of Decision Making Processes in Administrative Organisations*, Macmillan, New York.

Simon, H.A., 1960. *The New Science of Management Decision*, Harper and Row, New York.

Smith, J.D., 1984. "Expert systems for landscape/land use planning: its potential utility for METLAND V," in *Proceedings, Second Symposium on Computer-Aided Land Use Planning*, pages 93-100, Massachusetts Agricultural Experiment Station Bulletin 693, Amherst, MA, June.

Sprague, R.H. and E.D. Carlson, 1982. *Building Effective Decision Support Systems*, Prentice-Hall, Englewood Cliffs, NJ.

Stefik, M., J. Aikins, R. Balzer, J. Benoit, L. Birnbaum, F. Hayes-Roth, and E. Sacerdoti, 1983. "The architecture of expert systems," in F. Hayes-Roth, D.A. Waterman, and D.B. Lenat, eds., *Building Expert Systems*, Addison-Wesley, Reading, MA.

Tanic, E., 1986. "Urban planning and AI: The URBYS system," *Computers, Environment and Urban Systems 10*(3/4):135-146.

Waterman, D.A., 1986. *A Guide to Expert Systems*, Addison-Wesley, Reading, MA.

Winograd, T., 1975. "Frame representations and the declarative/procedural controversy," in D.G. Bobrow and A. Collins, eds., *Representation and Understanding: Studies in Cognitive Science*, pages 185-210, Academic Press, New York.

Wischmeier, W.H. and D.D. Smith, 1978. "Predicting rainfall erosion losses—a guide to conservation planning," *US Department of Agriculture, Agriculture Handbook 537*, US Government Printing Office, Washington DC.

Appendix A

The allocation of controls to nine land use activities for planning unit 6 is illustrated in this appendix. The first two activities are prohibited: there is no conflict in the objectives referring to these activities. According to the rules, both extensive agriculture and small rural allotments should be either prohibited or permitted. The planner decides to permit the first of these activities, make the second subject to Council consent, and adds comments explaining these decisions. Controls for the next five activities are decided by ADAPT. Finally, there is conflict over the control to be exercised over extractive industries and the planner decides to prohibit this activity. The table of controls over activities on this planning unit is then printed out.

```
=====================================================================
                    A Decision Aid and Planning Tool

v 1.3    Ian Grant               CSIRO                      May 1986
                                 Division of Water and
                                   Land Resources

v 2.5    Chris Czarkowski        Griffith University        May 1989
                                 Division of Australian
                                   Environmental Studies
=====================================================================

Last project used was called gatton
Is this the project you want to use (Y/N)? y
Loading project gatton
Project gatton ready
Command > add allocation for unit 6 all

UNIT 6
allocated activity INTENSIVE-AGRICULTURE to category PROHIBITED de-
cided by rules

allocated activity MIXED-AGRICULTURE to category PROHIBITED decided
by rules

Trying to allocate EXTENSIVE-AGRICULTURE in unit 6 RULES conflict
Rule 3 ACTIVE
ASSIGN EXTENSIVE-AGRICULTURE TO PROHIBITED
   IF LAND-SUITABILITY GT 10
         GROUPS: COUNCIL
         COMMENT: Policy 12
Rule 23 ACTIVE
ASSIGN EXTENSIVE-AGRICULTURE TO PROHIBITED
     IF LAND-TENURE LT 5
         GROUPS: COUNCIL
         COMMENT: Exclusion Policy 6.3
Rule 53 ACTIVE
ASSIGN EXTENSIVE-AGRICULTURE TO PERMITTED
     IF UNCLEARED-FOREST GT 70
```

```
          GROUPS: COUNCIL
          COMMENT: Policy 16
Rule 133
ACTIVE ASSIGN EXTENSIVE-AGRICULTURE TO PERMITTED
     IF ROCK-AND-MINERAL EQ 1
          GROUPS: COUNCIL
          COMMENT: small areal impact of quarrying
Which CATEGORY for EXTENSIVE-AGRICULTURE in UNIT 6?>permitted
     Enter comment (if no comment just hit return):
Decided to permit extensive agriculture because,
   although the capability rating is poor, this activity
   is unlikely to damage the land.
allocated activity EXTENSIVE-AGRICULTURE to category PERMITTED rules
conflicted, user decision

Trying to allocate SMALL-RURAL-ALLOTMENTS in unit 6 RULES conflict
Rule 24
ACTIVE ASSIGN SMALL-RURAL-ALLOTMENTS TO PROHIBITED
     IF LAND-TENURE LT 5
          GROUPS: COUNCIL
          COMMENT: Exclusion Policy 6.3
Rule 54
ACTIVE ASSIGN SMALL-RURAL-ALLOTMENTS TO PROHIBITED
     IF UNCLEARED-FOREST GT 70
         GROUPS: COUNCIL
         COMMENT: Policy 22
Rule 134
ACTIVE ASSIGN SMALL-RURAL-ALLOTMENTS TO PROHIBITED
     IF ROCK-AND-MINERAL EQ 1
          GROUPS: COUNCIL
          COMMENT: Policy 50; Policy 24
Rule 182
ACTIVE ASSIGN SMALL-RURAL-ALLOTMENTS TO PERMITTED
     IF SERVICES LT 4
          GROUPS: COUNCIL
          COMMENT: Policy 27
Rule 192
ACTIVE ASSIGN SMALL-RURAL-ALLOTMENTS TO PERMITTED
     IF ACCESS LT 5
          GROUPS: COUNCIL
          COMMENT: Policy 23
Rule 192
ACTIVE ASSIGN SMALL-RURAL-ALLOTMENTS TO PERMITTED
     IF ACCESS LT 5
         GROUPS: COUNCIL
         COMMENT: Policy 23
Rule 213
ACTIVE ASSIGN SMALL-RURAL-ALLOTMENTS TO PROHIBITED
     IF VEGETATION-CHANGE LT 3
         GROUPS: COUNCIL
         COMMENT: Policy 22
Rule 192
ACTIVE ASSIGN SMALL-RURAL-ALLOTMENTS TO PERMITTED
     IF ACCESS LT 5
```

```
        GROUPS: COUNCIL
        COMMENT: Policy 23
Rule 221
ACTIVE ASSIGN SMALL-RURAL-ALLOTMENTS TO PERMITTED
     IF ALTITUDE GT 2 AND ALTITUDE LT 5
          GROUPS: COUNCIL
          COMMENT: Policy 21
Rule 271
ACTIVE ASSIGN SMALL-RURAL ALLOTMENTS TO PROHIBITED
     IF RIDGELINES EQ 1
          GROUPS: COUNCIL
          COMMENT: Policy 22
Rule 281
ACTIVE ASSIGN SMALL-RURAL-ALLOTMENTS TO PROHIBITED
     IF BUSHFIRE-RISK LT 4
          GROUPS: COUNCIL
          COMMENT: Policy 20

Which CATEGORY for SMALL-RURAL-ALLOTMENTS in UNIT 6>consent Enter
comment (if no comment just hit return):
Generally OK, but development applications should have regard to
bushfires and ridgeline prominence

allocated activity SMALL-RURAL-ALLOTMENTS to category CONSENT rules
conflicted, user decision

allocated activity HOBBY-FARMS to category PROHIBITED decided by
rules

allocated activity URBAN to category PROHIBITED decided by rules

allocated activity VILLAGE to category PROHIBITED decided by rules

allocated activity OPEN-SPACE to category PERMITTED decided by rules

allocated activity SPECIAL-PURPOSES to category PERMITTED decided by
rules

Trying to allocate EXTRACTIVE-INDUSTRIES in unit 6 RULES conflict

Rule 30
ACTIVE ASSIGN EXTRACTIVE-INDUSTRIES TO PROHIBITED
     IF LAND-TENURE LT 5
          GROUPS: COUNCIL
          COMMENT: Exclusion Policy 6.3
Rule 59
ACTIVE ASSIGN EXTRACTIVE-INDUSTRIES TO PROHIBITED
     IF UNCLEARED-FOREST GT 70
          GROUPS: COUNCIL
          COMMENT: retain forested areas
Rule 139
ACTIVE ASSIGN EXTRACTIVE-INDUSTRIES TO PERMITTED
     IF ROCK-AND-MINERAL EQ 1
          GROUPS: COUNCIL
```

```
          COMMENT: Policy 50
Rule 197
ACTIVE ASSIGN EXTRACTIVE-INDUSTRIES TO PROHIBITED
     IF ACCESS GT 3
          GROUPS: COUNCIL
          COMMENT: Policy 51
Rule 218
ACTIVE ASSIGN EXTRACTIVE-INDUSTRIES TO PROHIBITED
     IF VEGETATION-CHANGE LT 3
          GROUPS: COUNCIL
          COMMENT: No comment

Which CATEGORY for EXTRACTIVE-INDUSTRIES in UNIT 6>prohibited Enter
comment (if no comment just hit return):
Prohibited because of poor access and damage to sensitive vegetation
in this area.

allocated activity EXTRACTIVE-INDUSTRIES to category PROHIBITED rule
conflicted, user decision
```

```
Saving changes

Command> display allocation for unit 6 all
UNIT 6
```

PERMITTED	CONSENT	PROHIBITED
Extensive-Agriculture	Small-Rural-Allotments	
		Intensive-Agriculture
Open-Space		Mixed-Agriculture
Special-Purposes		Hobby-Farms
		Urban
		Village
		Extractive-industries

Appendix B

The planner can question ADAPT about reasons behind decisions regarding control categories. In the first question, the planner asks ADAPT why a particular control category has been assigned to small rural allotments in planning unit 6. The program replied that, because of a conflict between objectives, the planner elected to assign this activity to the "permitted with consent" category with the comment that, at the time of development application, the bushfire risk and the intrusion of the development on the ridgeline should be examined.

The planner then asks ADAPT to show the extent to which rules advocating an objective for extractive industry have been fired (for the decisions taken on unit 6). The reply shows that five of the 15 rules concerned with this activity fired on unit 6 and one of these, rule 139, was overridden when the planner resolved a conflict. The planner then asks for this rule to be printed out.

Finally, the planner asks ADAPT to show the extent to which the objectives of the National Parks and Wildlife Service have been met when making decisions about assigning controls over activities in unit 6.

```
Command> why small-rural-allotments in unit 6
    Do you require text of rules displayed (Y/N)? n

In unit 6 activity SMALL-RURAL-ALLOTMENTS has been assigned to CON-
SENT

Reason: Rules conflicted, user decision
     User comment: Generally OK but development
        applications should have regard to bushfires
        and ridgeline prominence

Command< display counters about activity extractive-industry
```

Rule	Fired	Overridden
30	1	0
42	0	0
46	0	0
59	1	0
97	0	0
139	1	1
157	0	0
197	1	0
198	0	0
209	0	0
218	1	0
269	0	0
299	0	0
312	0	0
314	0	0
316	0	0

```
Command>display rule 139
```

```
 RULE 139
ACTIVE ASSIGN EXTRACTIVE-INDUSTRIES TO PERMITTED
     IF ROCK-AND-MINERALS EQ 1
          GROUPS: COUNCIL
          COMMENT: Policy 50
Command>display counters about group National-Parks-and-Wildlife-
Service
```

Rule	Fired	Overridden
201	0	0
202	0	0
203	1	0
204	1	1
205	0	0
206	1	0
207		

```
Command>exit

Are you sure you want to exit (Y/N)? y

Closing files....

ADAPT finished

Stop
```

3

Using Expert Systems to Check Compliance with Municipal Building Codes[1]

Eric J. Heikkila and Edwin J. Blewett

Introduction

Han and Kim (1989) recently posed the question, "Can expert systems help with planning?" The planning functions they review are primarily complex decision environments with no single best or correct answer. Examples of multidimensional planning problems addressed by Han and Kim include landfill site selection, urban design, and bus route choice. These are decision contexts that clearly do not satisfy the criteria for expert system applications set out by Waterman (1986), Silverman (1987), and Goodall (1985), who advocate crisply defined problems with certain solutions and domain rules. As Han and Kim point out, these guidelines are seldom likely to be met in a planning context. Nonetheless, they conclude that expert systems can work well in such contexts, particularly when combined with management database systems, geographic information systems, or decision support systems.

Not all planning functions are so nebulously defined, especially those pertaining to routine administrative or regulatory practices. Administration of local government building or zoning by-laws is a case in point. These by-laws are the most concrete articulation of the planners' vision, and they are the most immediate focus of the continuing dialogue between city hall and the development community. While the process by which these by-laws evolves is complex and at times obscure, the by-laws themselves are a set of codified rules that govern the development process explicitly.

In residential areas, municipal by-laws typically focus on a range of concerns, including yard setbacks and building standards. This is a relatively high volume

---

1. Reprinted by permission of the *Journal of the American Planning Association,* Vol. 58, No. 1, Winter 1992.

area for building permits and is handled by a cadre of front line staff whose job is to check whether submitted plans for residential building renovation or construction conform to the written code of the building by-laws. This plan checking function is often a difficult one to perform well because the by-laws themselves are often highly complex, confusing, and mutually interactive. Despite the complexity and high public profile of their jobs, plan checkers are typically quite low in seniority, receive relatively low wages, and are the first to suffer the frustration exhibited by developers or homeowners whose building permits are denied. Not surprisingly, staff turnover among plan checkers is quite high, further perpetuating this cycle of inexperience and frustration.

This chapter assesses the feasibility of using expert systems as a plan checking tool. To this end, we have developed a prototype expert system based on the fire separation component of the City of Vancouver Building By-Laws. Fire separation in this context refers to the relationship between the distance of a dwelling unit from the neighboring property, and the percentage of the dwelling's exposing building face that can be devoted to windows, doors, and other unprotected openings. In the following two sections, we describe the Vancouver by-laws that govern fire separation and how plan checkers determine whether applications comply with these by-laws. This is followed by a discussion of why expert systems be an appropriate technology in this context, and a subsequent description of the prototype expert system developed for this purpose and an assessment of the potential for widespread applications of this type. A number of general issues emerge in this discussion, both technical and organizational. On balance, however, we believe the benefits of consistency, speed, and accuracy afforded by expert systems makes them a logical tool for checking compliance of submitted plans with a set of municipal by-laws.

Fire separation and the Vancouver building by-laws

The motivation for fire separation by-laws is to reduce the risk of fire spreading from one dwelling to another. While details of the by-laws and their application in practice can be quite complex, the basic principle is straightforward: unprotected openings close to an adjacent dwelling increase fire threat. If the *exposing building face* is a single plane, then the percentage of unprotected openings allowed by the by-laws increases monotonically as the distance of the exposing building face is further away from the property line. The precise relationship between this *minimum limiting distance* and the *allowable unprotected openings percentage* (AUOP) depends upon other considerations as well. For example, if either of the dwellings (the one being considered or the one adjacent) have built-in sprinklering systems, or if the dwelling in question has protective fire-resistant glazing on all openings, then an enlarged AUOP may be warranted. The total square footage of openings permitted on the exposing building face is obtained by multiplying the total area of the exposing building face by the AUOP for the minimum limiting distance.

Further complications may ensue as the exposing building face is broken up into planes of varying distances to the neighboring property line. In this case, a plan checker first examines whether the proposed openings exceed the AUOP prescribed for the closest plane. If not, the fire separation requirements are clearly satisfied, since the AUOP increases with distance from the property line. However, if this preliminary test fails, the plan checker proceeds with a more elaborate test to determine whether the proposed openings are allowed by building by-laws that govern fire separation. The principle underlying this elaborate test is as follows. Each opening on the exposing building face has associated with it a certain *limiting distance* which may be conveniently thought of as its effective distance from the property line, from the perspective of fire threat. The limiting distance for any opening depends upon its position relative to flanking walls and relative to openings on those flanking walls. The strengths of an expert system are best displayed in this search when the structure of plan checking rules and procedures becomes most complex.

Once the limiting distance has been determined for any opening, a procedure known as *ratioing* is undertaken. The purpose of this exercise is to calculate a *reduction factor* for each opening that is used to effectively reduce the actual opening area to some smaller number, where larger limiting distances will be associated with smaller effective openings (i.e., with smaller reduction factors). Of special interest is the minimum limiting distance over all openings on the exposing building face. For each limiting distance, including the minimum, there is a unique AUOP (subject to additional information on sprinklering and protective glazing). The reduction factor for any opening, which is never greater than one, is defined as the ratio of the AUOPs for the minimum limiting distance and the limiting distance, respectively. That is,

$$RF_j = (AOUP)^* / AUOP_j \tag{3.1}$$

where RF_j is the reduction factor for the jth opening, $AUOP_j$ is the allowable unprotected opening percentage for the jth opening, and AUOP* is the AUOP for the minimum limiting distance (and so $AUOP^* \leq AUOP_j$). The "reduced area" for each opening (ROA_j) is calculated as the product of the actual opening area (AOA_j) and the reduction factor. That is,

$$ROA_j = AOA_j \cdot RF_j. \tag{3.2}$$

The plan checker will approve the proposed plan for fire separation if the total reduced area for all openings on the exposing building face is within the allowable total for the minimum limiting distance.

Thus, to recapitulate, the fire separation portion of Vancouver's building bylaws are concerned with the risk of fire spreading between adjacent dwelling units. As the distance from the dwelling to the property line increases, so does the percentage of the exposing building that can be allotted to doors, windows, and other unprotected openings. In determining whether a submitted plan conforms to the fire separation by-laws, the plan checker first looks to see whether actual openings fall within the prescribed maximum. If so, then the fire separation by-laws are satisfied. If not, and if the exposing building face resides on multiple planes of varying distances from the property line, ratioing is used to reduce the effective areas of openings whose limiting distances exceed the minimum limiting distance. Much of the plan checker's attention is directed toward the search for each opening's limiting distance. The next section provides an overview of how one determines what the limiting distance is for any one opening.

Determining the limiting distances for unprotected openings

The limiting distance for an unprotected opening is the effective distance of that opening to the property line from the perspective of fire threat. Vancouver's fire separation by-laws are based on thermodynamic experiments conducted decades ago to assess how various building configurations affect heat and flame radiation. Several crucial factors are embedded in the by-laws, including: (1) distance of an unprotected opening from flanking walls, (2) existence of unprotected openings on flanking walls, and (3) length of any flanking walls. For the purposes of this study, it is irrelevant whether the fire separation by-laws in place in Vancouver are optimal, thermodynamically or otherwise. Instead, what is important is to ascertain is whether current practice for checking conformity with the fire separation by-laws can be replicated via expert systems.

Current practice is best described with reference to Figure 3.1, which shows an opening W on the exposing building face of a courtyard with two flanking walls, where each flanking wall also has an opening. In this example, there are five candidates for the limiting distance of opening W. One candidate is simply the actual distance of the opening to the property line. Another candidate for limiting distance is termed "projection left" (PL), the distance of the edge of the left flanking wall from the property line plus the distance of opening W to the left flanking wall. When an opening is adjacent to a flanking wall, as is the case here, the wall has the effect of concentrating the heat and flame, thereby increasing the distance of radiation. The closer an opening is to the flanking wall, therefore, the further forward is its effective distance projected, and so PL becomes smaller. A third candidate for the limiting distance of opening W is "flanking opening distance left" (FODL), defined as the distance to the property line of the edge of the opening in the left flanking wall that is nearest to the property line. If opening W is too close to the left flanking wall, then there is a danger that fire will spread to that wall and that the flanking opening will itself become a source of flame and heat radiation, and so FODL becomes a candidate limiting distance. Two additional candidate limiting distances are PR and FODR, based on the same concepts applied to the right flanking wall. In general, the candidate that yields the smallest limiting distance is the effective limiting distance.

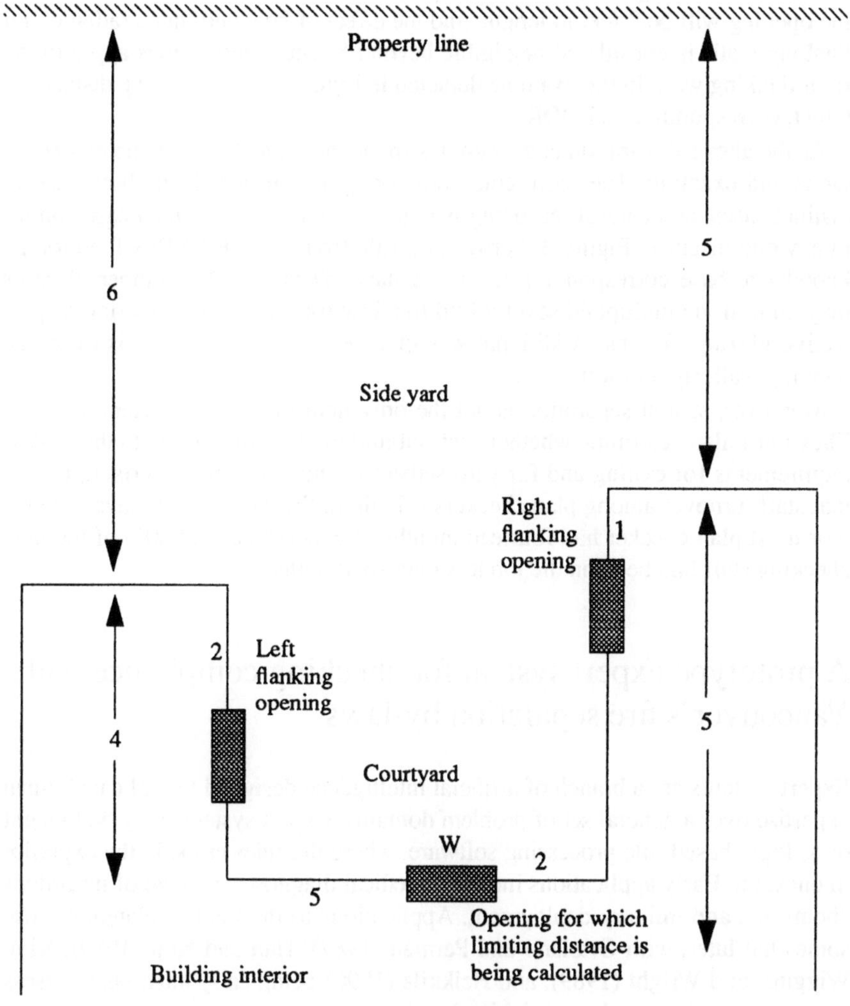

Figure 3.1. Courtyard with openings on each flanking wall.

Several exceptions apply. First, if a line drawn outward at a 45 degree angle from opening W does not intersect the left flanking wall, then neither FODL nor PL will be the limiting distance. The reasoning behind this rule is that flame radiates from an opening at a 45 degree angle, and so if the conditions described above are met, the flanking wall is not in danger from this opening. Another consequence of the 45 degree angle radiation is the "5 foot 8 inch rule," which states that neither FODL nor PL will be the limiting distance if the opening is more than 5 feet 8 inches from the left flanking wall. If the opening is just 5 feet 8 inches from the flanking wall, then a 45 degree line drawn to the flanking wall from the opening will be 8 feet in length, and the effect of heat and flame radiation on flanking walls is considered negligible beyond 8 feet. Similar rules apply to the right flanking wall. In the example depicted in Figure 3.1, the limiting distance is 6 feet, corresponding to FODR.

As the above illustration and examples show, the plan checker's job can be tedious and exacting. The basic courtyard configuration described above has 57 distinct cases, and careful reasoning is required to determine which case applies in any one instance. Figure 3.2 provides a rule from the VP-EXPERT prototype knowledge base corresponding to one of these 57 cases. The number of cases may, in turn, be multiplied several fold to allow for various sprinklering or protective glazing options. Additional sets of cases arise where there is only one flanking wall, right or left.

Moreover, spatial separation is not the only item that plan checkers look for. They must also determine whether each submitted plan complies with the by-law requirements for exiting and for yard setbacks. Thus, it is not surprising to find that staff turnover among plan checkers is high. In the case of Vancouver, the senior-most plan checker had fourteen months of experience, and 80% of the plan checking staff had been on the job less than six months.

A prototype expert system for checking compliance with Vancouver's fire separation by-laws

Expert systems are a branch of artificial intelligence designed to replicate human expertise over a general set of problem domains. Expert systems may be thought of as logic-based rule processing software, where the rules embody the expertise in question. Early applications included medical diagnoses, storage of hazardous chemicals, and military deployment. Applications to the field of planning arose somewhat later, with Ortolano and Perman (1987), Han and Kim (1989), Kim, Wiggins, and Wright (1989), and Heikkila (1990) being early attempts to assess expert systems' general potential in planning.

An expert system consists of three basic components: a knowledge base, inference engine, and user interface. What makes expert systems distinct from conventional programming is that the inference engine (the "thinking machine") can be disembodied from the rules which it processes (Ortolano and Perman, 1987). This permits development of expert system shells containing the inference engine and user interface, and indeed commercial development of these shells is a flourishing subindustry. Given an expert system shell, the task is to translate expert knowledge for the problem domain in question into precise rules that conform to logical and syntactical requirements of the shell.

For this research, we built a prototype expert system to check compliance with Vancouver's fire separation by-laws as they apply to single family residential properties within the city (those zoned RS-1). The prototype was developed over

```
RULE    COURTYD_WLEFT_PROJ_RIGHT_DOMINATES

  IF    single_plane = no
 AND    on_facing_wall = yes
 AND    length_of_flanking_wall_left > 0
        and length_of_flanking_wall_right > 0
        and opening_in_flanking_wall_left = yes
        and opening_in_flanking_wall_right = no
 AND    projection_right <> unknown
 AND    projection_left >= (projection_right)
 AND    flanking_opening_dist_left >= (projection_right)
THEN    limiting_dist = (projection_right);

!FPM-14
```

Figure 3.2 Sample code for one rule.

a three month period using the software shell VP-EXPERT and contains approximately 150 rules drawn primarily from sections 9.10.4 of the City of Vancouver Building By-law No. 6134 and from the Permit and License Department's Fire Protection Manual. No attempt was made to link the prototype to the City's other databases, although there is clearly potential for doing so. The prototype examines fire separation only, and does not address exiting requirements or yard setbacks. City of Vancouver staff from the Department of Permits and Licenses and from the Division of Computer Services were actively involved in the project, but the prototype itself was built by the authors alone. The prototype is now in use on an informal trial basis by plan checking staff.

There are several reasons for believing that plan checking is amenable to expert systems application. First, precedents elsewhere are encouraging. Hosking, et al. (1987) in New Zealand and Leary (1989) in England undertook early attempts to encode local government building or development by-laws, and both found that availability of a written code significantly reduces the effect required for development. While instances of researchers reporting that a formal written code was sufficient resource material for knowledge acquisition (Sergot, et al., 1986), it is more often the case that human experts are called upon as additional resources for interpretation of vague or incomplete by-laws. For the application reported here, the written by-laws and supplementary interpretive documents were indeed useful, but extensive and on-going consultations with plan checking staff were essential.

One clear advantage of the plan checking function over other planning applications is that relevant by-laws, while sometimes confusingly worded, are nonetheless definitive. That is, if the premises of the by-law are met, then the conclusion is valid with complete certainty. This avoids reliance on subjective "confidence factors" or other methods for dealing with decision making under uncertainty.

Additionally, the by-laws do provide a tangible focus for the transcription of facts and rules for the knowledge base.

Partial automation of the plan checking function suggests several other important benefits based on increased staff productivity. Accuracy, consistency, speed, and training are all likely to be enhanced with the aid of expert systems. On several occasions during the knowledge acquisition phase, we encountered instances where plan checking staff came to differing conclusions regarding an application. On other occasions a single plan checker would reach different conclusions on similar applications. Reduction or elimination of inconsistency between individuals or over time is one of the key benefits of a more automated plan checking operation. While the City of Vancouver did not undertake any formal tests to measure increased staff speed, accuracy, or consistency, the plan checking staff themselves attested vigorously to each of these benefits and are now using the prototype actively on a voluntary basis. The prototype is also being used as a training device for new staff, although it was not expressly designed as such.

Automation of the plan checking process also suggests linkages to other databases. Most local governments maintain large databases on real properties within their jurisdiction. These land-use databases may be part of a geographic information system, tax billing system, or general utility land-use information system. Data drawn from these databases may supplement the expert system consultation. For example, calculation of front and back yard setbacks may require data on adjacent parcels. Similarly, information on building footprints obtained via an expert system plan checking consultation may be used to gradually add detail to the agency's larger parcel files.

An additional benefit from a partially automated approach to plan checking is enhanced documentation. Currently, the calculations undertaken by plan checkers are scratched onto the plan itself in a highly unstructured format. If the builder should choose to appeal a rejection at a later date, the plan checking process is usually repeated in full for want of documentation. In contrast, most expert system shells feature detailed explanation capabilities. After a consultation is completed, the user can query the system on "how" a decision was arrived at, i.e., which rules were fired and in what order. Additionally, the user can search systematically through a list of variables to determine what their values are at any time during the consultation. Sophisticated expert systems also contain "what if" capabilities, allowing the user to ask how the conclusion might vary were certain variable values to differ. This capability is especially useful in plan checking contexts where the focus of the discussion shifts from screening to prescription. For example, upon having his plan rejected, a builder might seek to determine whether altering the size or location of a specific opening would result in a favorable decision. The entire consultation, together with related queries, can be kept in an ASCII, dBase, or spreadsheet file for further review at a later date. This is especially useful in the plan checking context where staff turnover is habitually a problem.

Expert systems have other advantages over conventional programming from a plan checking perspective. First, by-laws and other written regulations translate rather naturally into the rule syntax of expert systems. Moreover, one can rely on the expert system shell itself to undertake the logical processing of rules thereby permitting one to focus on construction of the rules themselves and various aspects of the user interface. Perhaps most importantly, however, expert systems are goal-oriented rather than process-oriented, and this is a decided advantage for plan checking applications. A conventional programming approach seeks to mimic step by step the process by which a decision is arrived at. In contrast, the logical sequence is endogenous in expert systems. The expert system is searching for an answer to a specific question ("Does this building permit application satisfy the fire separation requirements of the building by-law?"). The chain of reasoning it follows to determine a final answer depends entirely upon the intermediate answers it receives during the course of the consultation. For example, an entire set of questions may (or may not) be set off in response to a finding that the building is not sprinklered. It is this endogenous reasoning capability that qualifies expert systems as a branch of artificial intelligence.

Future prospects for expert plan checking systems

Our overall assessment of expert systems applications to the plan checking function is quite positive. Nonetheless, there are a number of issues, both technical and organizational, that should be addressed by planning agencies interested in following this avenue. The technical issues pertain primarily to the features one looks for in an expert system shell, while the general issues concern the manner in which the expert system fits into the organizational fabric of the plan checking unit.

Technical issues

Tello (1988) and Wiggins, et al. (1989) provide useful reviews of a range of expert system shells. Shells vary in terms of price, reasoning mechanisms, user interface and graphics capabilities, ability to link to external data files, methods for dealing with uncertainty and properties of inheritance. Prices of expert system shells vary from a few hundred to many thousands of dollars. These costs are quite small relative to plan checking budgets, so no one presumes that cost would not in itself be a major inhibiting factor. However, as the organization is learning more precisely what its requirements are, there is much to be said for purchasing a relatively inexpensive shell for prototyping purposes so that more seasoned judgement can then be applied to the selection of the shell to be used for full scale implementation.

Forward and backward chaining are the main types of reasoning processes found in expert systems. In either case, the basic rule structure is "if PREMISE, then CONCLUSION." Forward chaining begins with the known premises and attempts to infer conclusions implied by those facts. Backward chaining is more goal-oriented, beginning with the desired conclusion and working "backwards" to determine whether the corresponding premises are met. Both chaining methods may work through lengthy chains of rules before finally linking the initial premise with the final conclusion. Pedersen (1989) argues that forward chaining is more appropriate to situations where the objective is to flesh out all relevant facts associated with a given premise, and where the goal is not necessarily known. Backward chaining, in contrast, is more appropriate where the objective is to infer one key fact ("fire separation by-law requirements are met"). Forward chaining is typically more useful as an aid in configuration, planning, and interpretation while backward chaining is best for classification, selection, or diagnosis. For plan checking, backward chaining is clearly more appropriate for the limited objective pursued here, which is to determine whether or not a given plan conforms to the by-law requirements. If one wished to extend the expert plan checking system's scope to include suggested modifications to the plan, then elements of forward chaining might be usefully introduced.

A number of options exist for the use interface mechanism, with considerable potential for graphic interfaces. For this prototype, a simple query mode was adhered to. Most expert system shells come equipped with this capacity, and can be readily trained to prompt the plan checker with well-phrased questions eliciting information required during the course of a consultation. Another option is to make use of a spreadsheet template for inputting data. The data would then be accessed by the expert system directly from a spreadsheet file for each application. A data template input method has obvious advantages from a documentation perspective and may lead to even quicker processing speeds. One disadvantage of a data template is that much of the reasoning displayed by the system would then be less visible to the applicant and to the plan checker. The user interface can also be supplemented by a more graphic approach. The applicant might first select from a host of generic building types and then be prompted to provide information about each opening and flanking wall in a visual pattern that mirrors what the plan itself contains.

Some of the more sophisticated expert system shells feature an inheritance capability whereby objects of interest may "inherit" the characteristics of parent objects. For example, an exposing building face that resides in a courtyard may be a special instance of the parent, exposing-building-face. Then all rules that apply to exposing building faces in general would also apply to the specific case of an exposing building face found in a courtyard. The expert system shell used to develop the prototype reported upon here does not have an inheritance capability, but this did not appear to be a handicap for prototyping purposes. In contrast, Leary (1989) used software that allowed for one level of inheritance, and reported that it was a useful feature for his application.

Organizational issues

An expert plan checking system will only be useful if it conforms to and enhances existing practice. A critical factor is acceptance of the system by the plan checking staff, for it is their expertise being captured by the system. Any divergence by the system from established routine could lead to its abandonment by skeptical staff. An example of this tension between technical purity and contextual sensitivity arises in the use of backward chaining expert systems. Standard practice is to invoke as little procedural content as possible when using backward chaining. This way, the inference engine itself is unencumbered in its search for the desired conclusion.

In the plan checking context, we found that it was not always practical or desirable to avoid procedural programming. In a preliminary version of the prototype, staff would at times become confused by a sequence of questions that differed from the order they were accustomed to. A plan checker might ask, "Why is it asking me about the flanking window before asking me how far this opening is from the property line?" When the expert system is allowed to reason independently, there is little reason to believe the sequence of questions will correspond to established practice. As long as the rule base is complete and correct, this should not be a problem from the perspective of reaching the desired conclusion. However, it can affect staff's perception of the system and their willingness to use it. We found it useful, therefore, to delineate two distinct functions of the plan checking system: (1) to obtain pertinent data regarding the configuration of the proposed dwelling unit, and (2) to subject those facts to the logical processes of the expert system. The former function is more appropriately procedural, while the latter is best subject to pure backward chaining reasoning.

Another issue from an organizational context may be summarized by "do as I do, not as I say." It may happen on occasion that routine practice is based as much on an accepted interpretation of the by-laws as on the by-laws themselves. In such instances, it was our decision to base the prototype on existing practice to encourage its use. While this did not prove to be a problem in this particular application, we can envision situations where senior staff may not wish to formalize or document existing practice via the expert system for fear that controversy over interpretation of the by-laws may ensue.

Staff acceptance of an expert plan checking system will also depend upon how they perceive it will affect long term job security, where a favorable response will be engendered if they believe the system will enhance productivity and reduce tedium while in no way threatening to replace them as the plan checking staff. For the research reported on here, both the presence of a strong employee union and rapid turnover in plan checking staff mitigated overt concerns on this point. Moreover, the immediate gains in efficiency were welcomed by staff who are at times overburdened by the volume of plans to be checked.

The presence of a full scale expert plan checking system would have staff implications beyond the front line of plan checkers. As with any automated system, there is an ongoing need for maintenance and fine tuning. The by-laws themselves are hardly static, and existing practice evolves slowly but steadily in response to test cases that arise and whose outcomes are assimilated into future interpretations of the by-laws. It is essential that the system be able to evolve in a parallel fashion if it is to remain useful. Some in-house maintenance supplemented by outside expertise is probably in order, with the precise combination depending upon available resources. These considerations may also affect one's choice of software, with special attention on whether run-time only versions are available.

Finally, in designing an expert plan checking system one must consider the question, "Who is intended to run whom?" It would be a mistake to try and design a system that does it all. In addition to being practically infeasible, it is not desirable from a human standpoint. An expert system should be staff enhancing rather than staff replacing. To this end, the design focus should be on elimination of tedium and highlighting the logical path over which decisions are reached. The system can and should be designed to instruct as it infers, so that plan checkers become more rather than less actively involved in the process. Not only does this approach enhance quality of work and job satisfaction, it is also essential for enabling plan checkers to advise builders about their options in the event that the plan is rejected.

There is much to be said for first developing a prototype based on a relatively narrow set of by-laws, particularly where an expert system is new to the organization. A prototype, while smaller than a full scale system, can be quite wide ranging in its functionality. Numerous experiments can be undertaken with the coding of rules or with the user interface. As with most endeavors, it is only through active involvement that one comes to appreciate the subtle differences underlying seemingly innocuous choices about how to proceed. Even if, on the basis of having developed a prototype, one concludes that a different choice of shell is called for or that some other fairly radical changes are needed, there is often much that can be salvaged. For example, the prototype rules may be thought of as an additional source of knowledge to be encoded into the new system or a data input mechanism may still retain its usefulness. Or, as was the case in this research, the prototype can be developed to stand alone so that its usefulness for a specific purpose is not dependent upon future development decisions.

Conclusions

Based on the prototype developed here, it appears that the plan checking function of local governments is a natural application for expert systems. Plan checking is complex yet relatively unambiguous. The relevant by-laws are an excellent initial source of written code for knowledge acquisition, particularly when supplemented by staff training manuals. In addition to the fire separation by-laws there is clear potential for similar applications to exiting and yard setback requirements.

Notwithstanding the above, there are numerous issues to consider when developing an expert plan checking system. Technical issues include the type of reasoning mechanism to employ, user interface specifications, dialogue versus graphics, and inheritance capabilities. Organizational issues include procedural programming versus endogenous reasoning, determining the appropriate scope for applications, and developing design guidelines to ensure the system is staff enhancing. These issues complicate the decision to employ an expert plan checking system, and how to proceed once that decision has been made. These issues notwithstanding, the results of this research must be viewed as encouraging.

References

Goodall, A., 1985. "The Guide to Expert Systems," *Learned Information*, Oxford, UK.

Han, S.Y. and T.J. Kim, 1989. "Can expert systems help with planning?", *Journal of the American Planning Association 55*(3):296-307.

Heikkila, E.J., 1990. "Modeling fiscal impacts using expert GIS: Theory and strategy," *Computers, Environment, and Urban Systems 14*(1):25-33.

Hosking, J.G., W.B. Mugridge, and M. Buis, 1987. "Firecode: a case study in the application of expert systems techniques to a design code," *Environment and Planning B*(14):267-280.

Kim, T.J., L.L. Wiggins, and J.R. Wright, eds., 1989. *Expert Systems: Applications to Urban Planning*, Springer-Verlag, New York.

Leary, M.E., 1989. "Expertise and expert systems in British land use planning: analysis and progress towards a prototype," paper presented to the *First International Conference on Expert Systems in Environmental Planning and Engineering*, Cambridge, MA.

Ortolano, L. and C.D. Perman, 1987. "A planner's introduction to expert systems," *Journal of the American Planning Association 53*(1):98-103.

Pedersen, K., 1989. *Expert Systems Programming: Practical Techniques for Rule-Based Systems*, Wiley, New York.

Sergot, M.J., et al., 1986. "The British Nationality Act as a logic program," *Communications of the ACM 29*:270-386.

Silverman, B.G., 1987. "Should a manager hire an expert system?", *Expert Systems for Business*, Addison-Wesley, Reading, MA.

Tello, E.R., 1988. *Mastering AI Tools and Techniques*, Howard W. Sams & Co., Indianapolis, IN.

Waterman, D., 1986. *A Guide to Expert Systems*, Addison-Wesley, Reading, MA.

Wiggins, L.L., et al., 1989. "Expert systems technology," in T.J. Kim, L.L. Wiggins, and J.R. Wright, eds., *Expert Systems: Applications to Urban Planning*, Springer-Verlag, New York.

4

Machine Learning, Expert Systems, and an Integer Programming Model: Application to Facility Management and Planning[1]

Tschangho John Kim, Sang-Yun Han, and Ilker Adiguzel

Introduction

Decision makers in both the public and private sectors often face the difficult problem of effectively managing facilities that range from utility plants to small offices. The task of facility management not only requires quantitative reasoning but also a great deal of qualitative reasoning. In developing the best possible facility management plan, one should have easy access to a great amount of facility inventory data and knowledge of experts on various domains, including diagnosis of facility condition and analytical mathematical models, to ensure efficient use of resources.

Despite the fact that separate approaches can provide only partial solutions to the problem, there have been few attempts to devise an integrated computer tool in the area of facility planning management. Recent advances in artificial intelligence (AI), particularly expert systems (ES), can provide a means of narrowing the gap between what traditional problem-solving methods can do and what decision makers want (Han and Kim, 1989; Kim, Wiggins, and Wright, 1989).

In order to shed light on how traditional programming models and expert systems can be combined, this chapter has two objectives:

---

1. Reprinted with permission from *Computers, Environment and Urban Systems*, Vol. 15:189-201, S.-Y. Han, T.J. Kim, and I. Adiguzel, "Integration of Programming Models and Expert Systems: An Application to Facility Management and Planning," 1991, Pergamon Press plc.

1. To review how an operations research (OR) approach to problem solving is different from an AI approach, and
2. To discuss how to develop an integrated computer tool which effectively aids decision-making in facility planning management.

For the second objective, this chapter presents the design and implementation plan of a knowledge-based decision support system XPlanner. XPlanner combines ES with a zero-one integer optimization model and a database system to create a comprehensive decision aid for the management and planning of military facilities.

Operations research and artificial intelligence

The field of operations research (OR) is large and includes many subareas. The problems that have been solved by OR include queueing, inventory, allocation, routing, scheduling, search, replacement, and competition (Wilson, 1985). AI, on the other hand, has focused on such areas as representation and use of human knowledge through logic, learning, and understanding of natural language and perception (vision and touch). Simon (1987) compares OR with AI by defining OR as "the application of optimization techniques to the solution of complex problems that can be expressed in real numbers" while defining AI as "the application of methods of heuristic search to the solution of complex problems that defy the mathematics of optimization, contain nonquantifiable components, and involve a large knowledge base."

Both ES and OR models are designed to help decision makers. But while ES normally incorporates qualitative knowledge, OR primarily uses quantitative knowledge. Table 4.1 briefly compares the different solution approaches of linear programming (LP) models and ES. The linear programming model, for instance, involves arithmetic computations to solve a goal function:

$$Minimize\ \Sigma_i a_i X_i$$

$$subject\ to\ \Sigma_i \beta_i X_i \geq A.$$

Using a simplex algorithm, it finds the optimum values for the objective function (goal).

On the other hand, ES involve symbolic processing in representing knowledge and finding a value of goal parameter. For a simple example of evaluating land development proposal, consider this knowledge base written in Prolog syntax:

```
Goal acceptability (X, development-proposal)
  Rule 1 zoning-change (needed, X) :- landuse (incompatible, X)
  Rule 2 acceptability (reject, X) :- zoning-change (needed, X)
Fact landuse (incompatible, development-proposal)
```

In finding the value of the goal (e.g., acceptability of the land development proposal), ES may use backward chaining (deduction) to find out if the proposal requires a zoning change[2]. This is a simple example of the knowledge base and inference engine. Actual problems normally consist of hundreds of rules and a long inference process.

As briefly reviewed here, OR and AI (including ES) have developed different solution techniques according to the type of knowledge they possess. The important question here is how and when OR and AI could be combined to take advantage of both their strengths. The rationale for the integration of these two fields is that most decision-making problems, particularly in urban planning, require both quantitative and qualitative knowledge and reasoning. The facility management problem, for instance, may be quickly and optimally solved by OR techniques once the nature of the problem is mathematically formulated. But the process of formally defining the problem requires significant expertise in many subject areas of facility management. In addition, implementing the solution provided by the OR models may be better handled by ES through its qualitative reasoning process. The benefits of coupling ES and OR in this regard will be evident as the design concepts of XPlanner are discussed later.

Table 4.1. Different Approaches of Linear Programming Models and Expert Systems

	LP models	Expert systems
Knowledge	Constraint equations and objective functions	Knowledge base: rules and frames and goals
Solution method	Simplex algorithm	Inference engine: backward and forward chaining

Expert systems and decision support systems

Although some authors see expert systems as special instances of decision support systems (DSS), many authors distinguish ES from DSS because of the former's unique structure, technology, and ability to solve different types of problems (see, for instance, Kroeber and Watson, 1987, and Liang, 1988). Table 4.2 sum-

2. In this example, the backward chaining process first fires Rule 2 because this rule contains the goal parameter *acceptability*. Then it checks if the condition part of Rule 2 is true, i.e., whether a zoning change is needed or not. This fires Rule 1 because this rule contains the parameter *zoning-change* and checks if the condition part of Rule 1 is true, i.e., if the land use is compatible. Upon finding that the land use proposed by the development proposal is incompatible, it sets the value of the parameter *zoning-change* to *needed* and then sets the value of the goal parameter *acceptability* to *reject*.

marizes major differences between ES and DSS in terms of types of computer processing involved and types of decision problems solved.

In DSS, the computer stores data and various decision models and the user interacts with the computer through the user-interface, providing the computer with judgments. In ES, the computer stores all expert knowledge, including judgments, into a program and recommends a solution by using appropriate analyses and its own programmed logic.

Table 4.2. Taxonomy of Decision Problems

Type of knowledge needed for problem solving	Nature of problems	
	Structured or routine	Unstructured or semistructured
Quantitative reasoning: numeric computation	Data processing systems	Decision support systems
Qualitative reasoning: intuition and rules of thumb	Expert systems	Human experts

In discussing the possible relationship between OR, DSS, and ES, Wynne (1984) and Tingley (1987) describe these techniques as a sequence of means for providing assistance to decision makers (Figure 4.1). OR models are developed into DSS to incorporate human judgment into the decision-making process through an effectively developed man-machine interface. DSS, in turn, can be further developed into ES, in which human judgment is coded into a knowledge base. The feasibility and desirability of these progressions from OR to DSS and from DSS to ES are, of course, dependent on the nature of decision problems.

The idea of integrating ES into DSS to create powerful computer-based systems has garnered much attention and created new terminology: "expert decision support system" (EDSS), "knowledge-based decision support system," and "intelligent decision support system." The possible contributions of ES to EDSS include (Turban and Watkins, 1986):

1. Helping users in selecting models
2. Providing judgmental elements in models
3. Simplifying building simulation models
4. Enabling friendlier interface
5. Providing explanation capability.

Of these, the most important role of ES in EDSS is in model selection and building. As Strauch (1974) points out, the process of problem analysis (or mod-

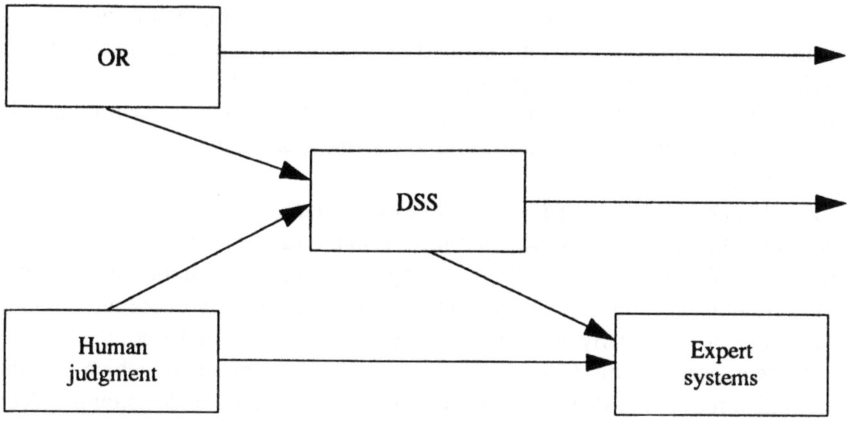

Figure 4.1. Progression of operations research to decision support systems and expert systems (Tingley, 1987).

eling) usually involves three interrelated components: formulation of the formal problem, mathematical analysis, and interpretation of the results. While the mathematical analysis is handled effectively with DSS through its embedded OR models, the formulation requires the subjective knowledge of the user. Further, the interpretation requires the personal judgment of the decision makers. The coupling of ES and DSS in this case is based on the assumption that subjective knowledge and personal judgment can be better made by experts than by decision makers and users of the system.

The sections that follow discuss the design concepts and implementation plan of XPlanner. XPlanner is an example of adding a rule-based system to the optimization model for facility management in order to aid users in the various tasks involved, such as formulating facility optimization models and diagnosing structural conditions of facilities. It is developed to stimulate planners to employ mathematical models more frequently and easily in their problem-solving processes.

Decision problems supported by XPlanner

Most decision problems may be broadly categorized into: (1) structured and (2) unstructured or semistructured. XPlanner is developed based on the notion that there are different types of computer-based information systems, each with its own unique ability to solve decision problems. Data processing systems are suited for structured problems that have standard operational procedures, decision rules, and clear output formats, such as identifying low income districts or determining the median income of a city. Decision support systems (DSS), on the other hand, are intended for unstructured or semistructured problems, such as evaluating land development proposals. DSS can be used to perform "what-if" type analyses estimating fiscal and other impacts of land development proposals based on different sets of variables and to provide quantitative support to the decision maker.

Kroeber and Watson (1987) define DSS as "an interactive system that provides the user with easy access to decision models and data in order to support semistructured and unstructured decision-making tasks." As the definition implies, the interaction between the decision maker and the system is very important in DSS. The interaction is usually achieved in the form of "what-if" dialogue.

While expert systems (ES) are very good at solving problems which require qualitative reasoning, they have the strict requirement that the decision problem be structured so that experts' knowledge on solving the problem can be captured in a computer program. As the characteristics of each computer system imply, only the integration of different types of computer-based systems can produce effective decision-making aids because most decision making in urban planning deals with a mixture of structured and unstructured problems.

The task of facility planning management has both structured and unstructured decision-making components. Cities in urban areas and military installations maintain a variety of facilities from utility plants to housing units and the decision tasks involved in facility management range from the task of determining the current physical and functional conditions of facilities to that of deciding which facilities to close or build. While the former example is a somewhat structured decision problem, the latter example can be regarded as an unstructured or semistructured problem.

XPlanner is targeted for both areas of decision making. In military installations, for which XPlanner was developed, the decision problems in facility planning management can be categorized as follows:

1. Project possible changes in planning constraints such as mission types and budget levels.
2. Estimate facility requirements based on the missions of installation.
3. Determine the current physical and functional conditions of facilities.
4. Estimate facility deficiencies based on 2 and 3 above.
5. To handle the facility deficiency or surplus, create a facility management plan for an effective and efficient utilization of existing and future resources.

Only an effectively integrated computer system can support all the decision-making tasks listed above. The task of projecting mission and budget changes can be handled by an effective man-machine interface of DSS, which incorporates the judgments of the decision maker and enables a series of "what-if" dialogue. The "what-if" interaction is critical in developing XPlanner, because decision-makers want to incorporate their judgment on the future of the Army (e.g., mission and budget changes) into the problem solution process of XPlanner. For example, when the user inputs different types of missions, the system produces new facility management plans.

The task of determining current *physical* conditions of facilities can be handled by a database management system that maintains comprehensive facility inventory data, whereas the task of determining *functional* conditions of facilities may better be handled by an expert system. Examining the functional conditions of facilities involves such heuristic tasks as determining whether a particular facility is suitable for accommodating a certain type of activity and assigning each activity (e.g., training or recreation) to a facility. Further, the task of developing a facility management plan can be supported by an optimization model in order to achieve efficient allocation of resources.

In short, the major decision problem involved in military facility planning management is developing an effective facility management plan that supports constantly changing missions. The facility management plan prepared by the Army planners should effectively utilize the current and future resources of the military installation and should also comply with the safety and welfare standards prescribed by the Army authorities. XPlanner is designed to support these tasks.

Structure and components of XPlanner

To effectively support all the tasks involved in facility planning management, XPlanner consists of several components (knowledge base, model base, database, user interface, and inference engine) as illustrated in Figure 4.2. The inference engine and knowledge base control the whole system.

The role of the knowledge base

The knowledge base of XPlanner contains two types of knowledge: (1) knowledge about the classification of functional areas of facilities and about the diagnosis of the physical conditions of facilities, and (2) knowledge of the formulation and interpretation of a zero-one integer optimization model for facility planning.

For the examples of the first type of knowledge, consider the following rules used in XPlanner for diagnosing physical problems of structure:

1. IF structure = wood-frame AND humidity-level >= 70, THEN termite-infestation may be high,
2. IF termite-infestation = high, THEN building-decay = serious AND wood-treatment by chemical A = required,
3. IF sagging-roof OR non-vertical-walls OR tilted-floors OR misalignment-of-doors, THEN excessive-settling,
4. IF excessive-settling, THEN foundation-wall = to-be-replaced.

These rules can be used to effectively estimate the cost and feasibility of each type of project (e.g., conversion and renovation) related to each facility. Based on this type of knowledge, the major function of the knowledge base is to estimate

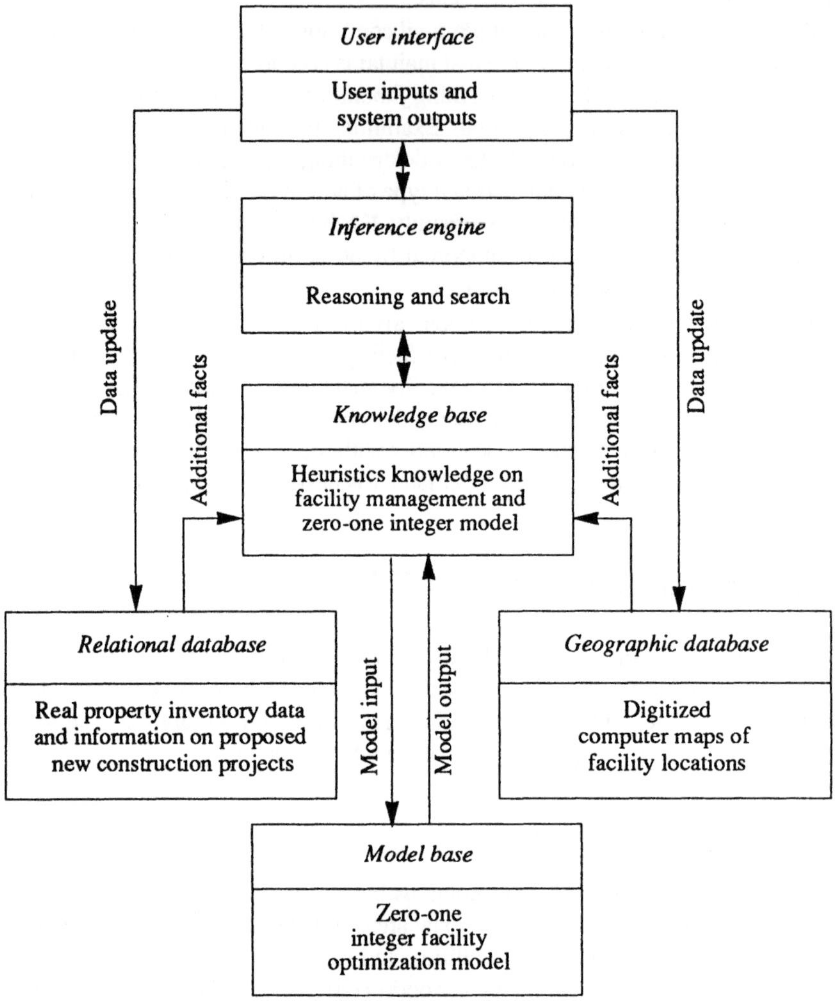

Figure 4.2. Schematic structure and components of XPlanner.

parameter values necessary to formulate the zero-one integer optimization model. The list of parameters to be estimated by ES (the knowledge base component) is described later under the model base.

In addition to estimating parameter values for the facility optimization model, ES screens existing facilities in order to eliminate the facilities that are irrelevant for consideration by the optimization model. For this purpose, ES maintains the knowledge base that classifies existing facilities into several condition categories. Machine learning programs with effective induction algorithms can be used to develop the knowledge base in this area. Many induction algorithms have been successfully applied to the knowledge acquisition problems, particularly to classification problems (Michalski, et al. (1985); Forsyth and Rada (1986); Shaw (1988); Liang (1989)). Because ES classifies and eliminates the irrelevant facilities, the facility optimization problem may be reduced to a size manageable by the zero-one integer optimization model (described later).

Another important role of the knowledge base is to interpret the results of the optimization model using its knowledge of a zero-one integer model. In this system, ES serves as an extra layer between the model and the user, translating qualitative criteria into the numeric input and also translating the model's numeric output to qualitative concepts that are more intuitive and informative to the user.

In addition, the knowledge base has the important function of controlling the whole system, accessing the database and the model base of XPlanner as necessary. In addition to the domain knowledge specific to the facility management, the knowledge base maintains control knowledge (meta-knowledge) to control the steps involved in reasoning. For instance, it initiates the forward chaining mode in the middle of backward chaining[3] or it forces some rules to be fired before other rules by using meta-rules. This type of control knowledge is necessary to control the flow among the knowledge base, the model base, and the database.

Knowledge acquisition through machine learning

As previously explained, the knowledge base for facility classification is obtained through a machine learning program which induces decision rules from sample cases. Military real properties are classified into several condition categories for management purposes. Assigning condition codes is very much a judgmental decision based on the rules-of-thumb about architectural and engineering knowledge. Because military facility managers could not clearly articulate their decision rules, a machine learning approach was adopted to induce their decision rules from sample cases. The condition codes used for XPlanner include:

1. *Adequate*: Generally meets criteria.
2. *Modest*: Upgrading required and practical.
3. *Substandard*: A facility that cannot practically be raised to meet the adequate level for housing the function for which it is currently designed.
4. *Unusable*: A facility which is no longer tenable for any purpose without having to upgrade it extensively.
5. *Disposable*: Disposal required.

To develop the knowledge base needed to classify facilities into these categories, a field expert in the real property management office of a military installation was asked to identify attributes that he normally considered for determining

3. In the ES shell Personal Consultant Plus, this can be accomplished by assigning ANTECEDENT property to necessary rules.

condition codes. The list of the decision factors identified at this stage includes:

CF—Current use of facility
1) administrative uses; 2) barracks; 3) warehouse

YR—Year of construction

CT—Type of construction
1) permanent facility; 2) semi-permanent facility; 3) temporary facility

MT—Major building materials used
1) concrete; 2) brick; 3) masonry; 4) wood

FS—Fire safety
1) fire protected; 2) fire resistive; 3) fire unprotected

HW—Availability of hot water
1) available; 2) not available

AC—Availability of air conditioning system
1) available; 2) not available

HT—Availability of heating system
1) available; 2) not available

CR—Presence of cracks on wall
1) none; 2) modest; 3) severe

WD—Signs of water damage to structural members or wall
1) none; 2) modest; 3) severe

ST—Signs of unusual settling in the foundation wall
1) none; 2) modest; 3) severe

SR—Signs of sagging roof
1) none; 2) modest; 3) severe

MD—Presence of misalignment of doors and windows
1) none; 2) modest; 3) severe

CI—Condition of interior finishes (ceiling, wall paint, etc.)
1) good (requires no repair); 2) average (requires minor repair or maintenance); 3) poor (requires major repair or replacement)

CE—Condition of exterior finishes (siding, gutter, etc.)
1) good; 2) average; 3) poor

CU—Condition of utility lines
1) good; 2) average; 3) poor

Fifty training cases were collected for the decision attributes selected (Table 4.3). The next step was to apply a machine learning program to produce generalization rules based on these examples. The ID-3 (Interactive Dichotomizer-3) algorithm developed by Quinlan (1979) was used for this purpose. ID-3 works by first finding the attribute which is most discriminatory and then by dividing the cases with respect to that attribute[4].

ID-3 generates a decision tree that classifies sample cases. The decision tree in Appendix A shows 21 branches, each with a condition code at the end of the

4. Quinlan used an information-theoretic measure of entropy for assessing the discriminatory power of each attribute. The entropy is calculated using the formula $-\Sigma_{i=1}^{n} P(X_i) \log_2 P(X_i)$ where $P(X_i)$ is the probability of event i.

Table 4.3. Training Cases Used to Induce Rules for Classifying Facility Condition Codes

CF	YR	CT	MT	FS	H	HT	AC	CR	W	ST	SR	M	CI	CE	CU	CD
1	40	3	4	3	2	1	2	3	3	1	3	2	2	2	1	5
1	40	3	4	3	2	1	2	2	2	1	2	1	2	2	2	4
1	40	3	4	3	2	1	2	1	2	1	1	1	2	2	2	4
1	60	2	4	2	1	1	1	2	2	3	2	3	2	3	2	5
1	60	2	4	2	1	1	1	2	2	1	1	2	2	2	1	3
1	70	2	2	2	1	1	1	2	3	1	1	1	2	1	2	2
1	70	2	2	2	1	1	1	2	1	1	1	1	2	2	1	2
1	75	1	3	1	1	1	1	1	1	1	1	1	2	1	1	2
1	75	1	3	1	1	1	1	1	1	1	1	1	1	1	1	1
1	75	1	3	1	1	1	1	1	1	1	1	1	2	2	2	2
1	79	1	3	1	1	1	1	1	1	1	1	2	1	1	1	1
1	79	1	3	1	1	1	1	1	1	1	1	1	2	2	2	2
1	80	1	2	1	1	1	1	1	1	1	1	1	1	1	1	1
1	82	1	2	1	1	1	1	1	1	1	1	2	1	1	1	1
2	40	3	4	3	1	1	2	2	3	2	2	2	3	3	2	5
2	40	3	4	3	1	1	2	2	3	2	2	1	3	2	2	5
2	40	3	4	3	1	1	2	1	3	1	2	1	2	2	2	4
2	40	3	4	2	1	1	2	1	2	1	1	2	2	1	1	3
2	40	3	4	2	1	1	2	1	2	1	1	2	1	2	2	3
2	40	3	4	2	1	1	2	1	2	3	3	1	2	2	2	5
2	55	2	2	2	1	1	1	1	1	2	2	1	2	3	1	3
2	55	2	2	2	1	1	1	1	1	2	2	1	2	2	2	3
2	60	1	2	2	1	1	1	2	2	1	1	1	2	2	1	2
2	60	1	2	2	1	1	1	2	2	1	1	1	2	2	2	2
2	60	1	2	2	1	1	1	1	1	1	1	2	2	2	1	1
2	70	1	2	2	1	1	1	1	1	1	1	1	2	2	2	3
2	70	1	2	2	1	1	1	1	1	1	1	2	1	1	1	1
2	70	1	2	2	1	1	1	1	1	1	2	1	2	1	1	2
2	70	1	2	2	1	1	1	1	1	1	1	2	1	2	2	2
2	70	1	2	2	1	1	1	1	1	1	1	1	2	2	1	2
2	79	1	2	2	1	1	1	1	1	1	1	1	1	1	1	1
2	79	1	2	2	1	1	1	1	1	1	1	1	2	1	1	1
2	82	1	1	1	1	1	1	1	1	1	1	1	2	1	1	1
2	82	1	1	1	1	1	1	1	1	1	1	2	1	1	1	1
3	40	3	4	2	2	2	2	3	2	2	1	3	3	3	3	3
3	39	3	4	2	2	2	2	2	2	3	2	3	2	1	1	4
3	39	3	4	2	2	2	2	2	3	3	2	2	2	2	1	4
3	40	3	4	3	2	2	2	2	3	2	2	2	3	3	2	3
3	50	3	4	2	2	2	2	2	2	2	3	2	3	3	2	3
3	50	3	4	2	2	2	2	2	2	2	1	2	3	3	2	3
3	50	3	4	2	2	2	2	2	1	2	1	1	3	3	2	2
3	60	2	4	2	2	1	2	1	1	2	1	2	2	2	2	2

Note: See previous page for the legend and description of the codes used in this table.

branch. These branches can be easily translated into production rules. For instance, the first two branches in Appendix A are written in rules that can be used by the ES shell GURU (DSS generator used for XPlanner) as follows:

1. IF (construction-year < 1945) AND (signs-of-sagging-roof = none) AND ((current-use = administrative) OR (current-use = barracks)) THEN (condition-code = substandard)
2. IF (construction-year < 1945) AND (signs-of-sagging-roof = none) AND (current-use = warehouse) THEN (condition-code = modest)

The role of the model base

The model base of Xplanner contains a zero-one integer optimization model that is designed to devise a facility management plan with efficient allocation of resources. Integer programming (IP) is a special case of linear programming with the characteristic that the values for all the variables in the solution must be integers. A further sub-class of IP is zero-one programming, in which all variables in the solution have a value of zero or one. Such a formulation is needed when the decision variables represent a binary decision. An example of the set of binary decisions involved in XPlanner are *renovate* or *don't renovate*. IP problems are computationally much more difficult to solve than linear programming problems (Greenberg, 1971).

All the parameter values contained in this model are supplied by the knowledge base. In return, the model base supplies model output to the knowledge base for the interpretation (see Figure 4.2). The structure of the model is described here without further discussion on zero-one integer programming.

Objective function

The objective function is to minimize troop stationing facility costs, which include new construction, renovation, conversion, and maintenance costs. To allow proper comparison of alternatives in equivalent dollars, present and future costs for each alternative are brought to a common point in time by being converted to an annual series of payments. The objective function is formulated as below:

$$\text{Minimize} \sum_i [\mu_i^S S_i] + \sum_i [(a_i + \mu_i^R) R_i] + \sum_i \sum_j [(\beta_{ij} + \mu_{ij}^C) C_{ij}] + \sum_l [(\gamma_l + \mu_l^N) N_l] \tag{4.1}$$

where:

$$S_i = \begin{cases} 1 \text{ if facility } i \text{ is to stay as it is} \\ 0 \text{ otherwise} \end{cases}$$

$$R_i = \begin{cases} 1 \text{ if facility } i \text{ is to be renovated} \\ 0 \text{ otherwise} \end{cases}$$

$$C_{ij} = \begin{cases} 1 \text{ if new construction project } i \text{ is to be converted to } l \\ 0 \text{ otherwise} \end{cases}$$

$$N_l = \begin{cases} 1 \text{ if new construction project } l \text{ is to be approved} \\ 0 \text{ otherwise} \end{cases}$$

$a_i =$ annualized initial renovation costs of facility i

$$= a_i \left[\frac{r(1+r)^{n_i}}{(1+r)^{n_i} - 1} \right], \text{ where}$$

a_i = initial renovation costs of facility i
r = discount rate
n_i = economic life of facility i

β_{ij} = annualized initial costs of converting facility i to facility j

γ_l = annualized initial costs of new construction project l

μ_i^S = annualized present value of total lifetime maintenance costs (recurring and non-recurring) of facility i when facility i stays as it is. Non-recurring cost is first converted to its present value before being annualized.

μ_i^R = annualized present value of total lifetime maintenance costs of facility i when facility i is renovated

μ_{ij}^C = annualized present value of total lifetime maintenance costs of facility i when facility i is converted to j,

μ_l^N = annualized present value of total lifetime maintenance costs of new facility to be constructed by project l.

As shown above, the cost figures used in the objective function represent life cycle cost which is the sum of the initial, recurring, and non-recurring costs expressed in equivalent dollars. The values of the parameter a_i through μ_l^N are endogenously determined by the knowledge base of XPlanner. When there are still facility deficiencies in the optimization model results, ES considers new construction (not already proposed) by checking the suitability of available space.

Choice constraints

The constraint shown below is used to ensure that only one option (e.g., renovate, convert, or use as is) is selected for each existing facility:

$$S_i + R_i + C_{ij} = 1 \quad \forall i, j \tag{4.2}$$

Budget constraints

The facility optimization model has the following budget constraints to ensure that the facility management plan is devised within given budgets:

$$\sum_i a_i R_i + \sum_{ij} b_{ij} C_{ij} + \sum_l g_l N_l \leq A \tag{4.3}$$

$$\sum_i (\mu_i^S S_i + \mu_i^R R_i) + \sum_i \sum_j \mu_{ij}^C C_{ij} + \sum_l \mu_l^N N_l \leq B \tag{4.4}$$

where:

- a_i = total initial costs of renovating facility i,
- b_{ij} = total initial costs of converting facility i to j,
- g_l = total initial costs of new construction project l,
- A = total (one time) budget available for renovation, conversion, and new construction projects, and
- B = annual budget available for minimal maintenance of facilities.

The knowledge base of XPlanner estimates the costs a_i, b_{ij}, g_l, and the users supply the value of A and B.

Mission constraints

The mission constraints are used to ensure that the solution by the model satisfies facility requirements generated by the mission. Through the user-interface, XPlanner aids users in developing scenarios regarding possible mission changes. The knowledge base maintains the rules that interpret scenarios developed by the user into the mission constraints (i.e., facility requirements). The mission constraints are formulated as below:

$$\sum_i (\delta_{ikp}^S S_i + \delta_{ikp}^R R_i) + \sum_i \sum_j \delta_{ijkp}^C C_{ij} + \sum_l \delta_{ikp}^N N_l \geq T_{kp} \quad \forall k, p \tag{4.5}$$

where:

- δ_{ikp}^S = facility i's capacity of accommodating troop type k in activity p when facility i stays as it is,
- δ_{ikp}^R = facility i's capacity of accommodating troop type k in activity p after renovation,
- δ_{ijkp}^R = facility i's capacity of accommodating troop type k in activity p after conversion to j,
- δ_{lkp}^N = new construction project l's capacity of accommodating troop type k in activity p, and
- T_{kp} = total number of troops of type k in activity p as given by new mission.

All the parameters in this model, (e.g., a_i through δ_{lkp}^N) are estimated by the knowledge base. An example of the parameter supplied by the knowledge base is the capacity of facility i (i = 1, ..., n) to accommodate troop type k (k = 1, ..., enlisted woman, ..., m) for activity p (p = 1, ..., recreational use, ..., l) after the facility is renovated. In addition, the values of the right-hand side constants (A, B, and T_{kp}) are also supplied by the knowledge base.

The task of the ES component of XPlanner to configure the zero-one integer optimization model requires much computer power and speed. Even with eight facilities, the number of decision variables in the objective function is forty, which is the maximum that can be solved by the integer program package called Lindo. The number of parameters to be estimated by the knowledge base is 472, with eight facilities, three conversion types, three activity areas (training, recreation, and housing), and four troop types (enlisted man, enlisted woman, non-commissioned officers, and officers).

The role of the database

As an additional component, XPlanner maintains a relational database to store and access factual data necessary in facility management and planning. The database basically maintains four types of data:

1. Real property inventory data which describe existing conditions of facilities, such as area, current use, age, type of building materials used, and structural conditions of each facility.
2. Detailed information on the new construction projects which have been proposed for the installation, which includes completion year, location, type, and expected capacity of facilities.
3. Data on construction and repair materials, including material types, costs, and durability.
4. Information on the space allowance standards compiled from the Army regulations and manuals.

In addition, XPlanner maintains a geographic database in which digitized computer maps are stored to provide spatial data such as location of roads and utilities (Figure 4.2). This information is used by the ES module of XPlanner in recommending candidate sites for new construction.

The knowledge base accesses the database to obtain additional facts. In XPlanner, the data flows from the database to the knowledge base and finally to the model base. The software dBase III Plus or Lotus 1-2-3 may be used as a database manager for XPlanner. The geographic database is developed by using AutoCad and is utilized through a graphic utility program of the ES shell, Personal Consultant Plus.

Workings of XPlanner

The workings of XPlanner involve frequent interactions among the user, the knowledge base, the model base, and the database. The user provides XPlanner with additional judgmental factors not encoded in the knowledge base and the database supplies additional facts to the knowledge base. The knowledge base generates inputs for the model base and the model base, in return, supplies model outputs to the knowledge base (see Figure 4.2). As depicted in the flowchart of XPlanner in Figures 4.3 and 4.4, the steps involved in the consultation of XPlanner can be summarized as follows:

1. As the consultation with XPlanner begins, XPlanner first helps users develop their own scenarios regarding the future of the Army. They can use their own judgment in deciding possible mission changes, budget levels, and some demographic changes (e.g., participation rate of the female labor force).
2. XPlanner's ES estimates facility requirements generated by the new mission, using its knowledge about space allocation standards set by Army regulations and manuals.
3. The database management system (DBMS) of XPlanner supplies information on the proposed construction projects, and the users further describe and modify the nature of the projects if they so choose.
4. XPlanner accesses the database (real property inventory data) to check current physical conditions of facilities.
5. Based on the physical conditions, the ES module classifies existing facilities into several condition categories ranging from Class 1, usable (meeting all criteria) to Class 6, disposable (no longer tenable for any purpose).
6. The ES eliminates facilities that are irrelevant for consideration by the facility optimization model[5].
7. ES estimates troop accommodating capacity of each existing and proposed facility. This information is needed later by the optimization model for its mission constraint equation (see Equation 4.5).
8. ES module estimates costs of renovation, maintenance, and conversion of facilities to be considered by the optimization model, using its heuristic knowledge in the knowledge base. The costs of new construction are given by the project proposal.
9. After obtaining all the necessary parameter values, ES formulates the zero-one integer facility optimization model. As described earlier, the decision variables included in the model are whether to: (1) renovate facility i, (2) convert facility i to j, (3) use facility i as it is, or (4) approve new construction project l.

---

5. For instance, it is worthless for the model to consider the facilities classified by ES as disposable.

10. The facility optimization problem formulated in an integer programming model is solved.
11. The ES components of XPlanner interpret the optimization model's results and explain them to the user.
12. XPlanner asks the user if the results given by the model are acceptable in terms of facility requirements. If yes, go to Step 13; otherwise go to Step 15.
13. XPlanner asks users whether or not another scenario should be developed. If yes, go to Step 1 and repeat the whole process; otherwise go to Step 14.
14. XPlanner displays the results (facility management plan) in computer graphic (map) format.
15. If the results given by the optimization model are not acceptable to the user in terms of facility requirements, XPlanner asks the user if XPlanner should develop new construction projects. If yes, go to Step 16; otherwise go to Step 13.
16. Utilizing its knowledge, the ES recommends candidate sites for new construction projects and the users finalizes the type of new construction projects.
17. XPlanner estimates the facility requirements that can be satisfied by constructing new facilities and returns to Step 7.

As explained above, the problem-solving process of XPlanner involves an *iterative* process between the optimization model, the expert system, and the user. The user may continue the consultation until an acceptable answer (facility management plan), which satisfies facility requirements generated by new missions, is found.

Summary and conclusions

As pointed out by O'Keefe (1985), as operations research has shifted away from pure optimization models, it is likely that expert systems will shift away from pure symbolic processing systems and will increasingly employ optimization techniques. The coupling of ES and DSS basically takes two different forms: (1) integration of ES into conventional DSS to provide qualitative reasoning capability and intelligent user interface, and (2) integration of DSS into the conventional ES to provide modeling capability. XPlanner chooses the first type of coupling, in which ES helps users input necessary parameters and interpret and modify the solutions by DSS. ES also provides judgmental capabilities to XPlanner in order to supplement qualitative tasks such as facility classification.

From the experience of designing XPlanner, we believe that the intelligent interface provided by expert systems for the modeling tasks in DSS may stimulate planners to employ mathematical models more frequently and more easily in their problem-solving processes. As often pointed out by planners, modeling components tend to be treated as black boxes, inadequately recognizing the need

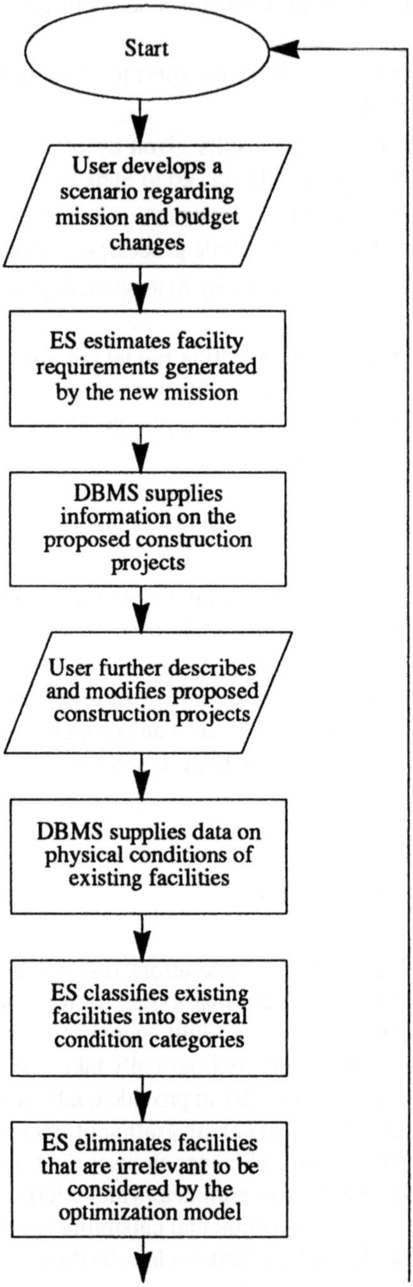

Figure 4.3. Flowchart of XPlanner (part one).

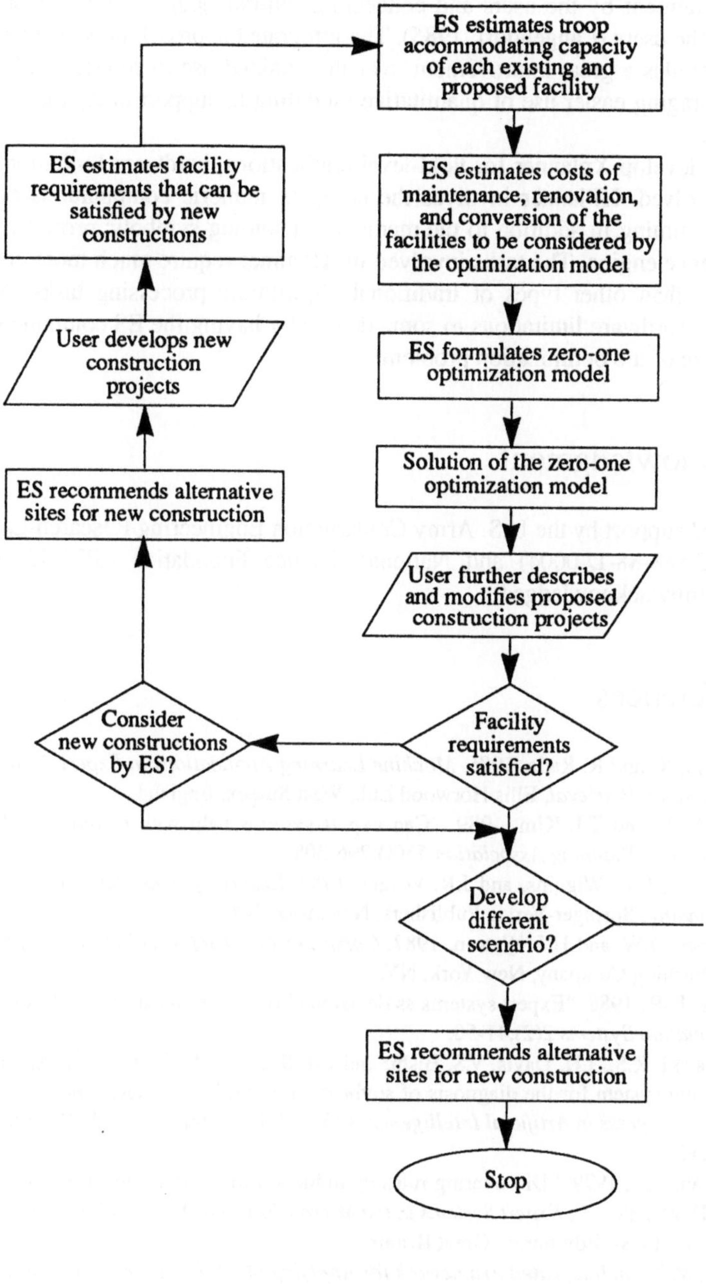

Figure 4.4. Flowchart of XPlanner (part two).

for judgment by the users and concealing implicit judgments and assumptions from the users (Langendorf, 1985). The integrated approach of XPlanner certainly provides a great improvement over the unaided use of modeling algorithms, encouraging easier use of quantitative modeling to support many planning decisions.

To develop Xplanner for field-level applications, hardware limitations should be resolved. XPlanner involves the complex numeric computations of integer programming in addition to the memory-consuming symbolic processing of the inference engine. The tasks involved in XPlanner require much more computing power than other types of traditional algorithmic processing tasks. XPlanner solves hardware limitations to some degree by having the ES component reduce the size of the optimization problem.

Acknowledgment

Partial support by the U.S. Army Construction Engineering Research Laboratory (DACA88-88-D-0004) and National Science Foundation (SES-87-18146) is gratefully acknowledged.

References

Forsyth, R. and R. Ráda, 1986. *Machine Learning Applications in Expert System and Information Retrieval*, Ellis Horwood Ltd., West Sussex, England.

Han, S.-Y. and T.J. Kim, 1989. "Can expert systems help with planning?", *Journal of American Planning Association 55*(3):296-308.

Kim, T.J., L.L. Wiggins, and J.R. Wright, 1989. *Expert Systems: Applications to Urban Planning*, Springer-Verlag Publishers, New York, NY.

Kroeber, D.W. and H.J. Watson, 1987. *Computer-based information systems*, Macmillan Publishing Company, New York, NY.

Liang, T.-P., 1988. "Expert systems as decision aids: issues and strategies," *Journal of Information Systems 2*(2):41-50.

Michalski, R.S., J.H. Davis, V.S. Bisht, and J.B. Sinclair, 1985. "Plant/ds: An expert consulting system for the diagnosis of soybean diseases," in L. Steels and J.A. Campbell, eds., *Progress in Artificial Intelligence*, 135-155, Ellis Horwood Ltd., West Sussex, England.

Quinlan, J.R., 1979. "Discovering rules by induction from large collections of examples," in D. Michie, ed., *Expert Systems in the Micro Electronic Age*, 168-201, Edinburgh University Press, Edinburgh, Great Britain.

Shaw, M.J., *An Integrated Framework for Applying Machine Learning in Intelligent Decision Support Systems*, faculty working paper no. 89-1534, Bureau of Economics and Business Research, University of Illinois at Urbana-Champaign,

Simon, H.A., 1987. "Two heads are better than one: The collaboration between AI and OR," *Interfaces 17*(4):8-15.

Strauch, R.E., 1974. *A Critical Assessment of Quantitative Methodology as a Political*

Analysis Tool, Rand, Santa Monica, CA.

Tingley, G.A., 1987. "Can MS/OR sell itself well enough," *Interfaces 17*(4):41-52.

Turban, E. and P.R. Watkins, 1986, "Integrating expert systems and decision support systems," *MIS Quarterly*, 231-245, June.

Wilson, J.M., 1985. "Classification of models in operational research," *Journal of Operational Research Society 36*(3):253-256.

Wynne, B.E., "A domination sequence—MS/OR, DSS, and the fifth generation," *Interfaces 14*(3):51-58.

Appendix A

Decision tree for facility classification rules

```
year
<1945 : roof sagging
             <2 : current use
                  <=2 : 3
                   >2 : 2
            >=2 : current use
                   <3 : door misalignment
                         <2 : 4
                        >=2 : 5
                  >=3 : year
                              >1940 : 4
                             >=1940 : 3
>=1945 : year
    <1980 : settling
             <3 : construction type
                   <2 : door misalignment
                         <2 : wall cracks
                               <2 : utility line
                                     <2 : interior
                                           <2 : 1
                                          >=2 : current use
                                                 <2 : 2
                                                >=2 : 1
                                    >=2 : 2
                              >=2 : 2
                        >=2 : wall cracks
                               <3 : year
                                  <1974 : utility
                                           <2 : 2
                                          >=2 : 3
                                  >=1974 : 2
                              >=3 : 3
                        >=2 : year
                            <1965 : current use
                               <3 : 3
                              >=3 : water damage
                                     <2 : 2
                                    >=2 : 3
                      >=1965 : 2
            >=3 : 5
    >=1980 : 1
```

PART TWO
GIS and spatial modeling
Introductory remarks by Jeff R. Wright

Geographic information systems (GIS) have been among the most widely embraced software technologies of the past decade. GIS began as the simple (albeit expensive) automation of two dimensional boolean overlay operations on crude graphic devices for purposes of attractive information display, and has become a powerful database technology for the management of data having a spatial dimension. GIS applications are possible in extremely diverse areas, including environmental and land-use management, public works planning and operations, facilities location, marketing, agri-business, mineral and resource exploration, and even military logistics.

GIS has also become a sort of aphrodisiac for public officials responsible for the management and analysis of spatially distributed data. Encouraged by the availability and affordability of a wide variety of social, economic, environmental, and physical geo-referenced data, virtually every state, county, and municipality around the world has at least considered adopting the technology. While GIS and related spatial information technologies have indeed become powerful in their ability to manage spatial data, the promise of comprehensive and general purpose spatial analysis systems has yet to be realized, and true spatial decision support systems are only now starting to emerge. Not surprisingly, the integration of GIS and expert systems technologies has become a central focus for the development of these systems.

The four chapters that follow present four different perspectives on GIS/expert systems integration, yet each chapter shares a common goal: to provide a non-procedural modeling environment that will help decision makers make better decisions. One perspective is given by Dundee Navinchandra in Chapter 5, who provides insight as to the conceptual as well as practical problems one encounters in trying to integrate GIS and expert systems technologies. His discussion focuses around four projects developed at MIT under sponsorship from the U.S. Army Corps of Engineers, Construction Engineering Research Laboratory: (1) CERES; an early attempt to develop a rules-based reasoning structure to ascertain likely environmental impact by utilizing an early object oriented map representation of environmental attributes, (2) LANDEX; a multidisciplinary environmental assessment system, (3) GEODEX; a system for layout planning, and (4) G; a fourth generation GIS system that allows external programs to query attribute maps. Based on his experience with each of these four systems, Navinchandra concludes that the integration of these technologies is presently limited by the lack of effective mechanisms for communication between system components.

Another perspective is given by Wright and Buehler in Chapter 6, who present a generic system for determining land suitability when circumstances suggest that Baycsian analysis is the appropriate analysis technique. B-infer is a Bayesian reasoning shell that facilitates the indexing of land units based on long-term (map attribute data) and short-term (contextual attribute data) through Bayes Theorem. An indexing system can be developed by a user for a specific application by preparation of a script using the B-infer program syntax, which is then translated into a custom application for use in a production environment. The authors suggest that this modular approach to the integration of GIS and non-procedural analysis techniques provides a framework that would prove ef fective for the integration of other inferencing mechanisms with spatial data.

Groundwater quality assessment is the focus of the work presented by Marc Armstrong in Chapter 7. The basic premise of this research is that even though state and federal agencies attempt to collect vast amounts of data about groundwater quality, the quality, consistency, and uniformity of those data tend to be questionable for a number of reasons. Armstrong presents one approach to the integration and management of such data and describes a series of programs that allow decision-makers to formulate and embed logical decision rules in a knowledge base which are then used in retrieving data from the database. This approach is effective in building and searching databases consistent with the style of a particular user and has applications in other domains as well.

Finally, the integration of two well-known applications development systems (the Arc/Info GIS and the Nexpert Object expert system tool) is presented by Maidment and Evans in Chapter 8. The goal of this research is the design of a system for use by land planners and municipal zoning officials to evaluate development requests. The authors demonstrate that the benefits of improved regulatory efficiency and effectiveness resulting from the integration of these technologies is significant in the domain of building regulation, and suggest that the approach is applicable to other environmental regulatory processes.

5

Observations on the Role of Artificial Intelligence Techniques in Geographic Information Processing

Dundee Navinchandra

Introduction

Since the advent of expert systems in the early 1980s, there has been a growing interest in applying computers to domains such as medical diagnoses, curing crop diseases, and marketing. Research groups around the world have jumped on the "artificial intelligence" (AI) bandwagon by producing thousands of toy systems. The ability to program with heuristics made it possible to introduce computers in various domains. Computer applications suddenly moved out of the realm of number crunching into symbolic processing and heuristic reasoning. It appeared as if a research paper was written for every toy system built, and a professional conference or workshop was started for each new domain of applications. As we enter the 1990s, a definite slow down can be observed. Numerous journals have folded, a majority of AI companies have gone out of business, and conferences have been abandoned.

These trends do not mean that expert systems are useless, or that AI is a dying field, however. Some AI application areas such as engineering design have gone through a life-cycle of phases, starting from the initial phase with its characteristic frenzy of activity, and settling down with small but intense research efforts aimed at answering fundamental questions about design theory, human problem-solving, and intelligence. There are many important lessons to be learned from other domains in which AI techniques have been applied and tested. This chapter presents some of the issues that are faced and recounts some experiences (1983 to 1986) in building AI-based environmental applications on top of geographic information systems (GIS). I feel that AI is still new to the GIS world and that we should be careful about how we approach expert system development.

GIS applications and paradigms

An AI application can be viewed as having three parts: (1) an inference engine, (2) a knowledge base, and (3) a working memory (or data). In this section, we examine applications that can be developed using geographic systems. One issue to consider is how and where AI methods are introduced in a GIS. As I view it, a GIS is essentially a spatial database management system. The system's data access capabilities can be used by the expert system to access working memory (spatial data). The inference engine performs some task such as site selection, layout, or environmental impact assessment. Let's examine these aspects in more detail.

Data representation

A map can be represented in three basic forms: a cellular grid, a vector database, or an object-oriented representation. I have found object-oriented representation to be most suited to rule-based programming and other AI-based techniques. In an object-oriented database, each entity on the map is identified as belonging to some class of objects. The map is then represented as a semantic network of relations. For example, objects on a map might include roads, settlements, clearings, farms, etc.

Grid and vector-based systems tend to be too detailed for high level heuristic reasoning. Running OPS5 type rules on any large map is beyond the capabilities of modern computers. We need ways in which a grid or vector-based map can be converted into an object-oriented form. This requires object segmentation, extraction, and recognition capabilities.

The knowledge base

For simplicity, we assume the system knowledge is stored as rules. In this subsection, we examine the rules used for three major map representation methods.

1. *Local cell rules.* An interesting and very fast approach to applying rules to grid-based systems is provided by a GIS system called GRASS. In 1985, Jim Westervelt (U.S. Army Construction Engineering Research Laboratory) developed a rule-based approach which operates locally on individual cells. For example, the system can run rules on a map to "find all sites which are south facing and have a ground slope of either less than 5% slope if the ground is loamy, or less than 7% if it is rocky." The rules are set up to operate on corresponding grid cells in several maps at once. The rules are interpreted, translated into C, and then compiled into the application. This approach works only for applications which do not require extensive intercell comparisons. There are, however, some neat "hacks" one can pull off. For example, if one was trying to isolate regions which were close to water, one could generate a map which labels each cell with a number corresponding to its distance from water. The generated map can then be thresholded

and used as a filter (by overlaying).

2. *Object-oriented data.* For applications which require high level reasoning about regions on maps and their interactions, object-attribute-value-based representations are best suited. With an object-oriented map, any production system-based tool can be easily used for problem-solving. The CERES system is an example of this paradigm (Navinchandra, 1984).
3. *Vector-based databases.* Vector-based systems which use relational databases are probably as good as object-oriented representations. I don't like vector-based systems and have not really tried building AI applications on or around them.

Inference engine and system organization

I have found two types of inference strategies relevant to GIS applications. For interpretation, diagnosis, and site selection type applications, production systems seem to work well. For more complex applications, such as layout planning and configuration problems, one needs some kind of search-based tool that can generate alternatives which satisfy constraints (Navinchandra and Goran, 1986). For example, a planner might be faced with the problem of locating a variety of land uses such as housing, shopping, schools, etc. on one site so that criteria such as access, cost, and compatibility are satisfied.

Major problems in applying AI to GIS

This section presents some issues one should be aware of. Also listed are areas which I feel are important and need more research.

1. *The initial success syndrome.* We often see papers which, in essence, read something like this "I built an expert system for XXX. It has twenty rules. I think the potential for this new technology is unbounded... ." This leads to the issue of scaling up.
2. *Scaling up.* A lesson to be learned from mature domains is that we should hold back on large rule-based expert system development projects. Expert systems with thousands of rules are difficult to maintain. When the expert system idea was originally suggested, many professionals believed that the knowledge base was an amorphous set of heuristics in which new rules could be painlessly added at any time. These professionals believed that updating the knowledge base just meant adding new rules and letting the inference engine use them. On the contrary, for any moderately complex application, rules interact with one another and have to be carefully programmed. It is now accepted that use of production rules is another programming methodology. There is no special relationship between heuristics and the syntax of rule-based systems. A serious knowledge acquisition and en-

coding problem is at hand.

3. *Knowledge acquisition.* Expert systems should be able to learn. Recent advances in machine learning show a lot of promise; however, these techniques are very new and cannot be expected to provide robust solutions to any real world problem. One of the new reasoning paradigms, case-based reasoning (CBR), is particularly relevant to environmental problems. CBR is a problem-solving paradigm where previous experiences are used to guide problem-solving (Carbonell, 1986; Kolodner, 1985; Simpson, 1985; Sycara, 1987; Hammond, 1986). Cases similar to the current problem are retrieved from memory, and the best case is selected from those retrieved and compared to the current problem. The precedent case is adapted to fit the current situation based on the identified differences between the precedent and the current case. Successful cases are stored so they may be retrieved and reused in the future. Unsuccessful cases are also stored so that they will alert the problem-solver to potential difficulties and help recover from failures. If a current case has features similar to a past failure, then the problem-solver is warned not to attempt the failed solution. After the problem is solved, the case memory is updated with the new experience. Thus, learning is integrated with problem-solving. An application of this paradigm to landscape design may be found in Navinchandra (1988).

4. *Interpretation and recognition.* When approached by the U.S. Army Corps of Engineers to examine the scope of AI in GIS and environmental engineering, I started out by building a prototype system. I worked with a simple database, represented as objects in a working memory, and wrote rules about environmental assessment. Initial results were very impressive (Navinchandra, 1984). When it came time to make the expert system work on real satellite data and SCS maps, I soon realized that geographic information systems were not amenable to this new paradigm. A grid cell-based map is a large matrix (with millions of data points), where each cell has characteristics such as soil, slope, vegetation, salinity, etc. One cannot just load a map into the working memory of an expert system shell.

   Consider, for example, an expert system for evaluating the environmental impact of building coal-fired plants. Such a system might have the following rule: IF the plant is within 10 miles of a large residential area, with sparse vegetation in between, and if the pollution isopleth is greater than 100 micrograms/cubic meter, THEN consider a mitigation strategy which includes adding precipitators/scrubbers to the plant. There are no expert system tools which will support such rules. The IF part of the rule requires the following capabilities: (1) finding residential areas, which may be done by using a population/development map, (2) determining if a settlement is large, which requires the ability to find clusters of development and finding their sizes, (3) finding the distance between clusters and finding the type of terrain between them, and (4) finding the intersection of isopleth maps with other maps. Tools which allow such actions to be taken are needed. The GIS should be able to provide such data manipulation capabilities to an external program (the inference engine's pattern matcher). To address this problem, a fourth generation language for geographic information systems was developed. The language implements a variety of algorithms taken from image processing and machine vision. The language, called G, allows developers to program with maps as if they were objects. G makes it possible to encode the type of operations required for complex rules such as the one listed above.

5. *Intra- and inter-representation operations.* A corollary to the issue of interpreting maps is that of allowing reasoning across various map types. For example, the G language can be used to convert bitmaps into an object-oriented form in a semi-automatic way. Consider the task of converting a given bitmap of development into an object database of large settlements. The bitmap has zeros and ones where a one represents development. The task involves finding clusters of ones and eliminating the small clusters. Following is a procedure for the task at hand: (1) *grow* all the regions in the map to make close, but unconnected, regions merge with one another, (2) *run* a segmentation algorithm assigning each cluster a number, (3) *thin* the segments back to their original size, (4) *shrink* all regions by the minimum size (which determines largeness), (5) *grow* back the regions that are left after thinning (the combination of steps 4 and 5 eliminates small regions), (6) for each category, create an object and record information such as location, size, etc. The above algorithm can be easily built in G language. The language is described in detail in the next section.

An overview of four AI in GIS projects

The following sections provide an overview of four AI-based GIS systems built jointly at the Massachusetts Institute of Technology (Cambridge, MA) and the U.S. Army Corps of Engineers, Construction Engineering Research Laboratory (Champaign, IL).

The first system, CERES (Navinchandra, 1984), was an early attempt at seeing how rules relating to environmental impact can be incorporated in expert systems. CERES was not attached to a GIS, but used a hand-crafted, object-oriented representation of a map. The second system, LANDEX (Navinchandra and Lacey, 1985) was aimed at developing a multidisciplinary environmental assessment system. The system was based on the blackboard architecture. LANDEX was not originally integrated with a GIS. It was only after the G language was developed that the integration was carried out.

The GEODEX system (Navinchandra and Goran, 1985) is a layout planner. It is based on the constraint directed search paradigm (Fox, 1983). The idea is to specify a solution in terms of constraints and have the system search out solutions. For example, in locating a house one might specify constraints such as: "good access," "not noisy," "close to shopping," etc. Finally, we will discuss a fourth generation language for GIS systems called G (Navinchandra and Shapiro, 1986). The language allows external programs to query maps to get complex information about proximity, access, intervisibility, etc.

CERES: An expert system for environmental impact analysis

Environmental impact analysis is an involved, multidisciplinary process which requires input from several experts who draw upon huge bodies of knowledge that have been garnered over years of experience. The expertise level of the impact analysts determines the quality, completeness, and validity of the impact statement. This section describes the design aspects of an environmental impact analysis expert system. By automating the analysis process, we intend to place the knowledge of experts at the user's fingertips. This will give the developers of EIS/EA the capability of preparing formalized, high-quality, and complete documents.

The following example involves the location of a coal-fired plant ("plant XYZ") which is found to have an impact on some cities in the vicinity. The system is based on rules. Here is what a rule from the functional area of air quality looks like:

```
Rule Airqual123:
    IF        <source under consideration is coal fired plant>
              <source is within 10 miles of residential area>
              <source emissions are of criticality greater than 5>
              <wind rose component is greater than 20 mph>
    THEN      <apply Pasquill-Gifford's equations>
              <check isopleths>
              <store the possibility of impact as: 'high'>
```

This rule involves air pollution and makes the expert system go to the map and look for the problem. On finding the potential for the problem, the system accesses the wind rose charts. Layering input keeps the system from requesting unnecessary information. The wind rose data need be input only when the computer senses an impending hazard. The rule above goes on to update the global database with inferences in the THEN part of the rule. These inferences are used by other rules to find overall impacts and mitigations. Mitigation rules look something like this:

```
Rule air-mits122
    IF        <isopleth through residential area is greater than
              250 micrograms/cubic-meter>
              <distance is within 10 miles of the source>
    THEN      <mitigation: call for plant relocation>
Rule air-mits120
    IF        <isopleth through residential area is greater than
               100>
              <isopleth through residential area is less than 250>
              <distance is within 10 miles of the source>
    THEN      <mitigation:add precipitators/scrubbers to source>
```

The representation

In the previous example, it was stated that the computer had to check for pollution source criticality. This would indicate that the computer needs access to data on the nature of the pollution source, which involves accessing a database of critical characteristics of the various pollution sources. Such data is called "elemental data." The database contains information such as: types of gases emitted by a thermal power plant, sound level of a M1 tank's shell, emission patterns of traffic, etc. This data requires continuous updating and support.

The other data set is the baseline data, which is the status quo data about the area in question. This site-specific database contains demographic data, water resources data, weather data, maps, etc. Maps are in the form of digitized matrix-based images which can be queried for checks like distance, juxtaposition, density of habitation, etc.

The next data set is the impact and mitigation data (IMD). The IMD is the global database created by the rule mechanisms. All inferences created by the rule mechanisms are stored in the IMD.

All these data sets are expressed as semantic networks. A semantic network is a method of representing symbolic knowledge and inter-relationships among the objects in the problem domain. The database is stored and updated each time a rule is fired. As each rule adds more and more to the IMD semantic network, it grows and comes closer and closer to becoming a store of all the impacts and mitigations.

Related systems

Several EA/EIA developers are currently using a form of computer aided environmental analysis and/or analysis tools to help them in this work. It follows that some users would have developed a style of EIS/EA preparation based on these tools. Our expert system emulates the human expert preparing the EIS/EA, we are not replacing any of the existing tools. One can quite easily conceive of the existing tools as inputs to the CERES expert system.

We can add the user's favorite forecasting models directly to the expert system and let the computer interpret the outputs of the model. This prevents the expert system from conflicting with current practice.

Limitations

The system had some serious limitations. Like many other expert systems, CERES was not able to collect knowledge from the user. Besides simple range checking, the system was not able to protect itself from erroneous data inputs.

In the preparation of the EIS, there are several activities (for example, correspondence, liaison, identification of new and unconventional factors, meeting with local interest groups around the site, etc.) which are tasks the developer will still have to carry out in order to prepare the EIS/EA.

Constant upgrading is required in order to maintain the completeness of the system. Whenever new equipment, construction techniques, tanks, etc. is introduced, the data banks should be upgraded with this new information.

LANDEX: An expert system for land use planning

This section describes the workings of a system called LANDEX, which is a rule-based system for site evaluation. Given a map and a list of proposed sites for a particular land use situation, LANDEX evaluates the sites from a multidisciplinary point of view. The system is currently configured to evaluate sites for use as maneuver areas for army training.

The rules in LANDEX are of an IF-THEN format, for example: "IF site has slopes greater than 20% and the soil is silty clay loam, THEN the site is severely unsuitable for off-road vehicles." Through the use of several heuristic rules and numerical routines, LANDEX is able to make deductions about a given site. A section of the program in operation is included in this chapter.

LANDEX was among the first efforts on an intelligent geographical data analysis project at the U.S. Army's Construction Engineering Research Laboratory.

Motivation

Land use planning is a multidisciplinary task that involves the integrated application of disciplines like noise abatement, ecology, soils and erosion science, etc. Each of these disciplines are complete fields in and of themselves and are ever expanding. It is becoming increasingly difficult for a planner, or one planning office, to keep abreast of new developments in relevant disciplines. To address this problem, experts continually prepare comprehensive guidelines.

When a set of guidelines is prepared, the users must read and understand the new techniques, then apply the techniques to the problem at hand. It is not clear why a planner should get involved in the details of some new noise prediction technique which he/she is going to use only once or twice a year!

Our proposal was that the computerization of guidelines would alleviate much of the users' burden. A system could be developed centrally and then distributed to planners nationwide. The goal was a system which would act as a planner's assistant and would handle the "nitty-gritty," leaving the planner free to plan and make decisions.

In order to substantiate this proposal, we put together a system for evaluation of army training sites. On studying existing guidelines, we found the analysis involved was both numerical and heuristic. The noise heuristics was calculation intensive, while ecology was heuristic knowledge intensive and terrain analysis was a combination of the two approaches.

LANDEX background

The LANDEX system is built to interpret maps and to reason about the suitability of a given site as an army maneuver training area. It has rules about several disciplines (for example noise assessment, soil problems, and land use conflicts).

The rules were initially acquired from a technical report by R. M. Lacey and W. D. Servinghaus (1981). After an initial, crude system was built, the authors of the report reviewed the system and added more rules. We quote from the report:

> *There is "...lack of a simple, quick and acceptable method by which to identify natural resource and environmental concerns....there is a need for a system or set of procedures which can be used to evaluate the land and natural resource suitability of areas proposed for training. This system should provide simple, illustrative material..." for decision making.*

LANDEX was a first attempt at such a system. We found it easy and natural to convert the report guidelines into rules for the expert system which strongly indicated the applicability of AI techniques to this domain. In retrospect, the encoding process was easy because we were working with well documented, predigested heuristics prepared by experts. If we had had to interview experts ourselves, the process would have been much more difficult.

System architecture

Site evaluation, with respect to land use planning considerations, has three major steps: (1) pre-evaluation assessment and data collection, (2) assessment with respect to eight disciplines: land use, noise, terrain, soil, air quality, water quality, vegetation, and wildlife, and (3) information integration and recommendations.

This process is reflected in the system's design. LANDEX consists of modules called knowledge modules. A knowledge module is a collection of heuristics (rules) about a discipline. The knowledge modules look at input data and make conclusions which are posted on a blackboard (Figure 5.1). The conclusions are available to the knowledge modules. The final step is to integrate conclusions and make a recommendation. LANDEX doesn't carry out this step, it is left to the user to make final decisions and judgments. We only implemented four modules of the proposed eight.

LANDEX: An extended example

The LANDEX system takes a description of the objects on a map and a description of the site under consideration as inputs. These descriptions usually contain the location, size, and physical characteristics of the objects.

In the process of evaluating the given site, LANDEX asks the user questions. The questions are dynamically generated; that is, they depend upon the user's answers to previous questions. At the end of the session the user can look at the

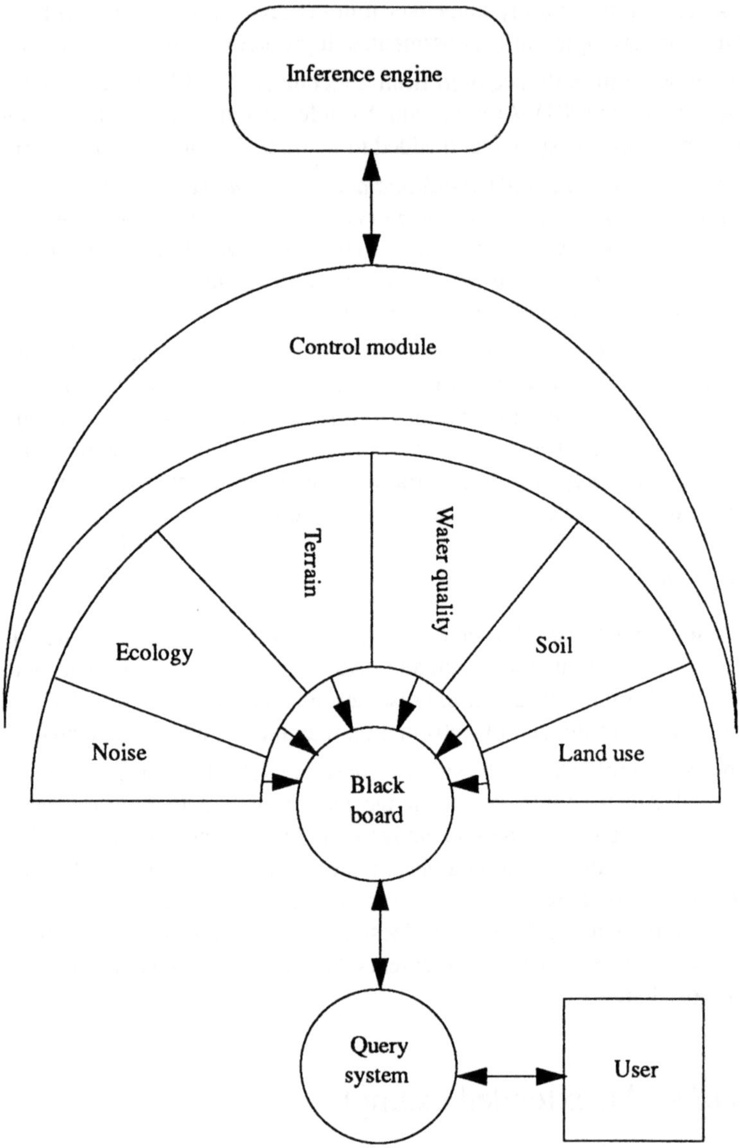

Figure 5.1. LANDEX architecture.

conclusions made by LANDEX. The user can also ask for explanations.

The map shown in Figure 5.2 is stored in LANDEX as a network of frames. A frame is a type of data structure well suited to object-oriented representations. For example: the goudard_lake can be represented as shown below:

```
goudard_lake:
    location: x_location y_location
    shape: size, orientation
    is_a: recreation_area
```

The frame above gives us the location and shape of the object called goudard_-lake. It also says that it is a recreation_area. By indicating this object is a recreation_area, we are telling the inference engine that we are ready to inherit the properties of a recreation_area. The corresponding frame is:

```
recreation_area:
    is_a: object
    acceptable_noise_level: 65 dBa
    distance_buffer: 2000 ft.
    (distance to non-natural land use areas)
```

The lake consequently inherits the properties of a general recreation_area. Most of the objects tend to be interconnected by links (as shown above) and are referred to as being in a network. Figure 5.3 shows such a network (partial) for the map in Figure 5.2. The other data included is the information about the site under consideration. The site is called "site1" and is indicated on the map as a dashed-line box. The frame for site1 is:

```
site1:
    is_a: maneuver_area
    location: xloc yloc
    shape: size, orientation
    vegetation: moderate
    slope: 5 degrees
```

The rules

Here are some examples of rules in the system. For example, a rule in the terrain analysis knowledge module reads:

```
RULE NAME: wet_land
      IF     <the site is lowlying>
             <the site gets heavy_rains>
      THEN   <it can be concluded that the site is prone to be wet>
```

The conclusion made by this rule will be deposited on the blackboard. It is then available to other rules to use. Here are two rules from the soil module:

```
RULE NAME: soil_loamy_sand1
    IF        <the site is_a maneuver_area>
              <the soil type is loamy_sand>
              <the terrain is prone to be wet>
    THEN      <it can be concluded that there is a SEVERE limitation
               due to soil conditions>
RULE NAME: soil_loamy_sand2
    IF        <the site is_a maneuver_area>
              <the soil type is loamy_sand>
              <the terrain is NOT prone to be wet>
```

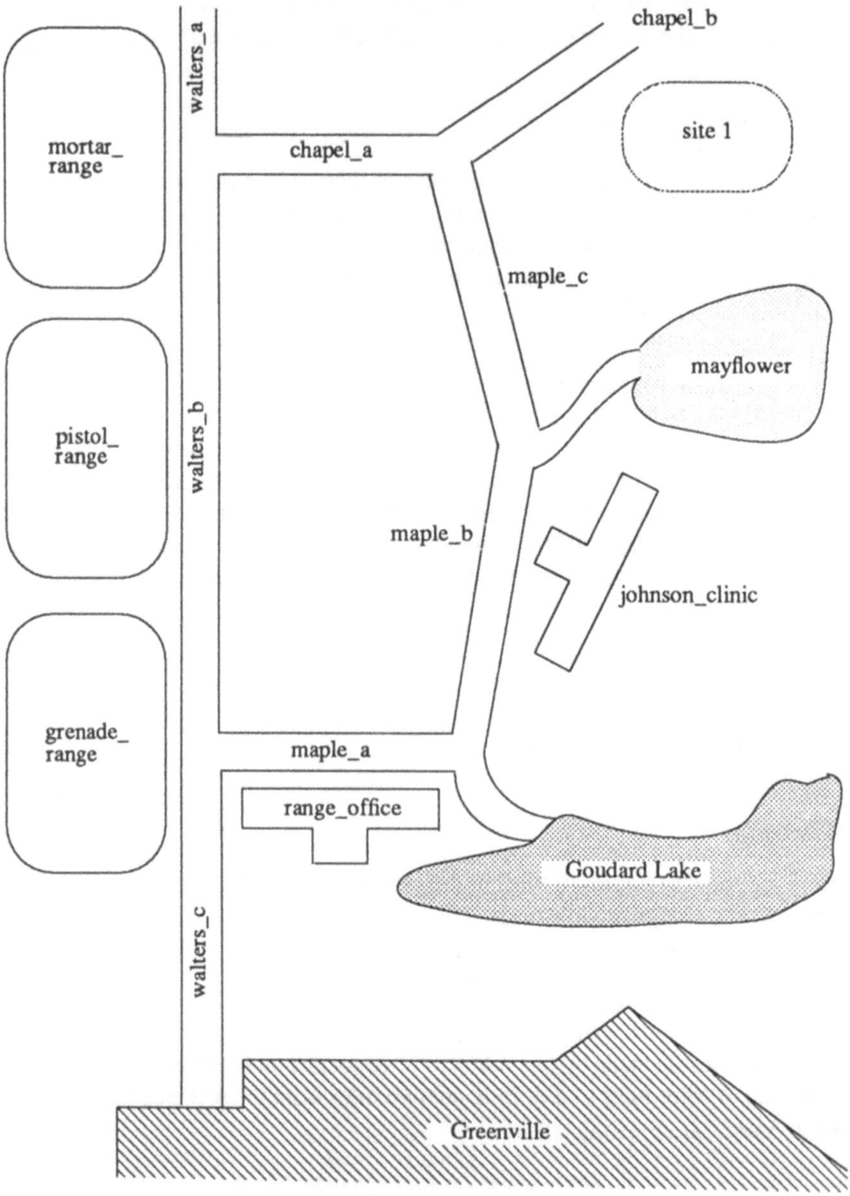

Figure 5.2. Input map.

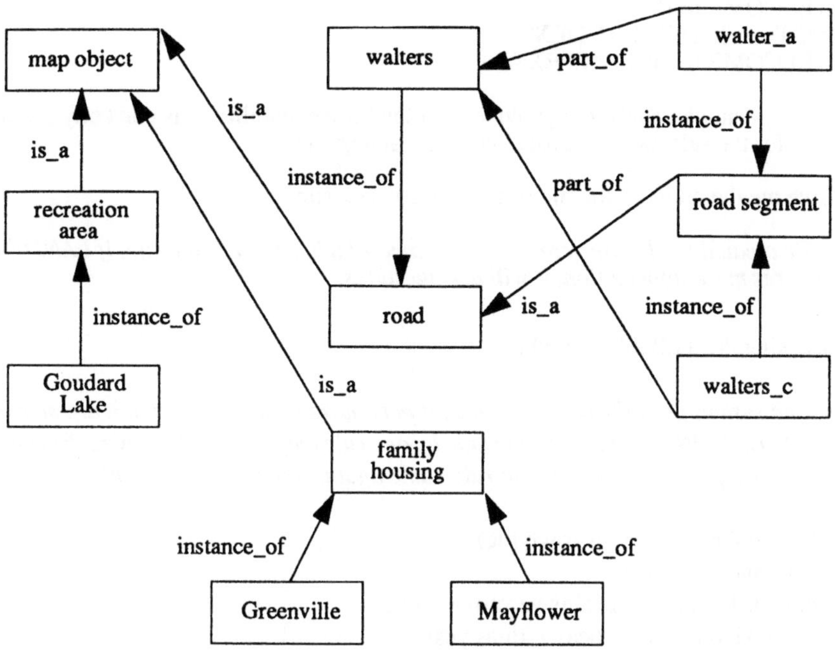

Figure 5.3. Frame representation of the map (partial).

```
THEN     <it can be concluded that there is a SLIGHT limitation
          due to soil conditions>
```

If the terrain rule: "wet_land" fires (i.e., is found to apply) and the site has loamy_sand, then the rule "soil_loamy_sand1" will fire while "soil_-loamy_sand2" will not.

A trace of LANDEX in operation

In this section we attempt to give the reader an idea of the system's operation. The reader is reminded that the system is written in the IMST environment (Navinchandra, 1985). The IMST package has a prompt "IMST++>." The text after the prompt is that which is input by the user.

Annotations are shown in italics. In places where LANDEX asks a question, the "==>" prompt is displayed. The user inputs data at this juncture.

---

unix% imst2.0

We type imst2.0 at the unix level. On getting the IMST==> prompt we ask the system to call up LANDEX.

WELCOME to IMST
IMST==> loadfile LANDEX
WELCOME to the LANDEX

This system can analyze a given site for land use suitability. It is currently set up only for the selection of tactical vehicle training sites.

Give me the name of the file with site info: ==> site1

A file created by the user contains the frame which describes the site. If LANDEX requires more information, it will ask questions.

Invoking the TERRAIN module

As the system goes through the rules, it may make conclusions which are shown as ASSERTIONS. The assertion made by one rule may be used by others. Further, the system prints out the current rule name against the flag "trying rule":

ASSERTING > (site1 is lowlying)
trying rule: lowlying
Does site1 get a lot of rains yes/no? ==> yes
ASSERTING > (site1 heavy_rains yes)

The lowlying rule asks a question. Rains, a data element, was not available in the original input. Notice the assertion made as soon as we said "yes."

trying rule: wet_land
ASSERTING > (terrain site1 is lowlying and is prone to be wet)
trying rule: drainage
How is the drainage at site site1 good/moderate/poor? ==> moderate
ASSERTING > (site1 drainage moderate)
trying rule: drainage1
ASSERTING > (terrain site1 will be water-logged as the drainage is moderate and the water table is moderately high

Invoking the NOISE MODULE

Once the terrain module has been run, LANDEX picks up the next module. In theory all these modules should be running in parallel.

Please answer a few questions about the traffic. I understand the following vehicles:
weapon_vehicles
heavy_trucks
automobiles
medium_trucks

Please give me the number of vehicles in each category which will be going to the site in question. Give me four numbers in a row. For example: 4 30 1 0 number of vehicles: ==> 45 5 0 1

trying rule: ASSERTING_vehicles
ASSERTING > (site1 weapon_vehicles 45)
ASSERTING > (site1 heavy_trucks 5)
ASSERTING > (site1 automobiles 0)
ASSERTING > (site1 medium_trucks 1)

Now I would like to know the access route to the site in question. Please list out the road segments that will be traversed. Route:==> walters_c maple_a maple_b maple_c.

ASSERTING > (walters_c is enroute)
ASSERTING > (maple_a is enroute)
ASSERTING > (maple_b is enroute)
ASSERTING > (maple_c is enroute)

The system asks us for the route to be taken for troops to go from the main garrison to the site. In this case we are going from Greenville (refer to map) to site1.

The shortest route is walters_c, maple_a, maple_b and maple_c.

trying rule: basic_dnna_weapon_vehicles1
ASSERTING > (maple_c dnna 65 203.5)
ASSERTING > (maple_c dnna 70 101.75)
ASSERTING > (maple_c dnna 75 50.875)
ASSERTING > (maple_c dnna 80 25.4375)
ASSERTING > (maple_b dnna 65 209.5)
ASSERTING > (maple_b dnna 70 104.75)
ASSERTING > (maple_b dnna 75 52.375)
ASSERTING > (maple_b dnna 80 26.1875)
ASSERTING > (maple_a dnna 65 209.5)
ASSERTING > (maple_a dnna 70 104.75)
ASSERTING > (maple_a dnna 75 52.375)
ASSERTING > (maple_a dnna 80 26.1875)
ASSERTING > (walters_c dnna 65 263.5)
ASSERTING > (walters_c dnna 70 131.75)
ASSERTING > (walters_c dnna 75 65.875)
ASSERTING > (walters_c dnna 80 32.9375)

The system uses site1's traffic information and existing traffic information for the roads in question and it runs routines which calculate the distances to noise attenuation (dnna).

The assertion (walters_c dnna 70 131.75) says that "the distance required for the attenuation of noise due to traffic from road segment walters_c for 70 decibels will be 131.75 feet."
Having done this, LANDEX will look for conflicts in noise levels. It has knowledge about the acceptable noise levels for the different objects in the database.

trying rule: basic_dnna_weapon_vehicles2
ASSERTING > (site1 dnna 65 184.0)
ASSERTING > (site1 dnna 70 92.0)
ASSERTING > (site1 dnna 75 46.0)
ASSERTING > (site1 dnna 80 23.0)

trying rule: isa_to_isa_inheritance
ASSERTING > (range_office acceptable_dba 70)
ASSERTING > (greenville acceptable_dba 65)
ASSERTING > (mayflower acceptable_dba 65)
ASSERTING > (johnson_clinic acceptable_dba 65)

The system figures out that Greenville's acceptable noise level is 65 decibels. This comes from the inheritance rule. Greenville, being a residential area, inherits the properties of residential areas.

trying rule: conflict_detection_road
ASSERTING > (noise walters_c conflict range_office)
ASSERTING > (noise maple_b conflict johnson_clinic)
Invoking the SOIL MODULE

I know of the following types of soil types:
sand
sandy_loam
loamy_sand
silty_clay_loam
sandy_clay
clay_loam
clay
loam
sandy_clay_loam
silt_loam
silty_clay
silt

Please give me the soil type at the site site1: ==> loamy_sand
ASSERTING > (site1 soil_type loamy_sand)
trying rule: soil_loamy_sand1
ASSERTING > (soil site1 limitation severe)

DONE

LANDEX has finished its consultation. We are now back at the IMST command level (LANDEX is built using the IMST development system). The user can now look at conclusions, ask for explanations and save the conclusions in a file. Saved conclusions for different sites can be evaluated by the user.

IMST==> describe terrain
Description of object: terrain
1 site1 slope is flat
2 site1 is lowlying and is prone to be wet
3 site1 will be water-logged as the drainage is moderate and the water table is moderately high

Having seen the three conclusions made by the terrain module, the user now asks the system HOW it arrived at the second fact in the terrain description.

IMST==> how terrain 2
The fact that terrain site1 is lowlying and is prone to be wet was deduced based on the wet_land rule.
Which says that BECAUSE
site1 is lowlying....8
AND site1 heavy_rains yes....9
THEN it can be concluded that:
terrain site1 is lowlying and is prone to be wet

Asking for the noise module's assessment

IMST==> describe noise
Description of object: noise
1 walters_c conflict range_office
2 maple_b conflict johnson_clinic

The noise module has come up with two conflicts. Referring to the map, we see that these conflicts are the range_office and the johnson_clinic.
If the noise module is re-run, using a new route for getting from Greenville to site1, the noise module comes back with only one conflict: the range_office (this is for route: walters_c walters_b chapel_a maple_c).
One may consequently decide to accept higher noise levels at the range office and adopt the second route.
Having the computer choose the best route automatically is a somewhat tougher problem.

IMST==> describe soil
Description of object: soil
1 site1 limitation severe

IMST==> how soil 1
The fact that soil site1 limitation severe was deduced based on the soil_-loamy_sand1 rule.
Which says that BECAUSE
site1 is_a maneuver_area....1
AND site1 soil_type loamy_sand....19
AND terrain site1 is lowlying and is prone to be wet....2
THEN it can be concluded that: soil site1 limitation severe

IMST==> how site1 1
The fact "site1 is_a maneuver_area" was given initially and was not deduced.

IMST==> bye
goodbye

unix% logout

Conclusions on LANDEX

The size of the LANDEX rulebase was too small to be used in the field. Building a full sized expert system requires a lot of care and time. Even after the deployment of a system like LANDEX, the knowledge base will continually undergo changes. A support staff will have to maintain the knowledge and keep it up to date.

LANDEX was an experimental attempt to examine the feasibility of having computer-based consultants at a site. It is a step towards a family of systems which could make the application of complex guidelines as simple as inputting maps and answering well structured questions.

GEODEX: a constraint-based layout planning system

This section describes the workings of an experimental knowledge-based geographical data analysis program called GEODEX. The program was built to assist planners in locating sites for a specified land use situation. GEODEX accepts as input maps of the site and a description of the land use application in question. The maps start out as grid cell-based matrices, but are converted into an object-oriented form using a parcel-finding program written in G. The objects are stored with a statement of the inter-relationships between them. Such a data structure allows us to include "real-world" information about the objects to use for automating the application of heuristics to the map. For example, a lake may be represented as a map object along with information about ecological and water-quality considerations. The land use requirements are represented as a set of constraints and heuristic rules and the constraints are simple bounds on certain parameters of a selected site. For example, a constraint may read: "the distance of the site shall be at least two miles from a major highway." The rules are heuristics drawn from expert land use planners which help in making inferences about maps and thus make access to expert knowledge much easier. An example rule may be: "IF the site has clay type soil and the site is lowlying and heavy rainfalls are expected THEN it may be concluded that the site will be waterlogged." The program starts off by applying the constraints to the maps. This step involves standard overlay techniques; the resulting map is enhanced using some machine-vision algorithms. The rules are then applied to the overlain map. At this stage, GEODEX returns a list of alternate sites along with a list of merits and demerits of each site. If no site is found, GEODEX can be asked to retract the constraints in an attempt to "open-up" new areas for potential sites.

Unlike LANDEX, GEODEX is a land use site selection system. It accepts as input a set of maps, a set of locatable objects, and a set of patterns/constraints. It produces, as output, a map with all the objects located so that none of the constraints are violated and all the patterns are satisfied. The algorithm used is constraint directed search (Fox, 1983; Chandra and Marks, 1985).

The domain

GEODEX views site planning as allocation of land to land uses while adhering to certain patterns/constraints. Taking an abstract view of the domain, problem-solving consists of three major stages: problem definition, suitability analysis, and synthesis.

At the problem definition stage, data is collected, goals are set, and a set of patterns/constraints is assembled. Data often change as a result of feedback from some later stage. Solution development consists of suitability identification and synthesis. Suitability identification consists of creating site suitability maps for each of the proposed land uses. This step uses constraints to produce site suitability maps and this is done quite easily on existing GIS systems. Synthesis is harder and requires search.

The constraints/patterns

Static constraints

Site planning problems have three components: (1) proposed land uses, (2) criteria, and (3) interrelating constraints. It is possible to produce maps for each land use with respect to several of the criteria (Tomlin, 1985).

Figure 5.4 shows how several maps can be created by criteria filtering. The maps are combined to give a final answer and the combination of maps (vertically or horizontally) is accomplished by using overlay functions. For example, vertical combination of all the "wetness" maps produces a map which characterizes water bodies, streams, and wetlands across all land uses. Horizontal combination

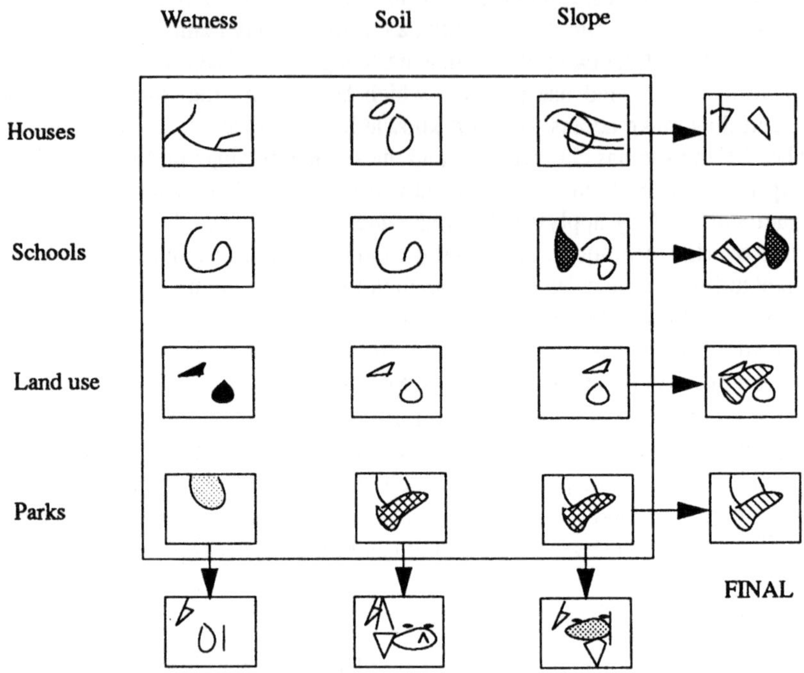

Figure 5.4. Using criteria as filters.

of maps yields a map which represents suitable sites for the land use in question. The constraints used are of the following type:

```
IF      the land use is schools
THEN    the sites near primary highways shall be avoided
IF      there is any land use x
THEN    x shall not be near a dumpsite
IF      there is a house x
THEN    x should be on a south slope
```

The constraints above are expressed as patterns. From this point, we use the words constraint and pattern interchangeably. We call the above constraints STATIC because they refer only to one proposed land use at a time. They can, however, refer to as many existing land uses as needed. A static constraint is one which is stated with respect to at most one proposed land use and several existing land uses and/or land characteristics. In contrast to this we have the dynamic constraints which require to be dealt with more judiciously.

Dynamic constraints

These constraints are used after all the static constraints are satisfied. The input to this process are the site-suitability maps. The problem now is to perform a synthesis using algorithmic techniques. A dynamic constraint is one which is stated with respect to at least two proposed land uses and several existing land uses and/or land characteristics. A typical example is a constraint which requires that schools and homes be close to one another. This constraint, innocuous as it may seem, can be hard to deal with because, at the beginning of the synthesis step, neither the schools nor the homes have been located.

GEODEX example

We shall now work through a simplified, but illustrative example of how constraint directed search is used.

Following is a list of some constraints in the system. The word "defconstraint" means define constraint:

---

defconstraint: shopping-convenience
There shall be one shopping center within each cluster of houses and nowhere else.

defconstraint: housing-clusters
There shall be two clusters with at least 3 sites on each. The sites within a cluster shall be within 2 units of distance from each other.

defconstraint: dump-sites
There shall be one dumpsite only.

defconstraint: dump-distance
There shall be no construction within one distance unit from a dumpsite.

defconstraint: cluster-distances
The housing clusters shall be separated by at least 2.0 distance units.

Three maps are given as inputs: legal housing sites, legal dump sites, and legal shopping center sites. These sites are generated by using the static constraints. The problem is to find two clusters of housing (three sites to a cluster) which satisfy the given constraints. Current GIS systems cannot handle the dynamic constraints. We use a search paradigm to find an acceptable layout. The three input maps are shown in Figure 5.5. For explanation purposes, an overlay of all three maps is also shown.

The search process involves making decisions, propagating constraints, and checking for solutions. For example, assume the system is at a stage where it has chosen two shopping areas: #3 and #4 as possible locations. Using the "shopping-convenience" constraint, housing sites #1, #2, and #3 are eliminated. Further, using the "dump-distance" constraint, dumpsite #2 is eliminated. This is shown in Figure 5.6.

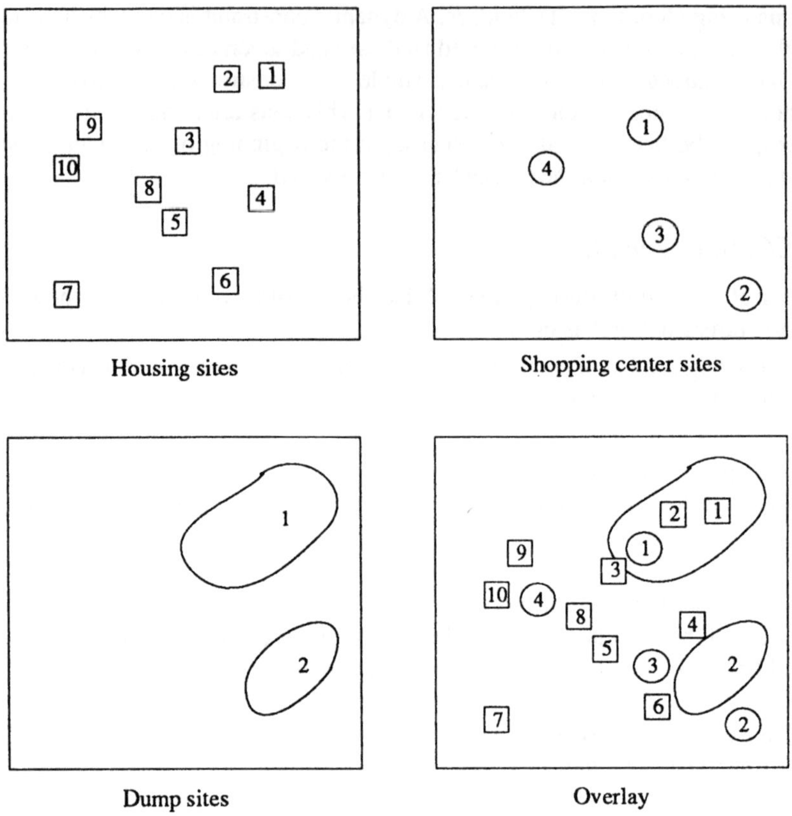

Figure 5.5. Input maps to GEODEX.

The solution, however, violates the "housing-distance" constraint because sites #5 and #8 are too close. This conflict will cause GEODEX to backup (backtrack), retract its choices for shopping areas #3 and #4, and try a different combination. In this case, the combination, #1 and #4 solves the layout problem, Figure 5.7.

Implementation of GEODEX

GEODEX is implemented in LISP and communicates with grid cell-based GIS. The program is reliable but doesn't use dependency-directed backtracking (Sussman, 1977), and is slow. An intelligent backtracking facility needs development.

Efficiency

Another limitation of GEODEX is its inability to handle area facilities. It can only choose between some predetermined set of legal sites and cannot find a cor-

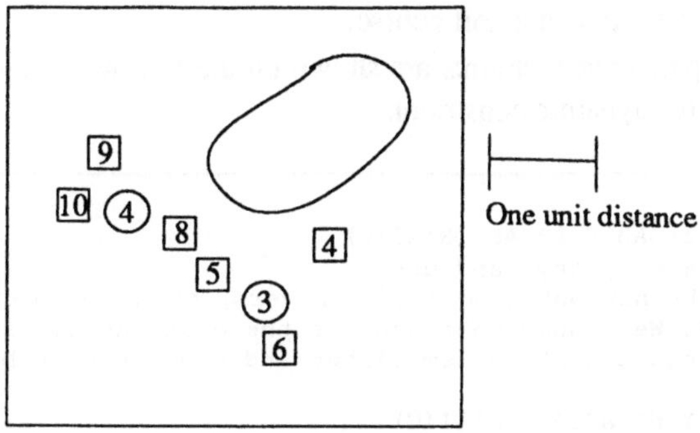

Figure 5.6. Two clusters are found, but are they legal?

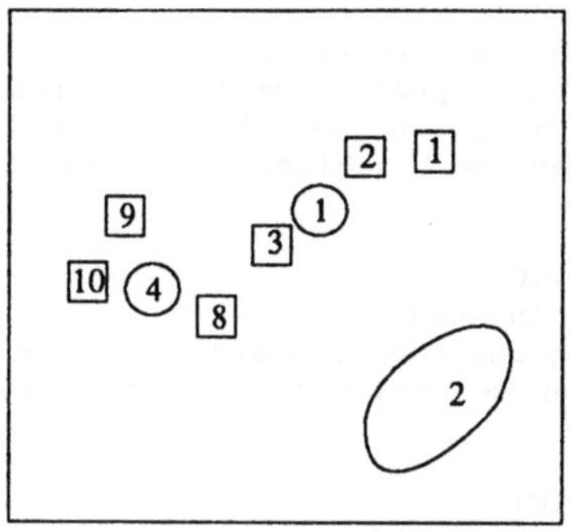

Figure 5.7. Two clusters which satisfy all the constraints.

ridor or a road. It can choose between several alternatives, but cannot choose one from scratch. Further, there is no measure of optimality; the program only looks for a satisfying solution and stops as soon as a solution is found. We are currently working on a computational model of trade-offs in symbolic optimization.

Constraint definition language

Constraints for layouts are called patterns. The idea of using patterns is taken from Christopher Alexander's 1968 book, *A Pattern Language Which Generates Multi Service Centers*. Some of the patterns are from *Site Planning* by Lynch, Kenin, and Mark (1984).

Example input

Following are the constraints which were input to a real world layout problem. The database used was that of an underdeveloped area around a small New England township. The program was instructed to locate: 100 houses in 5 clusters, a school, a shop, and a golf course.

The patterns/constraints are shown on the following pages. The actual code is shown for dynamic constraints.

```
1 UNBUILDABLE AREAS (STATIC)
IF: There is any land use x
THEN: Do not put x on roads, rivers, ponds, or existing structures.
REASON: We assume that none of the existing sites can be changed.
Buildings cannot be demolished and ponds cannot be filled.

2 LOWLYING AREAS (STATIC)
IF: There is any land use x which has to be located
THEN: Avoid lowlying areas
REASON: There is a problem of flooding in some parts.

3 GENTLE SLOPES (STATIC)
IF: There is any land use x which has to be located
THEN: Avoid all slopes 15 percent or more in grade
REASON: This escalates erosion and there is the problem of build-
ability. If the slopes are too high, then there is a danger of land
slides.

4 SCENIC VALUES (STATIC)
IF: A house is to be located
THEN: Locate it among the trees. Avoid open lands.
REASON: Open lands can be used for agriculture and commercial build-
ings.

5 SOUTH SLOPES (STATIC)
IF: Any facility x is being located
AND: x is among (housing, school, church)
THEN: Choose only South facing slopes
REASON: South facing slopes provide good winter sunshine and is es-
sential for dwellings and places people spend a lot of time. Offices
are not included as they tend to be air conditioned.

6 WATER PROXIMITY (STATIC)
IF: A house is being located
THEN: It should be close to water bodies
AND: It should be far from wetlands, marshes, etc.
REASON: Ponds and lakes and rivers are good spots to build. However,
wetlands and marshes should be avoided as they are ugly.

7 SHORES (STATIC)
IF: A facility x is being located
AND: It is to be close to a water body
THEN: It should not be right up against the water body. It should be
set back 3000 ft.
REASON: To build a house in a scenic setting it is important to
build such that the house commands a good view. However, the house
should NOT become part of the scenic spot.

8 CLUSTERS (DYNAMIC)
IF: There are to be n clusters of houses
THEN: The clusters shall be reported by at least 2000 feet.
REASON: Access to clusters may be complicated but a visual sense of
unity can be achieved without forcing social intercourse. There are
```

some savings on roads and facilities. The landscape is better conserved (Lynch and Mark, 1984).

```
(defpattern clustering_diff
     (house hous cluster loco)
     (not* instance clustering_diff hous loco cluster)
     (house hs clust lo)
test (selected? loco)
     (not (equal hous hs))
     (not (equal cluster clust))); diff cluster
 --> (set_value lo
     (loop for x in (get_value lo)
        when (not
               (member x
               (create_box(car(get_value loco))6)))
        collect x))
assert instance clustering_diff hous loco cluster))
```

9 CLUSTERS SMALL (DYNAMIC)
IF: There is a cluster x of n sites
THEN: n _ 5 sites
AND: The size of x is _ 25 acres
REASON: We assume that will be 5 clusters of 5 sites each. This gives 25 one acre sites. If 4 houses go on a site, then 100 houses can be placed on the given map.

```
(defpattern clustering_same
     (house hous cluster loco)
     (not# instance clustering_same hous loco cluster)
test (selected? loco)
     (greaterp (count (subtract (create_box(car(get_value loco))2)
        house_parcels)) 19_
--> (set_value loco '())):force backup
```

10 GOLF COURSE SIZE (DYNAMIC)
IF: There is a golf course to be located
THEN: It should cover a contiguous plot of land that is 1800 feet by 1800 feet.

```
(defpattern golf_course_size
     (golf_course golfn cluster loco)
     (any thing in there)
test (selected? loco)
     (not (equal thing golfn))
--> (set_value there
     (subtract (get_value there)
     (create_box(car(get_value loco)) 4)))
     (assert instance golf_course_size golfn cluster loco))
```

11 COMMERCIAL AREAS (STATIC)
IF: A shopping center is being sited
THEN: It should be close to the center of the town. It should be close to a primary road.
REASON: A lot of development occurs around primary roads.

12 SHOPPING AREAS (DYNAMIC)
IF: A shopping area is being located
THEN: It shall be at least 1000 feet from any school
AND: It shall be at least 1000 feet from any golf course
REASON: Shopping areas busy; avoid schools and houses.

```
(defpattern shopping_is_ugly1
     (shopping_center shopn global loco)
     (not* instance shopping_is_ugly1 schopn global loco)
     (school sn global schloc)
test (selected? loco)
-->  (set_value schloc
     (subtract (get_value schloc)
               (create_box (car (get_value loco)) 5)))
     (assert instance shopping_is_ugly1 shopn global loco))

(defpattern shopping_is_ugly2
     (shopping_center shopn global loco)
     (not* instance shopping_is_ugly2 shopn global loco)
     (golf_course sn global schloc)
test (selected? loco)
--> (set_value schloc)
     (subtract (get_value schloc)
     (create_box(car(get_value loco))5)))
     (assert instance shopping_is_ugly2 shopn global loco))

13 SCHOOLS ACCESS (STATIC)
IF: A school is being located
THEN: It should not be close to a primary road. It should be close
to a secondary road.
REASON: Primary roads are too busy. However, the school should be
close to some secondary road so that it can be accessed easily.

14 BAD AREAS (STATIC)
IF: Any facility is being located
THEN: It should be at least 2000 feet from dump sites and gravel
pits.
REASON: Avoid unsightly areas. Use distance as a buffer.
```

The program takes about a day to run on a dedicated time-sharing Vax 750.

The G language

This section describes a fourth generation language for building geographic analysis models. The G language provides the much needed bridge between a GIS and an external expert system. The expert system can query the GIS for patterns in the rules. For example, a pattern might need to check if a terrain map has flat, long areas for siting an airstrip.

The ideas used in the design of G are based on the map-algebra and map language concepts developed by Professor Dana Tomlin of Harvard University. G is now part of the U.S. Army's GRASS system. GRASS is a general purpose, grid cell-based geographical modeling and analysis package which includes both image processing and map analysis capabilities. G is written in both C and G itself. It is based in UNIX and uses the C-shell as it's control structure. In this section, we present our philosophy about end-user programming and its relevance in geoprocessing.

Introduction

In the past two decades cartographic data processing has evolved from early special purpose programs to modern general-purpose geographic information systems. A GIS gives us the ability to carry out complex analysis and modeling processes on maps.

Typically a GIS has three parts: (1) data representation and handling functions, (2) a tool-kit of analysis functions, and (3) a control structure.

The data handling module usually consists of low level functions that access maps, manage storage, and handle data conversion and graphics. Two data types have gained popularity among GIS developers: the raster data organization and the vector-based schema.

We feel that the actual choice of the data scheme is irrelevant to the ultimate goal of modeling and analysis. It is true that the choice of a data organization technique will impact the speed and memory requirements of the computer. However, in light of parallel processing architectures and optical data storage technologies, this issue is being rendered less and less important.

The analysis tool-kit constitutes the core of the GIS. These analysis tools include:

- Map reclassification and surgery.
- Functions to perform arithmetic on and among maps.
- Functions for image processing operations, i.e. segmentation (clump), edge-detection, enhancement, etc.
- Functions for proximity analysis.

Finally, a GIS provides a control structure that helps the user utilize the tools in the analysis tool kit. Typically, there are three ways of controlling analysis: (1) a menu driven approach, (2) a command driven approach, or (3) a structured-procedural command driven approach. The menu driven approach is the simplest, easiest to use, and the most inflexible. It is not suitable for professionals involved in model building. Command driven approaches are more flexible, but are not suitable for professionals involved in model building. Command driven approaches are more flexible and are amenable to a involved analysis process. In order to place the command driven approaches in perspective, we examine a typical modeling process. Assume that a user is trying to generate a suitability map for golf courses. The analyst might take the following steps:

1. Identify the relevant map layers to be used.
2. Reclassify (recode) map layers to indicate good or bad areas.
3. Perform a weighted analysis on the map layers; the resulting map has several areas of varying suitability ranging from., for example, 1 to 10 (10 is best).
4. If assuming the user wants areas greater than 100 acres, then display the map and look for suitable areas.

By using reclassify functions, weighting functions, and display functions, the user will be able to perform the above simple operations in a file. In a non-procedural command driven system, it is possible to encode the first three steps. Encoding the fourth step is not possible because query and choice are required. For this reason, we make a step towards structured-procedural command driven approaches. The result is a high-level, procedural language that lets us control the use of analysis tools.

Yet another programming language?

In the early days of geo-processing, computer programs were written to serve particular cartographic models. For every new model, a new piece of code had to be written. Over the years, basic operations that were common among models were recognized. These operations were identified and coded into modules. The collection of such modules constitutes a modern analysis tool-kit. These modules are coded in some base language like FORTRAN or C. The user of a GIS is expected to be able to solve a large number of his problems using the tools provided. However, as users grow more sophisticated, they "outgrow" the current tool-kit and demand new tools. In response to this, GIS developers write more code in the base language to form some new tool. It often happens that the new tool is some moderately complex combination of existing tools with a control code to guide the process.

Therefore, a GIS developer needs to be available to the user. We argue that such modeling processes should be left to the end-user (human or expert system). This allows the user to get better turn-around time on an application, and have better control over his/her program.

There is an interesting anecdote about telephones and telephone networks which illustrates the case for end-user programming. In the 1930's if one wished to make a telephone call, one picked up the receiver and asked the operator to connect them to a particular person. However, this was slow and inefficient as it required a lot of operators. The telephone company realized that, in a few decades, the number of operators needed would be a fair proportion of the total population of the country! The irony is the telephone company did just that: they made everybody a telephone operator. By providing telephones with dials, one did not ask an operator for a connection, but directly dialed the desired number. This radical idea was a success. In much the same way, we intend to let the end-user perform simple and moderately complex applications programming. All complex, low-level programming could still be done by the GIS development team. This is analogical to the fact that we still have telephone operators to help us with complex problems like international or third-party calls.

Fourth generation languages and G

In the beginning there were first generation machine languages. A 1GL consists of computer instructions in a numeric form. For example, 1010 110101 10101 may be a computer instruction that adds the contents of two registers. Then came assembly languages (2GLs). In AT&T-3B2 assembly the add instruction may look like this: adw2 %r1 %r2. Today, 3GLs are most commonly used. In C the add instruction may look like this: x + = y. A 4GL goes further by incorporating the following features (adapted form James Martin, 1985).

- It is friendly to the professional user (menu-based systems are best for casual users).
- It can be easily interfaced to a menu-based front end.
- Has a lot of domain-specific functions (tools) provided.
- Both procedural and non-procedural coding is supported.
- Operations like declaration statements, common blocks, etc. are eliminated.
- Intelligent default assumptions are made, where possible.
- Encourages structured code; GOTO statements are eliminated.
- Errors are usually high level; hardware or software dependent errors are rare.
- The programs in 4GLs are often interpreted with provision for compilation.
- Code is easy to read and is natural to the application domain.

The G language has the basic components of a GIS and 4GL. The data management and graphics for G are provided by the GRASS geographic information system. (GRASS is a UNIX/C based GIS developed by the U.S. Army's Construction Engineering Research Laboratory.) The G toolbox has been coded in both C and G. All the functions in G are based on those in the map analysis package (MAP) designed and developed by Professor Dana Tomlin. G's control structure is currently provided by the UNIX/C-Shell. We have chosen the C-Shell because it is easy to use and is fairly standard. In an earlier design the control was provided by a 5GL called LISP. This provides GIS access to AI applications written in LISP. As G is open structured, its control mechanism can be a menu system or any other program written in any language that is supported in UNIX.

Programming in G: some examples

Before introducing G functions, let us explain how our data organization works. The maps are stored in matrices and each cell in the matrix has a number. Each number denotes some category. The number zero is reserved to denote the absence of data. The most common function is the reclassification function called Grecode. It takes as input, a inputmap and an output map. It changes, as specified, the numbers in the cells of the map. For example,

```
Grecode <inputmap> <outputmap> changing 5 thru 10 to 1
and 25 thru 40 to 2
and 41 to 5
```

The above operation changes the categories between 5 and 10 to 1 and categories between 25 and 40 to 2 and category 41 to 5. The above statement can be entered and run as shown above. It constitutes a complete G program! The above function can be grouped with more functions into a longer G program. Each file of G operators in turn becomes a G function that can be called by other G program. This structure is directly inherited form the UNIX C-shell.

To carry our example further, we introduce two more functions. The Ggetcats <mapname> function returns a list of categories in the specified map. The Ggetarea <mapname> <category> operation returns the area (number of cells) occupied by the specified category in the specified map.

Example 1

Consider a situation wherein one has a map called development. Assume it is a bitmap of 0's and 1's, where a 1 indicates development. We are to write a program that will take such a map, find all clusters, and remove those which have less than N cells on the map (that is, remove all small parcels). Let us call this program clean.g.

```
#Program name is clean.g
#usage: clean.g <inputmap> <outputmap> <N>
#Note: the # sign marks comment lines in G
#Note: All variables have the $ sign preceding them.
#The $ sign is not used ONLY when the variable
#is being set.

#First, the three operands are bound to internal variables.
#The first, second and third operands are denoted by
#$1, $2 and $3 respectively.
set inmap=$1 # $1 is the first operand (by default)
set outmap=$2
set N=$3
#First find all clusters. Let a cell be part of a cluster
#if it is within 10 units of another
#Thin out the current regions to eliminate insignificant
#connections between clusters
Gthin inmap result 10
#grow back the map
Ggrow result result 10
#Find segments. Each segment is given a different category name.
Gsegment result result

#Now to find all parcels smaller than N cells
#We accomplish the task by looping through
#all the categories, checking the areas as we go
#along.

#first find all the categories.
#find the area of 'category'
set cats="Ggetcats inmap category"

#Then loop through the categories while checking their areas for
each category ($cats)

    #find the area of a category
    set area = "Ggetarea result category"
    if ($area <N) then #if the parcel is too small

    #eliminate category
    Grecode result outmap changing category to 0
    Gcopy outmap result
endif
end
```

Example 2

We shall now write a program, called build.g, that takes some user input and makes appropriate G calls. This program is supposed to do a weighted analysis on three maps. Let the three maps be SOIL, WATER, and SLOPE. The program creates a resultant map that has the SOIL, WATER, and SLOPE maps weighted according to w1, w2, and w3, respectively. The values of the weights for various types of buildings is different. Let us say that the values of w for the different types of buildings is as shown below:

Building	w1	w2	w3
School	3	2	1
High_rise	4	9	1
Homes	1	1	3

We are to write a program that takes one argument, the argument being the name of the building type.

```
#The program build.g
#get the argument
set building=$1

#set up a switch statement.
switch ($building)

case School:
     set w1=3
     set w2=2
     set w3=1
breaksw

case High-rise:
     set w1=4
     set w2=9
     set w3=1
breaksw

case Homes:
     set w1=1
     set w2=1
     set w3=3
breaksw

default:
     set building=NULL #A flag
endsw #end of the switch

#Now the weighting operation
if ($building != NULL ) then
#Switch successful
#The Gcalc function treats maps likes variables
     Gcalc RESULT= (SOIL * $w1 + WATER * $W2
     + SLOPE * $w3)/($w1+$w2+$3)
```

```
else
    echo "Unknown building type. Please input one
    echo "Homes, High_rise or School"
endif
#End of build.g
```

G is an experimental programming language and is by no means complete. The GRASS group will continue to add to G's functional tool-kit. We are also pursuing the geoLISP programming environment. LISP is a good 5GL for performing complex, object-oriented map analysis. It also provides us with a good database capability.

Conclusions

Two major issues are at hand. We need better ways in which a map can be queried for information such as proximity, clustering, inter-visibility, etc. This will help in programming heuristics into the GIS-based expert system. For example, an expert system should be able to ask a GIS if there is a flat, long strip of land on a given terrain map. The G language provides good features for such queries, however it lacks feature recognition capabilities.

As knowledge acquisition is a major problem, new machine learning methods need to be given a closer look. In particular, I believe, case-based reasoning will make important contributions to GIS and environmental engineering.

Acknowledgments

The research reported in this chapter was conducted when the author was with the Intelligent Engineering Systems Laboratory at the Massachusetts Institute of Technology (1983-1986). The research was supported by the U.S. Army Corps of Engineers, Construction Engineering Research Laboratory (CERL) under the supervision of Dr. Ravi Jain. I thank Dr. Jain for his foresight in recognizing the importance of AI in environmental engineering at an early stage. I also thank Mike Shapiro and Jim Westervelt of CERL for their contribution to the ideas presented here and for help in developing the expert systems described in the chapter.

References

Alexander, C., 1974. *Synthesis of Form.* University of California, Berkeley, CA 94720

Bennett, J.S. and R.S. Enameler, 1979. "SACON: A knowledge-based consultant for structural analysis," in *International Joint Conference on Artificial Intelligence (IC-JAI)6*:47-49.

Buchanan, B.G., G.L. Sutherland, and E.A. Feigenbaum, 1969. "Heuristic DENDRAL: A program for generating explanatory hypotheses in organic chemistry," in B. Meltzner and D. Michie, eds., *Machine Intelligence*, Vol. 4., Edinburgh University Press, pp. 209-254.

Buchanan, B.G. and E.H. Shortliffe, 1984. "Rule-based expert systems, The MYCIN experiments of the Stanford heuristic programming project," Addison Wesley.

Carbonell, J. G., 1986. "Derivational analogy: A theory of reconstructive problem-solving and expertise acquisition," in R.S. Michalski, J. G. Carbonell, and T. M. Mitchell, eds., *Machine Learning: An Artificial Intelligence Approach*, Vol. 2, Morgan Kaufman.

Foderraro, K.J., K.L. Sklower, and K. Layer, 1983. "The FRANZ LISP manual," UNIX User's Manual, Computer Science Division, Department of Electrical Engineering and Computer Science, University of California, Berkeley, CA.

Fox, M., 1983. "ISIS: A constraint directed job-shop scheduling system," Ph.D. thesis, Department of Computer Science, Carnegie Mellon University, Pittsburgh, PA.

Gaschnig, J., 1979. "PROSPECTOR: An expert system for mineral exploration," *Machine Intelligence*, Infotech State of the Art Report 9, No. 3.

Hammond, K.J., 1986. "CHEF: A model of case-based planning," in *Proceedings of AAAI 86*:267-271, Philadelphia, PA.

Joy, W., 1980. "An introduction to the C shell," UNIX User's Manual, Computer Science Division, Department of Electrical Engineering and Computer Science, University of California, Berkeley, CA.

Kolodner, J.L., R.L. Simpson, and K. Sycara-Cyranski, 1985. "A process model of case-based reasoning in problem-solving," in *Proceedings of IJCAI 85*:284-290, Los Angeles, CA.

Lacey, R.M. and W.D. Servinghaus, 1981. "Natural resource considerations for tactical vehicle training areas," U.S. Army Construction Engineering Research Laboratory, Technical Report N-106, June.

Lynch, K. and G. J. Sussman, 1977. "Forward reasoning and dependency directed backtracking in a system for computer aided circuit analysis," *AI Journal 9*:135-196.

Martin, J., 1985. *Fourth Generation Languages, Volume I: Principles*, Prentice-Hall, Inc.

Martin, J., 1986. *Fourth Generation Languages, Vol II: Representative 4GLs*, Prentice-Hall, Inc.

Navinchandra, D., 1984. "CERES: An expert system for environmental impact analysis," working paper, U.S. Army CERL, Environmental Division, Champaign, IL.

Navinchandra, D., 1985. "IMST user's manual: A tool for building rule-based expert systems," *CCRE: 85-6*, Department of Civil Engineering, Massachusetts Institute of Technology.

Navinchandra, D. and W. Goran, 1985. "Engineering topographic laboratories," proceedings of December 1985 workshop on GIS.

Navinchandra, D., 1988. "Case-based reasoning in CYCLOPS, a design problem-solver," in J. Kolodner, ed., *Proceedings of the DARPA Workshop on Case-Based Reasoning*, pages 286-301, Morgan Kaufman.

Navinchandra, D. and R. Lacey, 1985. "Steps toward alternative land use planning guidelines: A knowledge-based approach," Intelligent Systems Laboratory, Technical Report IESL-TR-85-2.

Simpson, R.L., 1985. "A computer model of case-based reasoning in problem-solving: An investigation in the domain of dispute mediation," Ph.D. thesis, School of Information and Computer Science, Georgia Institute of Technology.

Sycara, K., 1987. "Resolving adversarial conflicts: An approach integrating case-based and analytic methods," Ph.D. thesis, School of Information and Computer Science,

Georgia Institute of Technology.
Tomlin, D., 1983. "Digital cartographic model in techniques in environmental planning," Doctoral Dissertation, Yale University.
Tomlin, D., 1985. Personal communication, Graduate School of Design, Harvard University.
Tomlin, D., 1985. "The MAP analysis package," User's Manual. Graduate School of Design, Harvard University.

6

Probabilistic Inferencing and Spatial Decision Support Systems

Jeff R. Wright and Kurt A. Buehler

Introduction

The premise behind the use of formal land use allocation models is that certain geographical regions are better suited for specific facilities and land uses than other geographical regions. The goal of land allocation models is to generate good strategies for the assignment of specific land use activities to the most suitable available land units, based on appropriate measures of suitability. As distinguished from the widely-studied facility location models that treat optimal locations as points in a plane or along a network, land use allocation models treat optimal locations as two-dimensional areas.

Consider the region of interest to be partitioned into an arrangement of N non-overlapping geographical units numbered from $i = 1, 2, ..., N$. Every basic geographical unit can be characterized by a set of economic, environmental, and spatial attributes. Full and complete knowledge about the attributes for each unit is assumed. A subregion is any contiguous grouping of these units. The economic, environmental, and spatial characteristics of any subregion are a function of the collective attributes of the units comprising that subregion. Thus, for example, the total area of a subregion allocated for a land use is the sum of the individual areas of the units comprising that subregion.

The cost of acquiring and developing unit i for a specific land use can be based on a number of market and non-market factors, including physical characteristics of the land, proximity to services, etc. The suitability of a geographical unit, s_i, is defined to be a "measure of the physical capacity of a specific location (i.e., geographical unit) to support a specific land use" (Anderson, 1987). This determination is usually based upon natural characteristics of the land such as slope, soil type, and other physical factors. A number of mathematical models have been proposed to calculate suitability ratings as functions of these factors and underlying natural processes. The output from these models is a map of information depicting the relative capability of any given geographical unit to support a given land use. These models are discussed at length in Hopkins (1977), Chapin and Kaiser (1979), Anderson (1987), and Diamond (1988).

The underlying assumption of this discussion is that the problem of selecting or allocating a particular set of geographical units as a subregion for a particular land use is appropriately partitioned into three sub-problems (see Diamond and Wright, 1988): (1) collecting, storing, and managing data to be used in the determination of suitability for the intended land use, (2) the evaluation and analysis of data about each geographical unit to make that determination of suitability, and (3) the development of a strategy for selecting the set of units to form the subregion. Furthermore, three traditionally disparate technologies are most appropriate to address these sub-problems: (1) spatial data management using geographical information systems (GIS), (2) expert systems (ES) for determining land suitability, and (3) numerical optimization for identifying dominant land allocation strategies. The focus of this research is the design and implementation of a shell development environment that can be used in the determination of suitability indices for a given land use allocation activity. Implementation results in a "tight coupling" of modern GIS technology (a popular raster-based GIS) and a Bayesian inferencing procedure for purposes of generating GIS maps displaying the probability of land impacts for a given allocation strategy. As a shell environment, this system can be used to develop application-specific spatial decision support systems (SDSS) when a Bayesian predictive model is an appropriate probabilistic inferencing mechanism.

Very recently, the deficiencies of modern GIS technologies to support more than only very simple data analysis and modeling activities have received considerable attention. Most notably, Densham and Goodchild (1989) offered the following position statement to participants at the specialists meeting for the 6th Initiative of the National Center for Geographical Information Analysis held in Santa Barbara in March, 1990:

"Definitions of geographic information systems often focus on the capture, storage, manipulation, analysis and display of spatial data—implying that geographic information systems implicitly are designed to support spatial decision-making. For many spatial problems, however, geographic information systems do not support decision-making effectively: analytical modeling capabilities are lacking and system designs are not flexible enough to accommodate variations in either the context or the process of spatial decision-making. One response to these needs is the development of spatial decision support systems. We draw a distinction between geographic information systems and spatial decision support systems in terms of system design, the types of problem to which each can be applied, and the decision-making process supported."

To provide a context for the development of a spatial reasoning shell, we begin with a discussion of the emerging technology of spatial decision support systems followed by a rationale for non-procedural and approximate reasoning applied to the determination of land suitability indices. We then present an overview of the B-infer SDSS, and conclude with a discussion pointing out future research directions.

Spatial decision support systems

Decision support systems (DSS) evolved in the late 1960s and matured through the 1970s in response to the needs of the business data-processing community for more complete modeling capabilities, as well as better interaction with these models than that offered by management information systems (MIS). This evolution is well documented (see, for example, Gorry and Scott-Morton, 1971; Little, 1970; Alter, 1980; Bonczek, et al., 1981; Sprague and Watson, 1986; Holsapple and Whinston, 1987), a substantial DSS theory has evolved (Sprague and Carlson, 1982; Sprague and Watson, 1986; Aldag and Power, 1986; Malakooti, 1988; Zachary, 1988), and successful applications have motivated the establishment of DSS sub-units in many academic disciplines including operations research (OR), industrial engineering, and business and management programs. While numerous definitions and characterizations of DSS have been proposed (see, for example Alter, 1980; Bonczek, et al., 1981; Keen, 1980; Moore and Chang,1980; Geoffrion, 1983), for purposes of this discussion, the characteristics of DSS presented by Turban (1988) are used. Turban offers several distinguishing characteristics that may be found in a given application:

1. DSS provides support for decision makers mainly in semi-structured and unstructured situations by bringing together human judgement and computerized information. Such problems cannot be solved (or cannot be solved conveniently) by other computerized systems, such as electronic data processing (EDP) or management information systems (MIS), nor by management science.
2. Support is provided for various managerial levels ranging from top executives to line managers.
3. Support is provided to individuals as well as to groups. Many organizational problems involve group decision-making. The less structured problems frequently require the involvement of several individuals from different departments and organizational levels. DSS assist integration among individuals whenever appropriate.
4. DSS provides support to several interdependent and/or sequential decisions.
5. DSS supports all phases of the decision-making process: intelligence, design, choice, and implementation.
6. DSS supports a variety of decision-making processes and styles; there is a fit between the DSS and the attributes of the individual decision makers (e.g., the vocabulary and decision style).

7. DSS must be adaptive over time. The decision maker should be reactive, being able to confront changing conditions and adapt DSS to meet these changes. DSS must be flexible so that users can add, delete, combine, change, or rearrange basic elements (providing fast response to unexpected situations). This capability makes possible timely, fast, ad hoc analyses.
8. DSS should be easy to use. Users must feel "at home" with the system. User-friendliness, flexibility, strong graphic capabilities, and an English-like dialog language can greatly increase the effectiveness of DSS. This ease of use implies an interactive mode.
9. DSS attempts to improve the effectiveness of decision-making (accuracy, timeliness, quality), rather than it efficiency (cost of making the decision, including charges for computer time).
10. The decision maker has complete control over all steps of the decision-making process in solving a problem. A DSS specifically aims to support and not to replace the decision maker. The decision maker can override the computer's recommendation at any time of the process.
11. DSS leads to learning, which leads to new demands and the refinement of the system, which leads to additional learning, and so forth, in a continuous process of developing and improving the DSS.
12. DSS should be easy to construct. End-users should be able to construct simple systems by themselves. Larger systems could be built in the users organizations with only minor assistance from the information systems (IS) specialists.

Rather than being a structured analytical procedure, DSS denotes a technology that is flexible and adaptive within a problem domain, and, as a result, has enjoyed increasing popularity among OR researchers and practitioners paralleling advances in computer hardware and software technology (Turban, 1988).

If there is weak agreement as to a definition of DSS, there seems to be a general consensus that, at least conceptually, DSS are comprised of three components as described schematically in Figure 6.1:

- *A data management subsystem* that includes data query and management routines and necessary internal information about data structures.
- *A models management subsystem* that includes an appropriate suite of systems models, a directory of models and their requirements, and utilities for integrating and executing model elements.
- *A dialog subsystem* consisting of the hardware and software that provides the interface between the user of the system and the system itself.

A rich collection of software tools is available for the development of these individual components, but only recently have efforts focused on general purpose DSS development environments. Indeed, the technologies supporting the development of data management subsystems and dialog subsystems are quite mature, while the technology for model-based management systems is still relatively young. However, recent major advances in object oriented programming systems

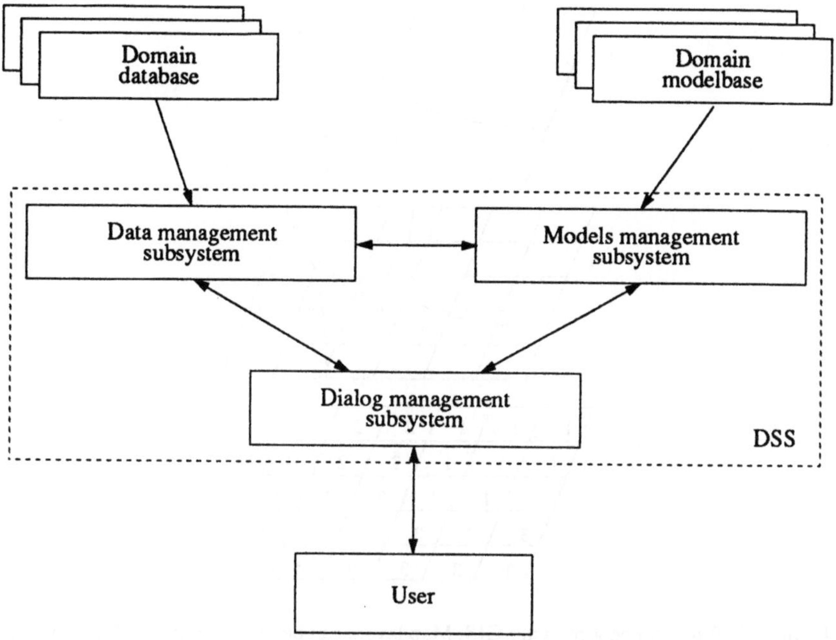

Figure 6.1. Logical framework for DSS applications showing the relationship between the three principal DSS subsystems (data management subsystem, models management subsystem, and dialog management subsystem). A given implementation may actually integrate all these elements into a single program module depending on the DSS generation technology employed.

(OOPS), hierarchical data structures, and compiler construction environments should speed the development of powerful model-based management systems.

When a DSS is developed for use with a domain database that has a spatial dimension (one or more data attributes having a distribution over space), or when the solution space generated by the DSS has a spatial dimension, and when most of the characteristics set forth by Turban are observed, the analytical system in aggregate may be referred to as a spatial decision support system (SDSS).

As suggested by Densham and Goodchild (1989), this description effectively describes spatial decision support systems (SDSS) and also serves to distinguish SDSS from geographic information systems (GIS), suggesting that some GIS functionality is an important component of any meaningful SDSS. From another perspective, the difference between GIS and SDSS is that SDSS deals with a solution space provided by one or more analytical (modeling) elements, as well as a problem (geographical) space manipulated by the GIS database. Thus GIS is an important component of any SDSS.

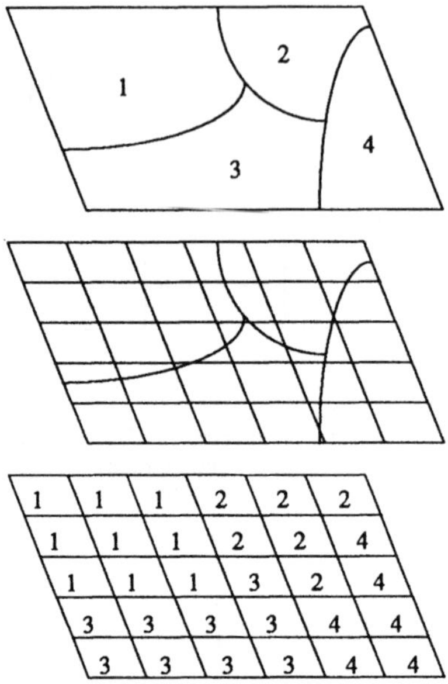

Figure 6.2. Basic concept of raster GIS. Map layers can be produced using this procedure for a wide variety of physical and logical spatial attributes.

Recent years have seen a dramatic increase in the use of GIS for managing spatial data. While there exist several fundamentally different types of GIS technologies, the most straightforward representation is that offered by Figure 6.2. Suppose we are interested in making a determination of the suitability of a particular region for a specific land use based on soil type. We might classify all relevant soil possibilities by assigning each a discrete category number (*1*, *2*, ..., *N*). Of course, there exists a continuous distribution of soil type across any region. By overlaying a grid cell referencing system over our soil map, we are able to assign a corresponding discrete value through some averaging process. Because the location of each grid cell is known, an approximation of our original soils map may now be stored and manipulated as a rectangular matrix of category values. By performing the same approximation to attributes such as slope, vegetation, land use, ownership, etc., we can subsequently pose questions such as "How much land in the region having sandy clay soils and average slope of 5° or less is presently being used for commercial purposes?"

Land use planning and design, site selection, and other problems requiring the manipulation of geographic information have benefited from recent advances in GIS. The incorporation of decision models into GIS has been discussed recently by several researchers, including Steinitz and Brown (1981), Tomlin and Tomlin (1982), Cederborg and Tosetti (1982), Gilbert, et al., (1985), and Elmes and Harris (1986). The role of the GIS in the proposed implementation environment would be to store, generate, and manage all of the data required for the land use allocation model. This would include the computation of distances between geographical units, the determination of areas and adjacency relationships, and the storage and calculation of acquisition and development costs for each unit based on, for example, the slope and soil characteristics of the associated land. All of this information could be conveniently stored and/or calculated by a GIS and passed to the land use allocation model without manual intervention.

The integration of GIS and DSS technology toward the development of SDSS applications would appear to be particularly useful in addressing problems requiring effective land suitability analysis. The determination of land suitability has been an important problem for a variety of planning activities. Examples of these activities include land use planning, site planning, facility location, corridor selection, and impact assessment. In all of the above examples, the suitability of a land area to a specific purpose or activity is must be determined in order to make a good decision about the allocation of an activity. For the purposes of this chapter, the term land suitability should be viewed as the relative ability or inability of a land area to provide the characteristics (factors) needed to support an activity or land use given the factors deemed important to the activity.

Methods of land suitability determination are based largely on identification of the factors that contribute to the suitability of the land for the purpose at hand, and the subsequent combination of these factors to produce some sort of suitability map. Hopkins (1977) provides a taxonomy of several methodologies for the generation of land suitability maps. Lyle and Stutz (1983) provides an excellent discussion of the computerization of this suitability map generation.

Since the late 1960s, GIS technology has been used to generate suitability maps. Earlier techniques were insufficient in that they did not allow for different weights on the factors affecting suitability and/or they ignored location of activities based on factors other than those of an ecological nature. Overcoming these limitations seemed to require the use of computerized systems. Many case studies exist that have taken advantage of GIS in the determination of suitability for later use in planning studies. Antenucci and Dangermond (1974) describes work done in the famous Maryland Automated Geographic Information (MAGI) system. Dangermond (1974) describes the use of computerized capability/suitability models in work ESRI did in the Zulia Regional Planning Study. In both cases, GIS technology aided planners in the development of solid land use plans. In the Maryland case, the MAGI system was used for many other planning studies. Another study on the computerization of suitability analysis was the implementation of the Land Evaluation and Site Assessment (LESA) system on a GIS (Williams, 1985). The prospect of using this system on studies covering large areas prompted the GIS implementation. The LESA system had been previously implemented by hand on a site-by-site basis. This study demonstrated that the "GIS approach extends the utility of LESA into area-wide modeling and allows a more flexible site analysis." In all cases the conclusion seems to be that GIS technology is extremely useful in aiding all types of regional and land planning activities.

The application of expert systems and proximate reasoning technologies to the analysis of land suitability is a relatively new concept. An even more recent concept is the idea of integrating expert systems with some database of geographic information (whether the database in question is a true GIS or not). The role of artificial intelligence (and especially expert systems) in solving spatial problems, however, has been well documented. Davis and Clark (1989) have provided an excellent bibliography of expert systems in natural resource management. A few notable papers are mentioned here.

Barath and Futo (1984) introduced a logic-based system for undertaking regional planning tasks. It was based on a hierarchical model of national, county, and district level activities. Every level has its own goals and the problem was to find an acceptable solution to these goals. Coulson, et al. (1987) provide a good description of possible applications of artificial intelligence techniques to problems in natural resource management and specifically landscape ecology. Topics discussed include:

- Integrated expert systems, which link management models with natural resource models.
- Intelligent GIS's, which permit interpretation of relationships within and among landscape data themes.
- Artificial intelligence modeling of animal behavior and interaction with the environment.

Davis, et al. (1987), describe an expert system shell, GEM-II, which has been developed specifically for classification problems in environmental management. Specifically, GEM-II has been used to model fire behavior and its effect on vegetation in Kakadu National Park in Australia. The incorporation of reasoning with spatial knowledge, as well as the combination of scientific models with heuristic models and the ability to handle taxonomic models within a spatial inferencing, is discussed in a later paper by Davis (1989).

Robinson, et al. (1987) provide an excellent discussion of several areas in which expert systems can and have been applied in the area of natural resource management including map design, geographic feature extraction, geographic database systems, and geographic decision support. These and many other references point to the desirability of the application of AI techniques to the manipulation and analysis of spatial data. A few prototype systems have been developed to try to solve spatial problems by integrating expert system and GIS components. Smith and his group at the University of California, Santa Barbara, have been working on the problem of integrating AI techniques into a GIS for several years. They have produced a couple of successful prototype systems, KBGIS and KBGIS-II (Smith,1984; Smith, et al., 1987). Han and Kim (1989) have developed an expert system prototype (ESSAS) which uses the Personal Consultant Plus interfaced with Lotus 1-2-3 for database files (pseudo-GIS) and AutoCad for graphic representation for the purpose of conducting site selection and analysis.

The emergence of expert systems shells has resulted from the need for software tools that can be used for rapid systems development. Shells are particularly useful when the system development must be done by non-programmers or engineers unfamiliar with traditional expert systems development in languages such as LISP and PROLOG, or when time constraints preclude exhaustive system specification methods of development. Shells are well adapted to rapid prototype development environments when the criteria for system performance are unknown or only generally understood.

A shell is a software module designed to promote the development of knowledge-based systems much the same as spreadsheets promote rapid development of accounting and business decision-support systems. The oldest shell systems grew out of the programming language LISP. While LISP has proven to be a very flexible tool for system development in the hands of expert programmers, it is relatively difficult to learn and use, and intolerably low levels of machine performance are frequently encountered with systems developed by less-than-expert programmers. While virtually any programming language can be used for developing expert systems, modern shell systems have proven to be an extremely cost effective and efficient alternative to lower-level languages.

Development tools have emerged at an astonishing rate over the past few years. Shells can be very narrow, application-oriented tools or broader more general-purpose tools. Because of the wide range of expert systems shells presently available, it is important that the proper development tool be selected for any given application.

Expert systems development tools are presently available on the full range of computer hardware from microcomputers to supercomputers (Bundy, 1984; Clancy and Rothenburg, 1987; Gevarter, 1987; Harmon and King, 1985; Hayes-Roth, et al., 1983; Maher, 1987; Richer, 1986; Waterman, 1986). Several conventions for grouping systems development tools have been espoused. Freedman (1987) categorized development tools by using their knowledge representation scheme and inferencing strategy. Whittaker, et al. (1986) classified development tools using knowledge representation and reasoning schemes. Gevarter (1987) classified tools by inference approach, object description, and actions. Waterman (1986) categorized expert system development tools in terms of knowledge representation and programming methods supported by the tool. Six categories of tools were defined: rule-based, frame-based, procedure-oriented, logic-based, object-oriented, and access-oriented. A more functional approach was used by Mullarkey (1987) who provides a refinement of the classification framework first presented by Maher (1984). Development tools are characterized in terms of required hardware support platform vs. software complexity showing the relative relationship between the major languages and shells. The interested reader is referred to one or more of these excellent reference materials for further study of the topology of these tools.

The preceding discussion of SDSS, its potential use in conducting land suitability assessment, and tools for efficient applications development, provide a framework for understanding how techniques for suitability analysis based on established analytical procedures may be integrated with GIS technology. In the next section, we provide a more detailed description of this type of integration.

Non-procedural and approximate spatial reasoning for suitability analysis

Most land use suitability assessment models are based on manipulations involving "factor maps." A factor is a unique geographical attribute such as slope of terrain, soil, or vegetative cover. Variations within a factor such as degrees of slope or soil classes are referred to as factor *types*. A factor map is a map showing the geographical location and distribution of factor types for a given factor and can be combined in various ways to generate a composite map depicting, either graphically or numerically, the relative capability of any given geographical unit to support a specific land use. A variety of different strategies for achieving these combinations have been suggested:

1. *Linear combination* is a common factor combination technique. This method relies upon the decision maker to supply both a rating schedule and a weighting schedule. The rating schedule depicts the relative suitability of the various types within a factor. For example, slopes greater than 20% are rated 0, between 10% and 20% a 5, and less than 10% a 10. The weighting schedule depicts the relative importance of the various factors in determining the suitability of a unit. For example, slope might be twice as important as depth to bedrock. The ratings for each type are multiplied by a weight for each factor and the result is summed.

2. The composite score for a unit might be conditioned upon the presence or absence of a factor or interactions between factors. If these interactions are well-defined, then it may be possible to define a non-linear function that explicitly captures these relationships. For example, a soil loss rating can be easily computed from standard equations given information on land cover, slope, and watershed shape. As such, *non-linear combination* methods can be very useful when applied to specific components of an overall suitability assessment. However, in practice the relationships between factors are rarely well-understood and it is difficult to represent most interactions as mathematical functions (Anderson, 1987).

3. The *direct assignment* technique involves enumerating all possible combinations of factors, and assigning a composite score to each unique combination when all factors are considered simultaneously. The advantage of this method is that it requires the explicit consideration of interactions between factors. The disadvantage of this method is that the number of composite score assignments that must be made increases exponentially with the addition of new factors.

4. The *rules of combination* method involves the use of explicit rules to assign suitabilities to *sets* of types rather than to single combinations. Rules of combination are expressed in terms of verbal logic rather than in terms of numbers and arithmetic. For example, "if type 1 of factor X and type 2 of factor Y occur together in a region, then the rating of factor X type 1 overrides that of factor Y type 2." These rules are based upon an understanding of the un-

derlying natural system being described rather than on a single set of relationships repeated for all combinations regardless of the specific types and factors being combined. If properly devised, rules of combination can handle interdependence between factors. Furthermore, they do not require an explicit mathematical relationship as do non-linear combination methods.

The integration of rule-based expert system technology and GIS, provides an excellent framework for automating suitability assessment via rules of combination methods. Suppose that an activity is to be assigned to a region and the suitability for any particular cell to support that activity has been judged to depend on slope, geology, and forest density. Suppose further that complete categorical information for the region of interest exists in a GIS as discussed above and as presented as Figure 6.3. An example of a combinational rule that might be used to determine suitability might be:

```
IFMAP slope 1-3
ANDIFMAP geology 4-6
ANDNOTMAP density 4
THENMAPHYP 4 suitability
```

This rule contains the judgement that if, for any particular cell, the average slope falls into slope categories 1, 2, or 3, and the average geology is of type 4, 5, or 6, and the forest density is not of type 4, then a suitability of 4 may be assigned.

Additional rules may be employed to reflect additional judgements:

```
IFMAP slope 1-3
ANDIFMAP soils 2-3 5 13 15-16 21 24
ANDNOTMAP vegetation 1-6
THENMAPHYP 4 suitability
!
IFMAP slope 6-7
THEN too steep
!
IFMAP soils 1 4 6-8 10-12 14 17-19 22-23
THEN unstable surface
!
IFMAP vegetation 1-6
THEN forest impact likely
!
IF too steep
THENMAPHYP 1 suitability
!
IFNOT too steep
ANDNOT forest impact likely
ANDIF unstable surface
THENMAPHYP 2 suitability
!
IFNOT unstable surface
```

```
ANDNOTMAP vegetation 1-6
ANDIFMAP slope 3-5
THENMAPHYP 3 suitability
!
IF too steep
ANDNOT forest impact likely
ANDNOT unstable surface
THENMAPHYP 3 suitability
```

These rules may then be imposed across the spatial extent of the region of interest and for all appropriate map layers producing a new map layer reflecting spatial suitability for the target land use as shown in Figure 6.3.

Three general observations can be made regarding factor combination methods:

1. Factor combination is often best achieved through the use of an explicit set of rules and verbal logic (Hopkins, 1977).
2. The selection of meaningful ratings, weights, and methods of combination is a complicated and tedious task to undertake manually. In short, decision makers are faced with at least three complex tasks: a) identifying the most relevant set of factors, b) deriving a meaningful set of ratings and weights, and c) identifying an appropriate functional form for combining factor maps that reflects the relationships between factors.
3. Expert judgement is an essential ingredient of the assessment process. For example, the rating assigned to each type of every factor must reflect all of the characteristics of that type and the potential impact of the proposed land use or facility on those characteristics and vice-versa. In describing an assignment technique, Lyle and von Wodkte (1974) state: "Although this sequence provides an orderly way of defining and articulating judgements, it relies heavily on the knowledge of those formulating the model. Usually, it is necessary to consult with experts in various disciplines during this phase to render the judgements as reliable as possible."

The problem characteristics outlined above are indicative of a "good" problem domain for a rule-based expert system. A rule-based expert system is a computer program designed to use domain specific knowledge gathered from human experts to perform problem solving tasks in a manner similar to humans. Rule-based expert systems are separated into an explicit rule base describing general problem solving strategies, and a control program to search the rule base. This programming framework allows the rule structure to apply domain specific knowledge to a problem and make "inferences" about problems in a more and flexible manner than traditional algorithm-based programs. It also facilitates the implementation of explanation procedures that can be used to justify the rationale behind a specific problem solving approach. Furthermore, most rule-based systems have ways of dealing with uncertainty, including certainty factors, fuzzy logic, Bayesian analysis, and a number of other statistically defensible procedures.

2	1	2	2	2
3	2	1	2	2
3	3	2	2	1
3	3	4	4	3
3	3	4	4	3

Slope category	Value
No slope data available	0
Flat slope (0-2%)	1
Very low slope (2-5%)	2
Low slope (5-10%)	3
Moderate slope (10-15%)	4
High slope (15-20%)	5
Very high slope (20-25%)	6
Severe slope (>25%)	7

1	1	10	15	15
10	10	1	10	10
13	13	10	1	1
13	13	15	10	10
13	9	9	15	15

Soils category	Value
No soils data available	0
Sandy loam	1,6,17
Clay loam	2,3,5,15,16,21,24
Silty clay	4,7,8,10-12,14,18,19,22,23
Clay	13
Rock outcrop	9

14	14	8	6	6
13	13	14	10	10
8	8	13	14	14
8	8	8	10	10
8	8	4	4	4

Vegetation category	Value
No vegetation data available	0
Dense forest	1,3,5
Sparse forest	2,4,6
Dense scrub	7,9,11
Sparse scrub	8,10,12
Grass	13,14
Barren	15

2	2	2	0	0
2	2	2	2	2
4	4	2	2	2
4	4	3	2	2
4	3	0	0	0

Suitability category	Value
No suitability data available	0
Poor suitability	1
Fair suitability	2
Good suitability	3
Excellent suitability	4

Figure 6.3. An example of rule-based inferencing across GIS map layers. A suitability raster map layer is produced from empirical data contained in other map layers through rule-based inferencing procedures that may include a reclassification of the base GIS categorical values achieved by proper structuring of rule antecedent conditions.

There has been limited research on the role of expert systems technology in the field of land use planning. Han and Kim (1989) have reported on the development of a prototype expert system for site selection and analysis. However, the system does not use suitability assessment techniques nor is it designed to interface with a geographical information system. Only a few researchers to date have proposed integrating GIS and rule-based expert systems. Among them, DeMers (1985) discusses the potential role of this technology in the classification of land according to agricultural economic viability, Smith and Pazner (1984) describe a knowledge based system designed to answer queries concerning geographic objects in large spatial databases, Jungert (1984) proposes a methodology for conducting efficient searches in spatial databases employing inference systems, and Coulson, et al. (1987) discuss the role of "intelligent" geographic information systems in natural resource management.

Another possible role of an expert system in the implementation environment is that of an interactive "front end" to a GIS. The expert system would query land use planners for the specific characteristics and requirements of the land use that is to be allocated. Based on this interaction, the expert system would identify the relevant geographical factors and prompt the user for the ratings and weights of each factor and the degree of "certainty" associated with each. If the user was relatively confident about these values, then no further questioning on this factor would be required. However, if the user was not confident about the values, the system would request additional information that might provide insight (either to the user or the expert system) as to the most reasonable set of ratings and weights. As the session progressed, the expert system would check for the consistency of ratings and weights based upon information available in the rule base and the users previous responses. At the same time, the expert system would serve as a knowledge acquisition facility by which all information input by the user or generated by the expert system would be stored in the rule base for future use as necessary. At the conclusion of this interactive session, the system would recommend a set of ratings, weights, and factor combination strategy reflective of an expert understanding of the underlying natural system. These recommendations could be accepted, modified, or rejected by the user on an ongoing basis until the process converged on an acceptable set of values.

For many problems requiring the determination of land suitability, applications are so diverse and unique that available knowledge is too incomplete or inexact for the use of formal rules of combination. Indeed, a domain expert may not be available or may not be able to articulate a rule set to a sufficient degree of precision. For these applications, a probabilistic inferencing procedure may be acceptable.

An "ideal" land allocation procedure using probabilistic reasoning to perform land suitability might be structured as shown in Figure 6.4. Such a two-level process might employ a formal procedure for computing the probability of an impact level that might be expected to result from the probabilities of various determining factors; level 1. Suitability would thus be determined in terms of the probability of experiencing one of a finite number of impact levels for a well-specified

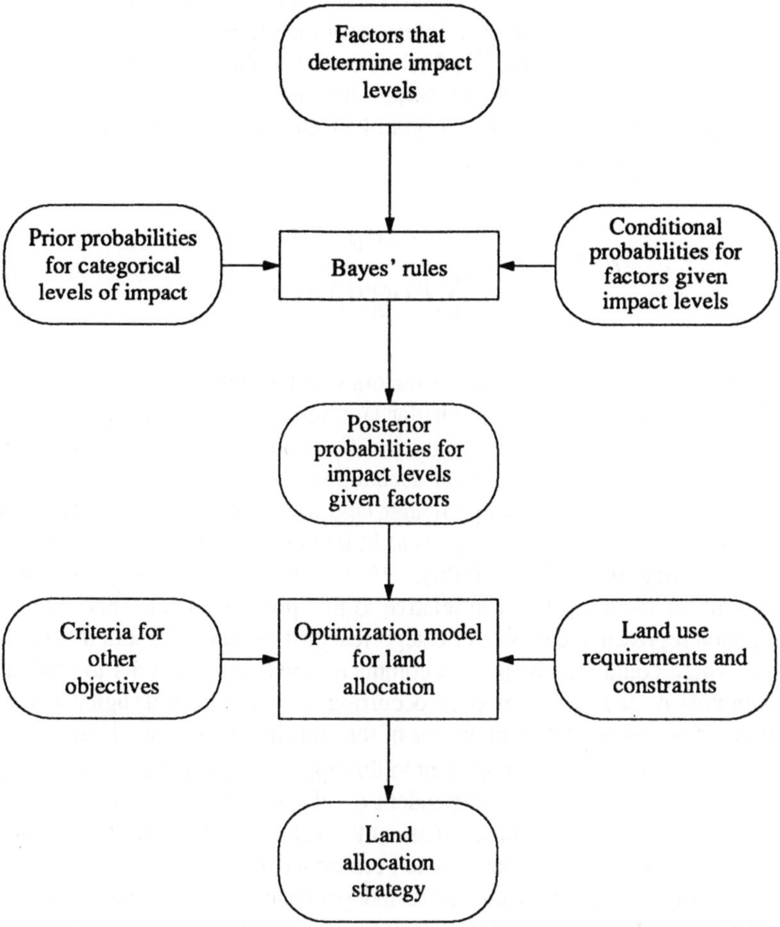

Figure 6.4. An idealized system for land allocation in which suitability assessment is accomplished by Bayesian analysis using a database of prior categorical probabilities, an input of determinant impact factors, and a set of conditional probabilities for factors given specified impact levels. The resulting posterior probabilities for impact levels are become suitability indices that are subsequently used by a formal optimization procedure in the assignment of land units to activities.

activity. An optimal allocation strategy, as measured by one or more sets of evaluative criteria (objectives), would be generated using some formal optimization procedure based on these probabilities; level 2. Instead of determining suitability based on a set of rules reflecting the subjective judgement of a domain expert, suitability in this instance is determined based on historical observations in the form of prior categorical probabilities for various impact levels.

One such procedure for determining these impact probabilities is the use of Bayes' Rule. Define H_i to be the ith member of the set of all possible outcomes of an event H, and E_j to be the jth subset of events in the set of possible events E that result in an outcome within H. Bayes' theorem may be used to compute the probability that the outcome H_i occurs given that the set of events E_j has occurred (Papoulis, 1984). In this case,

$$P(H_i | E_j) = \frac{P(E_j | H_i) P(H_i)}{\sum_i P(E_j | H_i) P(H_i)} \tag{6.1}$$

Assume that a set of appropriate factors can be identified that influence the suitability of a land area for a particular land use. Each possible combination of values of these factors make up the subsets E_j, and the set of all E_j's make up the set E. Furthermore, assume that an accounting could be made of these factors over many occurrences of this particular land use. Finally, assume that the suitability is judged (by a domain expert) to fit into one of a finite number (n) of appropriate categories of suitability, $H = \{H_1, H_2, \ldots, H_n\}$, and that an accounting of factors was made relative to this predefined set. This accounting would then represent a database of categorical past experience about that land use action, which could then be used in conjunction with Equation 6.1 to predict the probabilities of each member of H occurring for land areas designated for the same activity, provided that the values of the contributing factors, E, are known.

There are two assumptions inherent in this approach. First, the set of factors, E, are assumed to be mutually independent (exclusive). Independence means that any subset of the evidence must not affect the value of the rest of the evidence; a restriction that may prove troublesome because a comprehensive analysis of correlation should be done to guarantee independence (personal communication, David Gillette, December, 1989). Second it is assumed that the set of evidence, E, completely (exhaustively) covers the set of possibilities that effect the set of outcomes, H. This requires that a study of the factors that contribute (or do not contribute) to suitability be made, and that an all-inclusive set of factors be monitored over time to facilitate satisfaction of the first assumption. The validity of the Bayesian model for a given application requires satisfaction of these assumptions.

To illustrate the application of this concept, return to the example of rule-based inferencing discussed above and depicted in Figure 6.3. Suppose that we wish to develop suitability maps using probabilistic inferencing through Bayes' rule (Equation 6.1). A domain expert has determined that the factors most important in determining the impact of a given activity assigned to a region are again slope, vegetative cover, and soils type. For each of these factors, several categories are defined; for *slope*, flat, very low, low, moderate, high, very high, and extreme; for *vegetative cover*, barren, grass, sparse scrub, dense scrub, sparse forrest, and dense forest; and for *soils type*, sandy loam, clay loam, silty clay,

clay, and rock. For a given activity, our expert has determined a set of impact levels, none, low, moderate, high, and extreme.

An essential input to the probabilistic inferencing procedure discussed above is a historical profile of past impact resulting from previous instances of the activity of interest. An example of prior probabilities for these categorical impacts levels might be such as that presented in Table 6.1. Suppose that detailed information about 575 previous instances of that activity. Upon post-activity evaluation, impact was deemed extreme 38 times (prior probability 0.066), high 258 times (prior probability 0.449), moderate 206 times, etc. Prior to each activity, information was also collected according to each of the determining factors. The data presented in Table 6.1 reflect a complete profile of the conditional probabilities that each factor value will occur given that the activity was observed to have a certain impact level. For example, in this case the conditional probability of barren vegetation existing given that the activity resulted in a high impact is 163/258, or approximately 0.63. To the degree that all appropriate factors were considered during data collection, and independence among variables is maintained, Bayes' rule may be employed to generate the conditional probabilities in this data set. Further, as new observations are made, the conditional probabilities can be updated.

Using the categorical distribution of factors presented in Figure 6.3, probability maps for each impact level can be generated in which each cell of the map, the probability of a given impact level can be computed as shown in Figure 6.5. In addition, a *combined* map can also be generated by assigning that impact level having the highest probability to the respective cell in the matrix. For the output maps shown in Figure 6.5, computed using the data in Table 6.1, the cell in the upper left-hand corner of the map has a probability of a "high" impact of 44%. Because this is the highest impact category for that cell, the assignment of category 4 (for high) is assigned to that cell in the combined map.

To demonstrate that probabilistic reasoning as described by this example may be incorporated into a spatial decision support system, an SDSS application generator was developed for use in determining profiles for probable impact resulting from the assignment of military training activities to specific land areas (Buehler, 1990). The system is called B-infer—for Bayesian inferencing—and has been implemented as part of a well-known raster-based GIS: the Geographical Resources Analysis Support System (GRASS) developed by the U.S. Corps of Engineers (Westervelt, 1988). The following section of this paper presents an overview of B-infer. A more complete description of the system and its implementation may be found in Buehler and Wright (1991a and 1991b).

A Bayesian reasoning tool

The approximate reasoning procedure using Bayesian analysis, as discussed in the previous section, has been formalized as a shell development environment

Table 6.1. Database of Probabilities for the Bayesian Reasoning Example

	None	Low	Moderate	High	Extreme
Number of occurrences	21	52	206	258	38
Prior probability	0.037	0.090	0.358	0.449	0.066
Attribute categories	Number of occurrences/conditional probability				
Slope					
Flat	10/0.48	21/0.40	127/0.62	67/0.26	7/0.18
Very low	8/0.38	13/0.25	47/0.23	126/0.49	9/0.24
Low	3/0.14	9/0.17	16/0.08	54/0.21	11/0.29
Moderate	0/0.00	6/0.12	12/0.06	7/0.03	6/0.16
High	0/0.00	3/0.06	4/0.01	3/0.01	5/0.13
Very high	0/0.00	0/0.00	0/0.00	1/0.00	0/0.00
Extreme	0/0.00	0/0.00	0/0.00	0/0.00	0/0.00
Vegetation					
Barren	7/0.33	35/0.67	94/0.46	163/0.63	5/0.13
Grass	10/0.48	10/0.19	65/0.31	47/0.18	14/0.37
Sparse scrub	3/0.14	5/0.10	34/0.17	12/0.05	5/0.13
Dense scrub	1/0.05	2/0.04	13/0.06	32/0.12	8/0.21
Sparse forest	0/0.00	0/0.00	0/0.00	0/0.00	0/0.00
Dense forest	0/0.00	0/0.00	0/0.00	0/0.00	0/0.00
Soils					
Sandy loam	3/0.14	8/0.15	74/0.36	65/0.25	11/0.29
Clay loam	7/0.33	23/0.44	23/0.11	41/0.16	7/0.18
Silty clay	4/0.19	6/0.12	96/0.47	131/0.51	17/0.45
Clay	5/0.24	14/0.27	13/0.06	21/0.08	3/0.08
Rock	2/0.10	1/0.02	0/0.00	0/0.00	0/0.00

4	4	3	0	0
3	4	4	3	3
12	12	4	4	4
12	12	0	0	2
12	70	0	0	0

None (1)

3	3	2	0	0
2	1	3	2	2
28	28	1	3	3
28	28	39	5	3
28	30	0	0	0

Low (2)

41	74	48	0	0
25	33	74	48	48
20	20	33	41	74
20	20	33	61	37
20	0	0	0	0

Moderate (3)

44	16	41	61	61
51	55	16	41	41
26	26	55	44	16
26	26	9	12	39
26	0	0	12	35

High (4)

8	4	7	39	39
19	7	4	7	7
14	14	7	8	4
14	14	20	22	18
14	0	0	88	65

Extreme (5)

4	3	3	4	4
4	4	3	3	3
2	2	4	4	3
2	2	2	3	4
2	1	0	5	5

Composite

Figure 6.5. An example of the output of Bayesian inferencing across the same empirical data as the rule-based example. Probability maps are generated for each suitability level represented in the historical data in Table 6.1. In addition, a composite map can be generated where the index corresponds to the suitability level that has the highest probability of occurrence.

that allows a land management domain expert to develop a spatial decision support system for determining land suitability profiles. The system is properly viewed as a programming language for use in constructing program modules that accomplish level-1 analysis as presented in Figure 6.4. The B-infer development environment provides for direct input of GIS map layers as well as short-term *contextual* attribute values as factors for determining land suitability indices. System output is in the form of GIS map layers indicating posterior probabilities for various impact levels resulting from given ambient conditions. The general structure and function of B-infer is presented in this section, including a simple illustrative example of its use in determining a hypothetical suitability profile.

Consider the following scenario (adapted from Buehler and Wright, 1991a):

> *A judgement must be made as to the likely impact of assigning, to a specific land area, a specific activity. Complete and accurate information is available about the nature of the activity, the physical and environmental nature of the candidate land area, and the conditions present at the time*

the activity is to take place. The person responsible for making the assignment wishes to develop a decision support system that will aid that decision. Given that the Bayesian model is deemed appropriate as the reasoning engine for this decision support system the following steps are required:

1. *A person knowledgeable about the activity to be assigned, the candidate land area, and the nature of impact cause-effect, designs a Bayesian model that captures the most important aspects of the specific problem, adhering to exclusivity and exhaustivity requirements.*
2. *Categorical probability data are collected for that model and probabilities are computed. This information is assumed to reflect the cause-effect nature of the activity under consideration to the extent that future observations about impact may be used to update these probabilities.*
3. *A person knowledgeable about the mechanics of geographical information systems and the structure of the probability information acquired in step 2 produces a computer script program that reflects the model developed in step 1. This may require close interaction with the domain expert as well as the target user (if that is not the same person).*
4. *The script or program developed in step 3 is then used to control the execution of a system that will: 1) prompt the user for input about short-term conditions that might influence land impact, 2) access GIS map layers for long-term environmental information over time and space, and 3) produce information about likely impact in the form of probability maps as reflected by the Bayesian model.*
5. *The spatial impact maps are used, possibly with other data and analytical procedures, to make the activity assignment.*
6. *Following the activity, a judgement as to the resulting impact is made and short-term conditions are recorded. This becomes an additional observation upon which to update the probabilities for that activity.*

The short-term conditions mentioned in the scenario are important to this discussion because in the case that these conditions did not change there would be no need for a shell environment. The system could be executed one time and the output probability maps generated would suffice. The need to include contextual attributes (in some cases many of such attributes) demands an environment where the same application can be executed for any combination of these attributes. Consider as an example the same Bayesian reasoning application presented in the previous section with the addition of a soil moisture attribute that represents a judgement about the current relative moisture content of the soil. The categories defined for the soil moisture attribute are, dry, damp, wet, soaked, and drenched. Table 6.2 contains the historical profile of past impact resulting from previous instances of the activity of interest.

Table 6.2. Database of Additional Probabilities for the B-infer Example

	None	Low	Moderate	High	Extreme
Number of occurrences	21	52	206	258	38
Prior probability	0.037	0.090	0.358	0.449	0.066
Attribute categories	Number of occurrences/conditional probability				
Soil moisture					
Dry	15/0.71	21/0.40	0/0.00	0/0.00	0/0.00
Damp	5/0.25	17/0.33	64/0.31	18/0.07	0/0.00
Wet	1/0.04	14/0.27	74/0.36	85/0.33	9/0.24
Soaked	0/0.00	0/0.00	62/0.30	129/0.50	17/0.45
Drenched	00.00	0/0.00	6/0.03	26/0.10	12/0.31

Using the same categorical distribution of factors as the previous examples, probability maps of the given impact levels can be generated. The combined maps generated are presented in Figure 6.6.

Applications with many contextual attributes could be implemented quickly and easily with a shell tool such as B-infer. Furthermore, the applications would be easily executed to aid in the activity assignment decision with the current values of all contextual attributes, so that the decision would always be based in part on the current judgement of impact.

Figure 6.7 presents a schematic diagram of B-infer. The application developer prepares the input script (referred to in the previous section) and places it into a file using any system editor. The software is then executed within the GRASS environment. During execution B-infer reads the input script (arrow from input script to B-infer software in the figure), and translates the information into an internal representation (see Buehler and Wright, 1991b). B-infer then obtains the value for contextual attributes by prompting the user using menus constructed from the input script (the bidirectional arrows from B-infer software to the user in the figure). GRASS is then called upon to reclassify or recategorize the input maps corresponding to the ranges of layer attribute values given in the script[1].

1. Maps within the GRASS GIS are structured as a matrix of integer values that are used to categorize attribute information. To reclassify or recategorize these maps involves a reassignment of the information. In the case of B-infer, categories may be generalized so that more detailed categorical descriptions are grouped into more general categories without data preprocessing.

1	1	1	0	0
1	1	1	1	1
2	2	1	1	1
2	2	2	2	1
2	1	0	0	0

Dry

3	3	3	4	4
3	3	3	3	3
2	2	3	3	3
2	2	2	3	3
2	1	0	4	4

Damp

3	3	3	4	4
4	4	3	3	3
4	4	4	3	3
4	4	3	3	3
4	2	0	5	5

Wet

4	3	4	4	4
4	4	3	4	4
4	4	4	4	3
4	4	3	3	4
4	0	0	5	5

Soaked

4	3	4	5	5
5	4	3	4	4
5	5	4	4	3
5	5	5	5	5
5	0	0	5	5

Drenched

Figure 6.6. An example of the composite impact maps for each moisture level.

For example, a map of soils may have many specific soil types represented, but we really only need a general soil type. A reclassification operation would recategorize existing soil types into a new, simplified description. These reclassified maps are then used as input to B-infer (the arrow from reclassed maps to GRASS software and back to B-infer software in Figure 6.7). Once B-infer has all of this information it can begin the inferencing process.

A new map is created for each possible value of the inferred attributes (output maps in Figure 6.7). The value in each cell represents the probability of that inferred attribute value occurring in that grid cell. These probability maps are then ready for input to some other decision-making process or system. B-infer also determines which value is most likely to occur in each grid cell and writes that to yet another cell file (combined map in Figure 6.7). This map can be used to obtain a profile for the value of the inferred attribute over a land area.

Future directions

This research demonstrated that raster-based GIS technology combined with a formal Bayesian analysis system provides a SDSS environment for performing land suitability indexing in support of certain land allocation activities. In this context, the use of a Bayesian probabilistic inferencing model is only one possi-

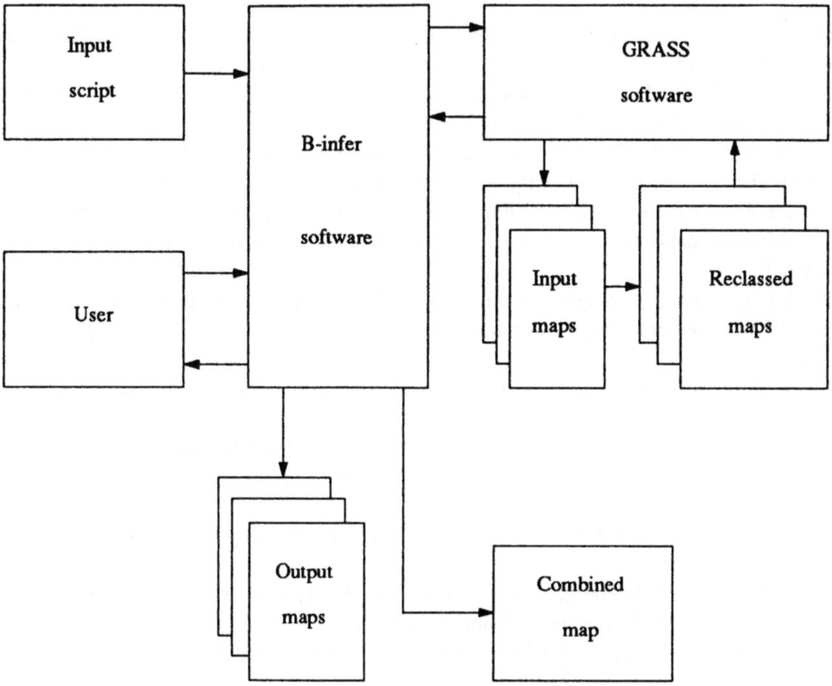

Figure 6.7. A schematic of the structure of the B-infer reasoning shell.

ble procedure for assigning suitability. Other meaningful suitability assessment procedures might include a formal rule-based expert system with uncertainty incorporated as certainty factors or fuzzy logic, neural network procedures, frame-based inferencing systems using formal hypothesis, and test algorithms such as minimal set cover, and other statistical models such as the Dempster-Shafer theory of evidence. Work is presently under way on the development of these model coponents.

References

Aldag, R.J. and D.J. Power, 1986. "An empirical assessment of computer-assisted decision analysis," *Decision Sciences 17*(4):572-588.

Alter, S.L., 1980. *Decision Support Systems: Current Practice and Continuing Challenges*, Addison-Wesley, Reading, PA.

Anderson, L.T., 1987. "Five methods for calculating capability suitability," in E. Crane, ed., *Papers from the 1987 Annual Conference of the Urban and Regional Information Systems Association, Fort Lauderdale*, Urban and Regional Information Systems Association, Washington, pp. 125-139.

Antenucci, J. and J. Dangermond, 1974. *Maryland Automated Geographic Information*

System (MAGI).

Barath, E. and I. Futo, 1984. "A regional planning system based on artificial intelligence concepts," *Papers of the Regional Science Association* 55(135-154.)

Bonczek, R.H., C.W. Holsapple, and A.B. Whinston, 1981. *Foundations of Decision Support Systems*, Academic Press, New York.

Buehler, K.A., 1990. "B-infer: A Bayesian Reasoning Shell for Land Management," Master's Thesis, School of Civil Engineering, Purdue University, 68 pages.

Buehler, K.A. and J.R. Wright, 1990a. "A Bayesian reasoning shell for land management. Part I: Structure & function," under review: *Journal of Computing in Civil Engineering*, ASCE.

Buehler, K.A. and J.R. Wright, 1990b. "A Bayesian reasoning shell for land management. Part II: Implementation," under review: *Journal of Computing in Civil Engineering*, ASCE.

Bundy, A., 1984. *Catalogue of Artificial Intelligence Tools*, Springer-Verlag, New York, 150p.

Cederborg, E. and R. Tosetti, 1982. "Site selection for low level radioactive waste disposal sites," in R. Post, ed., *Waste Management '82*, University of Arizona College of Engineering, Tuscon, Vol. 2(215-223).

Chapin, Jr., F.S. and E.J. Kaiser, 1979. *Urban Land Use Planning*, U. Illinois Press, Urbana, IL, pp. 289-315.

Clancy, W. and J. Rothenberg, 1987. "Tutorial TA1: Evaluating expert system tools," *Proceedings of AAAI-87*, Kaufmann Publishers, Inc., Los Altos, CA.

Coulson, R.N., L.J. Folse, and D.K. Loh, 1987. "Artificial intelligence and natural resource management," *Science 237*(262-267).

Dangermond, J., 1979. "A case study of the Zulia regional planning study, describing work completed," *Harvard University Laboratory for Computer Graphics and Spatial Analysis 10*(35-63).

Davis J.R., J.R.L. Hoare, and P.M. Nanninga, 1986. "Developing a fire management expert system for Kakadu National Park, Australia," *Journal of Environmental Management 22*(215-227).

Davis J.R., P.M. Nanninga, J.R.L. Hoare, and A.J. Press, 1989. "Transferring scientific knowledge to natural resource managers using artificial intelligence concepts," *Ecological Modeling 46*(73-89).

Davis J.R. and J.L. Clark, 1989. "A selective bibliography of expert systems in natural resource management," *AI Applications in Natural Resource Management 3*(3)1-17.

DeMers, M., 1985, "The formulation of a rule-based GIS framework for county land use planning," in Hanham, R., Vogt, W., and M. Mickle (eds.), *Modeling and Simulation: Proceedings of the Sixteenth Annual Pittsburgh Conference*, ISA, Research Triangle Park, NC, 93-97.

Densham, P.J. and M.F. Goodchild, 1989. "Spatial decision support systems: A research agenda," *Proceedings: GIS/LIS '89*, Orlando, Florida, Vol. 2(707-716).

Diamond, J.T., 1988. "A Multiobjective Discrete Optimization Model for Land Use Allocation," Doctoral Dissertation, School of Civil Engineering, Purdue University, 179 pages.

Diamond, J.T. and Wright, J.R., 1988. "Design of an integrated spatial information system for multiobjective land-use planning," *Environment and Planning B: Planning and Design 15*(205-214).

Elmes, G. and T. Harris, 1986. "Hierarchical data structures and regional optimization," *Proceedings of the Second International Symposium on Spatial Data Handling*, pp.

289-305.

Freedman, R. 1987. "27-product wrap-up: Evaluating shells," *AI Expert* 2(9)69-74.

Geoffrion, A.M., 1983. "Can OR/MS evolve fast enough?", *Interfaces 13*(10-25).

Gevarter, W., 1987. "The nature and evaluation of commercial expert system building tools," *IEEE Computer 20*(5)24-41.

Gilbert, K.C., D.D. Holmes, and R.E. Rosenthal, 1985. "A multiobjective discrete optimization model for land allocation," *Management Science 31*(12)1509-1522.

Gillette, D., 1990. Personal communication.

Gorry, G.A. and M. Scott-Morton, 1971. "A framework for management information systems," *Sloan Management Review 13*(56-70).

Han, S.Y. and T.J. Kim, 1989. "ESSAS: Expert System for Site Analysis and Selection," in T.J. Kim, L.L. Wiggins, J.R. Wright, eds., *Expert Systems in Urban Planning*, Springer-Verlag.

Harmon, P. and D. King, 1985. *Expert Systems: Artificial Intelligence in Business*, Wiley Press, New York.

Hayes-Roth, F., D. Waterman, and D. Lenat, 1983. *Building Expert Systems*. Addison-Wesley Publishing Company, Inc., Reading, MA, 444p.

Holsapple, C.W. and Whinston, A.B., 1987. *Decision Support Systems: Theory and Application*, Springer-Verlag.

Hopkins, L.D., 1977. "Methods for generating land suitability maps: A comparative evaluation," *Journal of the American Institute of Planners 43*(4)386-400.

Jungert, E., 1984. "Inference rules in a geographical information system," in *IEEE Workshop on Languages for Automation*, IEEE Comput. Soc. Press, Silver Spring, MD, pp. 32-37.

Keen, P.G.W., 1980. "Decision support systems and personal computing," *Data Base 12*(1-2), Fall.

Little, J.D.C., 1970. "Models and managers: The concept of a decision calculus," *Management Science 16*(8), April.

Lyle, J. and F.P. Stutz, 1983. "Computerised land use suitability mapping," *The Cartographic Journal 20*(1)39-49.

Lyle, J. and M. von Wodtke, 1974. "An information system for environmental planning," *Journal of the American Institute of Planners 40*(6)394-413.

Maher, M.L., 1984. "HIRISE: A Knowledge-Based Expert System for the Preliminary Structural Design of High Rise Buildings," Doctoral dissertation, Carnegie-Mellon University, Pittsburg.

Maher, M.L., ed., 1987. *Expert Systems for Civil Engineers*, American Society of Civil Engineers, New York, 148 pages.

Malakooti, B., 1988. "A decision support system and a heuristic interactive approach for solving discrete multiple criteria problems," *IEEE Transactions on Systems, Man and Cybernetics 18*(273-284).

Moore, J.H. and M.G. Chang, 1980. "Design of decision support systems," *Data Base 12*(1-2), Fall.

Mullarkey, P.W., 1987. "Languages and Tools for Building Expert Systems," in *Expert Systems for Civil Engineers: Technology and Application*, American Society of Civil Engineers, New York, 148 pages.

Richer, M., 1986. "An evaluation of expert system development tools," *Expert Systems 3*(166-183).

Robinson, B.R., A.U. Frank, and H.A. Karimi, 1987. "Expert systems for geographic information systems in resource management," *AI Applications in Natural Resource*

Management 1(1)47-57.

Smith, T.R. and M. Panzer, 1984. "Knowledge-based control of search and learning in large-scale GIS," in D.F. Marble, K.E. Brassel, D.J. Peuquet, H. Kishimoto, eds.,*Proceedings, International Symposium on Spatial Data Handling,* Geographisches Institut, Abteilung Kartographie, EDV, Zurich, Switzerland, pp. 498-519.

Smith, T.R., D. Peuquet, S. Menon, and P. Agarwal, 1987. "KBGIS-II: a knowledge-based geographical information system," *International Journal of Geographic Information Systems 1*(149-172).

Sprague, R.H. and E.D. Carlson, 1982. *Building Effective Decision Support Systems,* Prentice-Hall, Englewood Cliffs, NJ.

Sprague, R.H. and J.J Watson, eds., 1986. *Decision Support Systems: Putting Theory into Practice,* Prentice-Hall, Englewood Cliffs, NJ

Steinitz, C. and H. Brown, 1981. "A computer modeling approach to managing urban expansion," *Geo-Processing 1*(341-375)

Tomlin, S. and D. Tomlin, 1985. "Computer-assisted spatial allocation of timber harvesting activity," in *Proceedings Auto Carto 5,* American Society of Photogrammetry, 677-686.

Turban, E.,1988. *Decision Support and Expert Systems,* Macmillan, New York.

Waterman, D.A., 1986. *A Guide to Expert Systems.* Addison-Wesley, Reading, MA.

Westervelt, J., 1988. "An Introduction to GRASS," in *GRASS 3.0 User's Manual.* U.S. Army Corps of Engineers, Champaign, IL.

Whittaker, A.D., R. Thieme, D. Jones, and J. Barrett., 1986. "Getting started with expert systems," *ASAE Technical Paper 86-5037,* St. Joseph, MI.

Williams, E. and A. Massa, 1983. *Siting of Major Facilities,* McGraw-Hill, New York.

Zachary, W.W., 1988. "Decision support systems: Designing to extend the cognitive limits on the user decision processes," in M. Helander, ed., *Handbook of Human-Computer Interaction,* Elsevier Science Publishers, North Holland.

7

Database Integration for Knowledge-Based Groundwater Quality Assessment[1]

Marc P. Armstrong

Introduction

Groundwater is an important source of potable water. In the United States, for example, 30 states obtain over half of their drinking water from groundwater sources (Government Accounting Office, 1988), and in some areas it serves as the only viable source of household water. Because of this widespread dependency, many areas in the United States and other countries (Ademoroti, 1987) have experienced a decline in the quality of groundwater used for human consumption.

Although there are numerous causes for decreases in groundwater quality, many problems are attributable to human activities such as agricultural practices and leakage from waste disposal sites. Because of a growing awareness of the potential seriousness of groundwater contamination, this has become a visceral issue, attracting both media and legislative attention. As a result, several important pieces of legislation, each mandating the collection of groundwater information, have been signed into law. This legislation, along with growing scientific and public concern, has led government agencies to respond by developing and expanding operational groundwater quality monitoring programs.

Government agencies now regularly expend large sums of money to collect and warehouse individual databases that can be used to help solve complex interdisciplinary groundwater problems. The objectives of the monitoring activities have lacked a clear focus, however. The resulting diversity has made it difficult to use the collected data in decision-making because each database has been collected and organized without considering its relationship to other sources of informa-

1. Reprinted with permission from *Computers, Environment and Urban Systems*, Vol. 14, No. 3:187-201, M.P. Armstrong, "Database Integration for Knowledge-Based Groundwater Quality Assessment," 1990, Pergammon Press plc.

tion needed for effective decision-making. Moreover, problems have also occurred because of the irregular, and in some cases unknown, spatial manifestation of groundwater contaminants.

Yet the fact remains that environmental managers are often confronted with complex questions that arise in the course of making decisions about groundwater quality (Roberts and Butler, 1984). The questions are complex, in part, because their answers require the integration of data from numerous sources. Because of increasing uncertainty about the characteristics of drinking water obtained from ground supplies, it is imperative that strategies be developed to overcome difficulty encountered in the integration and use of different groundwater data sources, which in turn will decrease the likelihood that future decisions will be based on incomplete information. One way to overcome these problems is to increase the usefulness of existing databases and current monitoring programs. These existing information sources can be used to construct knowledge-based advisory systems that are designed to help decision-makers sort through the complexity inherent in groundwater information. Advisory systems assist groundwater quality managers in deciding whether changes are occurring in measured levels of harmful substances, and enable uniform enactment of strategies for detecting and assessing groundwater contamination.

Additional information can be provided to the user to help determine whether the amount of change in quality (areal or temporal) warrants remedial action. To achieve this goal a two-stage approach for assisting decision-makers is adopted. First, data from different sources are integrated to provide a synoptic view of groundwater quality. The second stage involves the development of procedures which employ artificial intelligence principles for querying the integrated database. Each stage is described in the following sections.

Scope of the knowledge-based advisory system

General design principles for the knowledge-based system were derived from research conducted on the development of decision support systems (DSS). These systems are designed to provide near-real-time feedback to decision-makers from results of computer models and database queries (Sprague and Carlson, 1982). Such systems must be flexible so that users can explore and, perhaps, reconceptualize each problem: information obtained at one step in a problem-solving process can be used to change the strategy for solving the problem (Hopkins, 1984). The knowledge-based DSS described in this chapter draws upon a fixed set of governmental regulations (rules) to govern the types of actions that may be required to assess groundwater problems; the set of rules embodied in the system concern invariant aspects of the problem (e.g., maximum contaminant levels for toxic substances). When incorporated into a solution framework, these rules lead to consistency in problem-solving across sites despite varying local conditions. The system, therefore, provides information about groundwater quality to decision-makers and contains rules to spell out appropriate strategies for intervention, if required. As implemented, the advisory system described here is formed from integrated software modules consisting of a database, a rule base, and a display and report generator. These modules provide the user with a powerful set of tools for producing and evaluating alternative strategies for solving locational and environmental problems (Armstrong, Densham, and Rushton, 1986). Although each component is important, in this chapter only the first two are described.

A strategy for groundwater database organization

The design of a suitable data structure is a crucial part of any system used to conduct spatial analyses. It is sometimes necessary, for example, to incorporate point, line, and area data types and their respective attributes into the formulation of a solution. Because human decision-makers have powerful capabilities for recognizing patterns present in spatial data, a graphic display capability should be present. The following general data types and relationships are included:

- Point, line, and area data.
- Topological relationships.
- Data of different levels of aggregation.
- Data describing attributes of entities.

When databases are designed several levels of abstraction are referred to, extending from the real world to the actual physical implementation of a model of the real world on the computer. Bestougeff (1984) specified three levels of abstraction which conform closely to those used by Peuquet (1988) in her recent work on spatial representations (external/conceptual, logical, and physical).

The lowest level of abstraction concerns data organization issues that are largely system-specific and therefore beyond the scope of this chapter. The external/conceptual and logical levels of abstraction are defined here through the use of two general models. At the highest level of abstraction (external/conceptual), the entity-category-relationship (ECR) model is used to describe entities and their allowable interactions (Navathe, Elmasri, and Larson, 1986). The relational model (Codd, 1979) is used at the second (logical) level of abstraction.

Conceptual level: ECR

The ECR model is an extended version of the entity-relationship approach (Chen, 1976 and 1983; Davis, Jajodia, Ng, and Yeh, 1983) which has been sucessfully employed in other spatial data modeling applications (Calkins and Marble, 1987; Robinson and Zhang, 1988). In the ECR model an entity can be an object, an event, or a concept within the general application domain of the database. Categories allow a database designer to clump similar entities or objects in hierarchies and allow the designer to impose structural constraints on the participation of entities in relationships.

Relationships exist among entities and provide a mechanism for creating associations among classes of object categories. Figure 7.1 illustrates these concepts in diagrammatic form. In this generalized representation, entity 1, which consists of two categories, has some (unspecified, in this example) relationship with entity 2, which has three unspecified attributes (e.g., color, length, and width).

Groundwater quality normally is evaluated on the basis of measurements obtained from sample wells. The well, therefore, is the basic conceptual unit of analysis. Figure 7.2 shows a simplified version of an ECR diagram for wells. In this case, the named relationship contain exists between wells and contaminants. Although this simplified representation is useful for the purpose of providing an example ECR structure, spatial databases generally contain other data types, which in this application must be related to well observations: locational, topological, and thematic.

Locational

In groundwater analyses, locational data are often represented by latitude and longitude, UTM, or state plane coordinate values. In some instances these coordinate references serve as the main identifier for a given well. Other locational data items such as state, county, and legal descriptions also are commonly specified to enable analyses at different levels of spatial aggregation (e.g., county) or to provide alternate forms of access to database entities (e.g., legal descriptions).

Topological

Topology can be used to form neighborhood relationships and networks which are especially useful in areas with high spatial sampling densities, such as around hazardous waste disposal sites. Topological linkages also allow gradient analyses and interpolation to be accomplished efficiently. Implementing a topological structure for point distributions is straightforward: the neighbors of observation wells designated by proximity are recorded and entered into a list.

Thematic

Thematic data for wells can be separated into two general categories: physical and chemical. Physical data items pertain to general characteristics of each well, such as depth and construction date. Chemical data refer to observations made at a given well and include: substance sampled, date, depth, value, and units used. In this classification, the physical data are nearly invariant, whereas the chemical data change as new observations are recorded. Change can occur by adding new observations to an existing temporal series, or by causing a recomputation of existing values in a database (e.g., causing a downward adjustment to a mean).

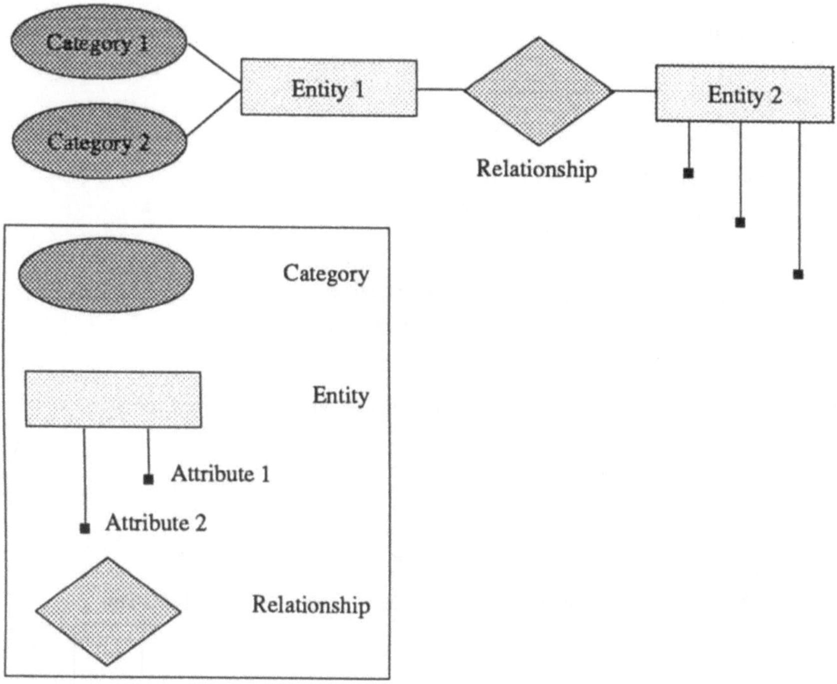

Figure 7.1. General format for ECR diagrams.

Combined view

Figure 7.3 shows the ECR diagram for groundwater observations derived from the above descriptions. The well, which is assigned an identifier, is the focal point for organizational strategy. Each well has a single location given in a 1:1 structure. The recursive relationship linked_to_other is used to specify the presence of topological linkages in the database. The relationship contains exists between wells and their characteristics. As described above, characteristics are broken into physical and chemical categories. Although this is a conceptual model, separation of characteristics into categories is done for reasons related to the implementation of the database model in software: when data are searched, if records are divided into categories, and if the user knows a category in which to search, then search times can be decreased because all records will not require examination.

Logical level - relational

At the second level of abstraction, several models are available for the logical design of the database module for the advisory system (Densham and Armstrong,

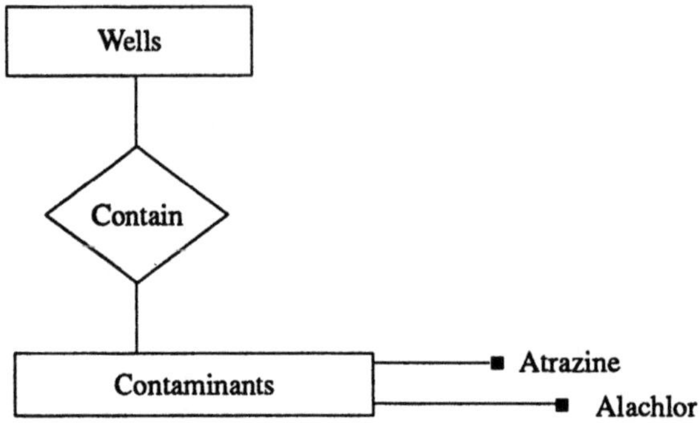

Figure 7.2. Simplified ECR diagram for groundwater quality data.

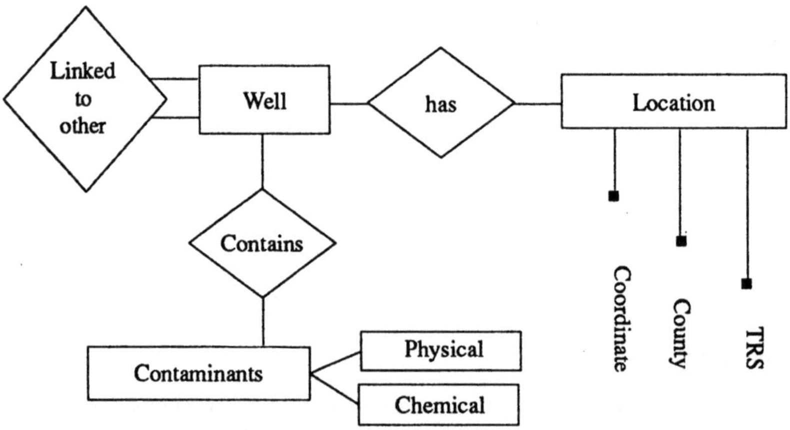

Figure 7.3. Completed general ECR diagram for groundwater quality databases.

1987; Armstrong and Densham, 1990). The relational model garnered much acclaim by database researchers (Codd, 1979; Dueker, 1985; Lorie and Meier, 1984; Morehouse, 1985), but it has been criticized (Calkins and Marble, 1987; Peuquet, 1988) because it is unable to handle complicated descriptions of spatial objects and their relationships in an efficient manner. For a database with points (e.g., wells) as its main focus, however, these limitations do not apply because complicated polygonal layers need not be encoded or modeled in the system.

The user's view of data in the relational model is one of its strongest advantages. The view simply consists of a series of linkable tabular structures called relations. Each relation can be culled by row or column to produce a subset relation from the original. Furthermore, two relations can be joined by matching (or in some implementations, using other Boolean operations) values drawn from a common domain in each; the resulting relation can thus have values that were originally contained in both original tables. The database can be augmented, or updated, by adding records to existing relations or adding new relations. If the new relations must be linked to existing ones, this is accomplished by ensuring that each new relation contains a domain already present within at least one existing relation. In practice, the database is formed from a series of record types which contain information about the location, topological relationships, and characteristics of groundwater samples for various contaminants (Figure 7.4). Each well is assigned a unique identifier (ID), which is used to form linkages with other relations. This is illustrated by an example shown in Figure 7.5, where two relations, one containing information about the physical characteristics of wells, and the other containing chemical characteristics, are joined on the basis of a common well identifier. Note that for each of these data types a unique identifier is normally used to enable the association of each data item with a particular well.

In Figure 7.4 the locational information consists of a state plane coordinate, county, and legal description for each well. Topological information is included to permit the efficient specification of groups of wells for analysis. Topology is defined by a list of wells associated with a given well based on proximity.

Thematic information is contained in two general record types for each well: physical and chemical characteristics. Physical characteristics pertain to features such as: (1) elevation above MSL of the sample, (2) well depth, (3) depth to bedrock, (4) lithologic units encountered, and (5) well date. The chemical records comprise the bulk of the database and contain information about the specific measurements obtained at the sampling wells. The chemical records may be stratified by substance, date, and sampling depth.

It must be noted that all groundwater databases do not conform to the general model. Most databases, however, contain enough elements to allow them to be incorporated into the general organizational framework. It is especially noteworthy that some measure of location is a common unifying theme in virtually all collections of groundwater data. In one form or another, location can be used as a basis for database integration.

Database integration strategy

An underlying premise in this chapter is that a single, unified repository of information facilitates decision-making. Because current data sources exist in a variety of formats, these data must be logically and physically integrated. In a sense, the different databases have been mandated, collected, and maintained in different forms because each represents a different view of groundwater problems. These differences arise from legal restrictions based on enabling legislation, and from a lack of unanimity in agency and disciplinary perspectives. For example,

LOCATIONAL

ID	Coordinate	County	Legal	Description

TOPOLOGICAL

ID	ID

THEMATIC

Physical characteristics

ID	Elevation	Depth	Bedrock	Lithology	Date

Chemical characteristics

ID	Substance	Date	Depth	Value	Units

Figure 7.4. Example record types and variables.

Physical relation

ID	Elevation	Depth	Date
abc	730	40	1983
def	705	48	1986
ghi	592	63	1984
jkl	821	50	1988
mno	689	105	1988

Chemical relation

ID	Substance	Date	Value	Units
abc	Lindane	010188	10	ppm
abc	Lindane	020188	11	ppm
abc	Endrin	010188	09	ppm
def	Lindane	010188	09	ppm
def	Aldrin	010288	01	ppm
ghi	Lindane	010188	09	ppm
jkl	Lindane	010188	09	ppm

Join

ID	Substance	Date	Value	Units	Elevation	Depth
def	Lindane	010188	09	ppm	705	48

Figure 7.5. Example join between physical and chemical relations (after Peuquet, 1988).

the U.S. Geological Survey's WATSTORE database contains mainly raw water observations, and most observations are for wells greater than 100 feet deep (Rajagopal, 1987). It is, therefore, a good source of information concerning deep aquifers. It is also useful for analyses involving inorganic substances and trace metals. On the other hand, WATSTORE was never intended for, and is therefore not usually useful for, analyses concerning exposure to human-induced problems related to land use and waste disposal practices. Another agency, with a different mission, the U. S. Environmental Protection Agency, has developed the MSIS database which is designed to manage provisions of the Safe Drinking Water Act. It deals largely with finished water and helps in the determination of maximum contaminant level (MCL) violations and the notification of violators (Rajagopal, 1987). It contains information about organic (e.g., Endrin), and inorganic substances such as lead and arsenic. The focus of each database reflects the mission of the agency charged with collecting and maintaining the information.

WATSTORE, MSIS, and other existing databases can be thought of as different, but complementary, sources for gleaning information useful in solving groundwater problems. Note that the intent here is not to argue that these databases are ideal candidates for merging, but rather to demonstrate the methodology for performing such a task. The databases can be made to function as a whole, thereby increasing our ability to obtain a synoptic view of groundwater quality. To generate such a database system a framework must be developed for resolving data definition and organization conflicts, and for tuning the design to the problem domain (Navathe, et al., 1986). This is a data engineering task, and in this case spatial linkages form the basis for the integration process.

A user's view about what is in a database reflects the user's perception about both the types of entities recorded and the relationships among entities (Navathe, et al., 1986). The view that a user holds of the database, however, may be quite different from what is actually contained in the database. The formal description of database content is contained in a schema definition. By combining schemas and user views we extend the usefulness of individual databases. Note that the ECR model described earlier has been shown to be suited for the task of view integration (Navathe, et al., 1986).

The process of view integration can be awkward when databases of very different content and design intent are involved. For example, when integrating databases in some applications, complexity occurs because the domains, or set of allowable values, of individual databases are disjointed and lack correspondence. In other situations, the domains may be identical, contained, or overlapping, and in those cases, there is virtual certainty that the entities or classes refer to the same concept. In the specific case of integrating groundwater databases, even though there are incompatibilities and fragmentation, the general intent of each database is similar; this greatly simplifies the integration process. Each database contains information about entities (wells).

The content and relationships among records varies only in form, and in many ways these differences are complementary.

The following steps based on the work of Navathe, et al. (1986) are followed during the groundwater database integration process:

1. Establish naming conventions.
2. Establish correspondence for entities, object classes, domains and attributes.
3. Establish candidate keys and domains for object classes.
4. Establish mapping of equivalent attributes of object classes.

Strong commonalities exist in the WATSTORE and MSIS groundwater databases (Table 7.1). For example, there are three ways to specify commonality in space, (place, county, and coordinate) and one in the temporal dimension. Note, however, that additional steps may be required to move from one domain to the other (e.g., county to county code). This process involves creating an explicit mapping from one variable to another by building a lookup table, or inverted list. An example of such a list is a correspondence table for counties and county FIPS codes. Additional work must be done, however, to form linkages on the basis of latitude and longitude. The U.S. Geological Survey uses a long integer identifier, which consists of latitude and longitude concatenated. MSIS uses a public water supply ID which can be linked to latitude and longitude. An additional table, therefore, must be constructed to permit retrieval on the basis of location. Similar processes are undertaken for place and for temporal stratification.

The ECR diagram for the integrated databases is depicted in Figure 7.3.

Bundling intelligence into query systems

The second problem addressed in this chapter focuses on ways to help decision-makers extract information relevant to their tasks from a database in a timely fashion. The programming environment that is used to support this task also allows artificial intelligence principles to be applied to groundwater problem-solving. The integrated database view created in the previous section can be implemented as a series of records, or facts, which serve as input for a query program written in Prolog. Prolog is a declarative language that uses facts contained in a database, and rules contained in a rule base, to provide a programming environment in which deductive logic is used to arrive at goals. Facts, in conjunction with rules governing temporal analyses, user interaction, and compliance levels, comprise the knowledge base of the groundwater quality assessment system. Goals normally are specified by a system designer, or they may be entered, and modified by a user in the course of arriving at a decision.

Inference and control are provided by constructs incorporated into the Prolog language (Sterling and Shapiro, 1986). Prolog implements a depth-first search algorithm through a process of forward chaining and subsequent backtracking. Forward chaining, which proceeds by checking rules for truth and adding knowledge to working memory, is well suited for querying the groundwater database, and for other monitoring and assessment problems, because many outcomes are possible. On the other hand, if only a limited number of outcomes is possible, then backward chaining (working from goals backward through sub-goals) may

Table 7.1. Identical and Overlapping Domains

WATSTORE	MSIS
Place	PWSMLC (place)
County	County code
Year and month	Date
USGSID (lat/long)	PWSID (inverted list)

be more efficient. The implementation of Prolog used for the system (Turbo Prolog) has a database subcomponent that enables the specification of record structures for a relational-like database (Turbo Prolog Owner's Handbook, 1986). In some ways the language can be thought of as an extension to the relational model (Sterling and Shapiro, 1986), and in general, allowable queries in Prolog databases resemble those available in relational database management systems. Relational operators are implemented and controlled using rules expressed in the logic program.

System implementation and design of the Prolog program

The following steps were followed in designing the advisory system:

1. Compile rules (e.g., exceedence levels).
2. Formulate predicates, clauses, and goals in Prolog.
3. Code program and correct errors.

The program, therefore, is developed in a fashion similar to that of an expert system, although the elicitation process is made simpler by relying, for example, upon legal restrictions (Armstrong, 1987).

An analyst initiates a query by selecting options including bounding (limiting) variables from a menu. Combinations of the following variables can be used to construct a query:

1. Site_id
2. Well_id
3. Contaminant
4. Contaminant_concentration
5. Date_of_observation

A series of general query types is now supported and others are being added. The generic queries are:

1. For a given site, list all wells.
2. For a given well, list all information.
3. Find all wells with a specified contaminant.

4. Find all wells exceeding any MCL.
5. Find all wells exceeding the MCL for a specified contaminant.

Additional queries have been developed to allow the user to examine the database for temporal trends (Armstrong and Bennett, 1988, in press).

A family of Prolog rules determines whether a candidate well observation, which is represented as a fact in the knowledge base, is consistent with a query constructed by a user. This query is implemented as a Prolog clause which possesses the same structure as the associated fact in the groundwater database. The variables within the query clause assume one of three states.

1. *Bound*: a single discrete value is specified by a user.
2. *Bounded*: values are restricted to a user specified range.
3. *Free*: no value restriction.

Facts in the database are tested for compliance with this pattern of bound and bounded variables. If the fact does not match the pattern prescribed by the query, the rule fails, Prolog backtracks up one level in the search tree, and tests the next fact for compliance. If the pattern match succeeds, then the bounded and free variables of the query clause are instantiated to the values of the associated fact, the display is updated, and the information is added to a dynamic database. The dynamic database records information about successful searches and is used for subsequent processing. To continue the search, a backtrack is explicitly forced by invoking the standard fail predicate at the end of the search rule.

The output format is determined by the specificity of the query. For example, a query constrained only by a single bound variable such as site_id generates a list of the structures it owns and/or is owned by, in this case, a list of the well_ids and contaminants found at the specified site. Queries which specify a well_id and a chemical can, at the analyst's discretion, generate a tabular report, a graphic display, and a temporal analysis. In its present form, unless the display is reset the system will display the results of up to six queries concurrently, thus allowing comparisons, for example, among wells at a site or among contaminants at a well during different time periods.

Rule base specification

The rule base for the system is derived in part to meet data quality objectives established by government agencies. General guidelines are provided by a series of maximum contaminant levels for 20 chemicals and health advisories for 22 contaminants in the Safe Drinking Water Act. The Clean Water Act also has specified national ambient water quality criteria for 95 chemicals in ambient (non-drinking) water systems, and many states have established their own more stringent standards. Other rules are available to assist the decision-maker in evaluating temporal trends in data, and in determining the number of samples that should be taken given a specified confidence level (Mar, et al., 1986; Rajagopal,

1986a, 1986b). Each rule can be altered and rules can be added or removed by the system user to provide flexibility and adapt the system to different locales and regulatory environments. Determination of suitable action, therefore, can be made general or can be tailored to individual sites.

Goal examples

In this section, example strategies are described for specifying and implementing Prolog for an advisory system. Assume the following relation has been formulated:

```
ID, DAY, MONTH, YEAR, SUBSTANCE, VALUE, UNITS
```

where ID is the well identifier, DAY, MONTH, and YEAR provide the time of sampling, and SUBSTANCE is the sampled contaminant which is measured as VALUE in UNITS of measurement. This is encoded by specifying the domain of each variable in the relation:

```
DOMAINS
   ID, SUBSTANCE, UNITS, MONTH = symbol
   DAY, YEAR = integer
   VALUE = real
```

Given that declaration, the following predicates and clauses can be defined:

```
PREDICATES
   WELL (ID, DAY, MONTH, YEAR, SUBSTANCE, VALUE, UNITS)
   CLAUSES
      WELL (AAAA, 30, 2, 1987, LINDANE, 0.002, MG/L).
      WELL (ABCD, 01, 1, 1987, LINDANE, 0.006, MG/L).
```

Goals are then specified in a similar manner:

```
GOAL
   WELL (ID, DAY, MONTH, YEAR, LINDANE, VALUE, UNITS) and VALUE >
   0.004.
```

The above goal will find observations with values of Lindane exceeding the threshold (0.004) specified in the goal. The goal was derived from the maximum contaminant level for Lindane under the Safe Drinking Water Act. This goal seeking process can be generalized in several ways to enhance the interactivity of the system. For example:

```
DOMAINS
   ID, SUBSTANCE = symbol
PREDICATES
   Findit (ID, SUBSTANCE)
GOAL
   Write ("Enter well ID:"),
   Readln (ID),
   Findit (ID, SUBSTANCE),
   Write ("Substance found is:", SUBSTANCE).
CLAUSES
   Findit ("AAAA", "Endrin").
```

```
Findit ("AAAA", "Lindane").
Findit ("BCBC", "Endrin").
```

In this example, the user is prompted for the identifier for a well. That ID is then used in a search of the database and any SUBSTANCEs found in the clause section are returned and written to the screen. The process can be refined to include maximum contaminant levels in the goal. In that case, only wells with observations that exceed the contaminant level are reported. This is illustrated below.

```
DOMAINS
   ID, SUBSTANCE = symbol
   AMOUNT = real
PREDICATES
   Findit (ID, SUBSTANCE, AMOUNT)
GOAL
   Write ("Enter substance:"),
   Readln (SUBSTANCE),
   Write ("Enter level:"),
   Readreal (VALUE),
   Findit (ID, SUBSTANCE, AMOUNT) and AMOUNT > VALUE,
   Write ("Well:",ID,"exceeds threshold of",SUBSTANCE).
```

In this example, defining goals with variable levels provides the user with a means to explore the information contained in the database. This flexibility is important in an advisory or decision support context. Note also from the above example, that the ID is returned as a result of the query. This ID then can be used to retrieve additional information about the well. For example, coordinate information could be used to produce a display of wells exceeding a user-specified threshold.

Conclusions

Groundwater contamination has become an important environmental problem, and in response, different government agencies have been empowered to collect groundwater data. As a result of variation in the goals and missions of these agencies, however, a patchwork of databases has resulted. One problem that can arise from this lack of coherence is a decreased ability to use the databases effectively during decision-making processes. In this chapter a two-pronged approach to dealing with database complexity in groundwater decision-making is described. In the development of a knowledge-based system for assessing groundwater quality, databases that were designed and implemented in settings that fostered restricted views of groundwater problems were integrated to demonstrate the utility of the procedure. The resulting integrated database provides a wider, more synoptic view of groundwater quality and presents increased value for decision-making.

The second task involves the use of a declarative language to construct a system that allows powerful queries to be formulated by users. Prolog, unlike procedural languages such as Pascal, does not require the programmer (or user) to control iteration and reassignment of variables as the program executes. Rather, Prolog uses facts and rules about objects and their relationships. This allows the description of a problem to take place, and Prolog uses its built-in inference mechanism to search for solutions from among the information contained in the knowledge base. In this way, the approach described here allows a user to sort through data complexities by using rules and query facilities available in the logic programming environment. This approach works well for building and searching databases in a flexible manner to support environmental decision-making processes.

Acknowledgments

I would like to thank Mr. David Bennett for his assistance in implementing the final versions of the Prolog programs and for providing insightful comments on earlier drafts of this paper. Lois Peterson and Rhonda Weaver provided invaluable assistance during the final formatting of this chapter. Finally, I would also like to acknowledge the support services provided by University House, the center for advanced study at The University of Iowa. Address for reprints: The University of Iowa, 316 Jessup Hall, Iowa City, IA 52242.

References

Ademoroti, C. M. A., 1987. "Contamination of shallow wells in Nigeria from surface contaminant migration," *Environment International 13*:491-495.

Armstrong, M. P., 1987. "A rule-based advisory system for groundwater quality assessment at hazardous waste disposal sites," *Papers and Proceedings of Applied Geography Conferences 10*:32-38.

Armstrong, M. P. and D.A. Bennett, 1988. "Design and implementation of a groundwater quality advisory system," *Papers and Proceedings of Applied Geography Conferences 11*:10-16.

Armstrong, M P. and D.A. Bennett, 1990. "A bit-mapped classifier for groundwater quality assessment," *Computers and Geosciences 16*:811-832.

Armstrong, M. P., P.D. Densham, and G. Rushton, 1986. "Architecture for a microcomputer based spatial decision support system," in *Proceedings of the Second International Symposium on Spatial Data Handling*, pp. 120-131, IGU Commission on Geographical Data Sensing and Processing, Williamsville, NY.

Armstrong, M. P. and P.J. Densham, 1990. "Database organization strategies for spatial decision support systems," *International Journal of Geographical Information Systems 4*:3-20.

Bestougeff, H. 1984. "Designing the database management system," in J. R. Rumble and V. E. Hample, eds., *Database Management in Science and Technology*, pp. 119-132,

Elsevier, New York, NY.

Calkins, H. W. and Marble, D. F. 1987. "The transition to automated production cartography: Design of the master cartographic database," *The American Cartographer 14*:105-119.

Chen, P. P., 1976. "The entity-relationship model: toward a unified view of data," *ACM Transactions on Database Systems 1*: 9-36.

Chen, P. P., 1983. "English sentence structure and entity-relationship diagrams," *Information Sciences 29*:127-149.

Codd, E. F., 1979. "Extending the database relational model to capture more meaning," *ACM Transactions on Data Base Systems 4*:397-434.

Davis, C. G., Jajodia, S., Ng, P. A., and Yeh, P. A., 1983. *Entity- Relationship Approach to Software Engineering*, Elsevier Scientific Publishing Company, New York, NY.

Densham, P. J. and Armstrong, M. P., 1987. "A spatial decision support system for locational planning: design, implementation and operation," in *Proceedings of the Eighth International Symposium on Computer-Assisted Cartography*, pp. 112-121, American Society for Photogrammetry and Remote Sensing, Falls Church, VA.

Dueker, K., 1985. "Geographic information systems: toward a geo-relational structure," in *Proceedings of the Seventh International Symposium on Computer-Assisted Cartography*, American Society for Photogrammetry and Remote Sensing, pp. 172-177, Falls Church, VA.

General Accounting Office, 1988. "Groundwater Quality: State Activities to Guard Against Contaminants," Report to the Chairman, Subcommittee on Hazardous Wastes and Toxic Substances, U.S. Senate, GAO/PEMD-88-5.

Hopkins, L. D., 1984. "Evaluation of methods for exploring ill-defined problems," *Environment and Planning B 11*:339-348.

Lorie, R., and Meier, A., 1984. "Using a relational DBMS for geographical databases," *Geo-Processing* 2:243-257.

Mar, B. W., Horner, R. R., Richey, J. S., Palmer, R. N., and Lettenmaier, D. P., 1986. "Data acquisition: cost effective methods for obtaining data on water quality," *Environmental Science and Technology 20*:545-551.

Morehouse, S., 1985. "ARC/INFO: A geo-relational model for spatial information," in *Proceedings of the Seventh International Symposium on Computer-Assisted Cartography*, pp. 388-397, American Society for Photogrammetry and Remote Sensing, Falls Church, VA.

Navathe, S., Elmasri, R., and Larson, J., 1986. "Integrating user views in database design," *Computer 19*:50-62.

Peuquet, D. J., 1988. "Representations of geographic space: toward a conceptual synthesis," *Annals of the Association of American Geographers 78*:375-394.

Rajagopal, R., 1986a. "Conceptual design for a groundwater quality monitoring strategy," *The Environmental Professional 8*:244-264.

Rajagopal, R., 1986b. "The effect of sampling frequency on groundwater quality characterization," *Ground Water Monitoring Review*, Fall, 65-73.

Rajagopal, R., 1987. "Large data bases and regional ground-water quality assessments—An Iowa case study," *Ground Water 25*:415-426.

Roberts, R. S. and Butler, L. M., 1984. "Information for state groundwater quality policy-making," *Natural Resources Journal 24*:1015-1041.

Robinson, V. B. and Zhang, G., 1988. "Modeling consistency preserving land database transactions in a logic programming environment," in *Proceedings of GIS/LIS '88: Volume II*, pp. 971-980, American Congress on Surveying and Mapping, Falls Church, VA.

Sprague, R. and Carlson, E., 1982. *Building Effective Decision Support Systems*, Prentice Hall, Englewood Cliffs, NJ.

Sterling, L. and Shapiro, E., 1986. *The Art of Prolog*, MIT Press, Cambridge, MA.

Turbo Prolog Owner's Handbook, 1986. Borland International, Inc., Scotts Valley, CA.

8
Regulating the Municipal Environment Using an Expert Geographic Information System

David R. Maidment and Thomas A. Evans

Introduction

One function of city government is to regulate the manner in which the city is affected by construction of new buildings, subdivisions, or other developments. City government performs this function through the mechanism of a regulatory code, which is a text reflecting an accumulation of ordinances passed by the City Council together with interpretation by the city staff. Architects and engineers present designs and plans for new developments to city staff who determine whether the plans comply with the code. If plans don't comply with the code, they must be altered or a waiver (or variance) from the provisions of the code must be obtained by application to the City Council or another panel of officials.

While, in theory, municipal regulation could function like a well-oiled machine, swiftly dispensing decisions on interpretation and compliance with city codes, in practice the process rarely runs so smoothly. Although delays are often blamed on the city bureaucracy, even with the most conscientious of city staff and a well-organized regulatory code, plan review and approval is an inherently complex and time-consuming process. The root causes of delays are two. First, there is naturally an adversarial relationship between city staff and the architects and engineers representing land development projects. It is the responsibility of city staff to guard the welfare of citizens throughout the entire city, while it is the responsibility of the developer's representatives to get the greatest return on a particular project. Second, the regulatory code that governs how this adversarial relationship will function is highly complicated by layer upon layer of rules and conditions which may or may not apply to the project in question. Rules from different areas of the code can overlap and even conflict with one another. Some sections of the code are vague and difficult to understand; in addition, the rules are constantly subject to change as new ordinances are passed.

The repeating cycle of design and review, design and review, may consume months or even years before producing the approved site plan for a new commercial building or subdivision. These delays carry a substantial cost which is initially paid by the developer, then passed on to the occupants of the buildings, who may in turn pass it on to their customers, and so on into the general population. This is a hidden cost of regulation. As an example, a piece of commercial property worth $5 million being held for a project with a loan at ten percent interest rate and a 20 year period, requires $1600 per day in payments just to satisfy the lender. Literally, time is money and anything that can be done to simplify and accelerate municipal review, while maintaining compliance with the city code, is beneficial to both the city and the developer. Automation, using computer technology, is one means of accomplishing this goal.

The intent of this chapter is to describe how the municipal regulatory process is transformed into a conceptual model for representation in a computer system, and how the prototypes of the model are programmed in an integrated computer technology called an expert geographic information system (EGIS).

Site Code concept

Site Code is a system of linked computer programs which automates some features of the regulatory process. Its implementation will allow city staff and applicants to find which regulations are applicable to a project and what requirements they impose without having to spend many hours poring over paper copies of the regulations.

At present, no operational Site Code systems exist, but a series of prototypes have been developed at the University of Texas at Austin using the land development regulations of the city of Austin, Texas as a test subject. Austin is a particularly appropriate test environment for this purpose because the city has made considerable efforts in recent years to simplify and streamline its regulatory code and its review and approval procedures. Thus, our research serves to show how automated tools can provide additional increases in consistency and accuracy of regulatory information beyond those attainable by thorough organization of the code and staff procedures.

An EGIS is a linkage between an expert system, used for describing the logic of municipal regulations in its rule-based programming environment, and a geographic information system which maps the spatial jurisdiction of the regulations and locates the site within those jurisdictions. Site Code uses a prototype expert GIS developed by linking the Nexpert Object expert system shell with the Arc/Info GIS using an Arc-Nexpert interface. This interface, operational in UNIX on a Sun 4 workstation, permits Nexpert Object and Arc/Info to interact by issuing executable commands to one another.

Site Code applies expert GIS to regulation processing by using the expert system to express the logic of the regulations, and the GIS to represent the maps normally used in applying those regulations, thus transforming a paper-based process into a computerized process. Nexpert searches its rule base to identify applicable regulations and activates routines in Arc/Info as required to supply spatial information such as the site's location, its zoning classification, proximity to principal roadways, position in sensitive watersheds, and so on (information presently found on paper maps kept in various city offices).

This concept is limited in practice by the inherent complexity of some municipal regulations and the requirement for case by case interpretation. It seems unlikely that it would be cost-effective to capture the regulatory logic completely, even if it were possible to do so. A compromise has to be drawn that only the most explicit and constraining regulations are actually programmed into the expert system and the remainder of the code is left for review and interpretation by city staff. Site Code's chief value lies in its ability to free staff from the mundane tasks of sorting through maps and volumes of regulations. It is intended to aid the city's staff in their work with applicants by allowing them to focus on questions and interpretations which require professional understanding and experience. Staff may be helped further by having access to a computerized form of the city code, which enables an applicant or city staff member to navigate the text of the regulatory code quickly and view its contents. This computerized form of the code, which uses hypertext technology, is an important adjunct to the Site Code EGIS and future Site Code work will incorporate hypertext directly.

Previous work in this field

The first version of Site Code was developed as part of a Ph.D. dissertation by Douglas T. Shaw (Shaw and Maidment, 1990), who studied the land development regulations and processes of three cities (Austin, TX, Fairfax County, VA, and San Diego, CA) to determine whether these processes are sufficiently similar to permit a generic regulatory processing system to be developed. He found that while each of these municipalities has special features concerning how its regulatory process functions, they all have the same basic structure. This similarity suggests that the techniques developed in automating the regulations of one city may be applied in others as well.

The first prototype version of Site Code (Shaw and Maidment, 1989), which we will call Site Code I, was programmed entirely within the Nexpert Object expert system shell and contained about 100 rules representing Austin's site plan regulatory process. Site Code I can be considered a concept demonstration system because it represented an image of what an ultimate system might look like without being complete in any aspect. There was no true GIS in Site Code I; its function was simulated with a set of scanned map images. All the regulatory facts were captured as objects within the expert system; they were not stored in an external database.

The design for Site Code I was influenced by similar efforts to represent other design processes or regulatory codes in expert system packages, notably the work by Garrett and Fenves (1987) on knowledge-based methods for structural design of buildings, Rosenman and Gero (1985) who computerized portions of Australia's uniform building code, Hosking, et. al (1987) who developed a knowledge-based computer model of New Zealand's building fire safety design code, and Leary (1988) who showed that the design guide employed by Essex County in England could be depicted as a goal tree, which lends itself to expert system programming. GIS can also be employed for planning decisions on a large scale such as selection of optimal site locations (Wright and Buehler, 1990).

It became obvious during the development of Site Code I that the location of the project has so much impact on determining which regulations apply to the project that an interactive link between the Site Code expert system and a GIS is required. An operating version of such a system was impossible to construct with the technology available at the time, because there was no direct linkage between the expert system and a GIS. This barrier to the advancement of Site Code prompted the first author of this chapter to organize a seminar on the subject "Integration of Expert Systems and GIS: Building an Arc-Nexpert Interface" at the University of Texas at Austin in July 1990. It became apparent rather quickly that the expert system/GIS linkage needed for Site Code is also desired for many other applications in which expert systems need a spatial data representation or in which it is desired to increase the decision support capabilities of GIS. Thus, a research initiative was launched in EGIS, the object of which is to develop a working prototype of this linked technology and explore its capabilities in various applications.

The first Arc-Nexpert interface was programmed by Dean Djokic as part of his Ph.D. dissertation studies (Djokic, 1991), where he used it to develop an intelligent network builder for analysis of stormwater drainage networks. The expert system sorts through the GIS database of stormwater pipes and inserts artificial pipes where necessary to account for the emergence of stormwater drainage from a pipe into a surface ditch from which it flows back into a pipe again. In the absence of these artificial pipes, the representation of the drainage network would be incomplete and portions of the contributing watershed area would be unaccounted for in the predicted flow at the outlet in the stormwater flow routing program (Djokic and Maidment, 1991).

Using this Arc-Nexpert interface, the second author of this chapter developed a new prototype version of Site Code which we will term Site Code II. In Site Code II, a different approach was taken than in Site Code I in the sense that instead of trying to take an overview of the whole regulatory process, we addressed only the first task normally faced in determining compliance of a site plan: does the site zoning permit the land use which the applicant wishes to place upon it, and assuming that it does, what are the design constraints such as maximum building height, front and side setbacks from the property boundary, and impervious cover, that are imposed by the zoning category? Site Code II is complete and exhaustive for the task it addresses; it captures all 32 of the zoning categories and the more than 1100 combinations of zoning category and land use type permitted in Austin, together with the design constraints pertaining to each zoning category.

Methodology

The mechanism for transforming the municipal regulatory process into a conceptual model can be considered in two dimensions. First, there is the permitting sequence which is followed from the initiation of a completely new project through a number of permitting steps ending with the receipt of a building permit and the construction of the facility. Second, there are the technical areas, usually represented by different city departments, such as those for land use, environmental, drainage and utilities, who must approve the project at various stages in the permitting process.

Permitting process

Figure 8.1 shows the four steps in the permitting process prior to construction. Each is a complex procedure in its own right: zoning, subdivision, site planning, and the building permit. The City of Austin has seven volumes of regulations, each several hundred pages long, which spell out the details of these four steps.

The first step is zoning. The symbols shown on the zoning map in the left panel of Figure 8.1 are zoning categories (LI stands for light industrial and SF-1 for single family low density residential housing, for example) and associated with each zoning category there is a list of permitted land uses which are legally defined, such as condominium, single family attached residential (a duplex), or general services retail. The land use proposed for the site must fall within one of the permitted land use categories or else the site must be rezoned to accommodate the proposed use. The zoning map is defined by the city's comprehensive planning process and compliance with it ensures that each new development is compatible with its neighbors and with an overall concept for growth of the city.

Once a tract of land has been zoned, it may be legally subdivided, which means that it can be broken down into individual lots whose extent is defined by the subdivision plat shown in the second panel of Figure 8.1. These lots inherit the restrictions imposed by the zoning category and they acquire additional constraints associated with access from streets, location of floodways, and water and electric utility lines. In addition, it may occur that restrictions applying uniformly to the subdivision, such as allowable percentage of impervious cover, may apply non-uniformly to individual lots within the subdivision so that more impervious cover is permitted on some lots than on others while maintaining the entire subdivision within the overall limit on average impervious cover. This is the case in mixed use commercial-residential developments. Planning of individual lots or sites within the subdivision must comply with constraints imposed by the subdivision plat.

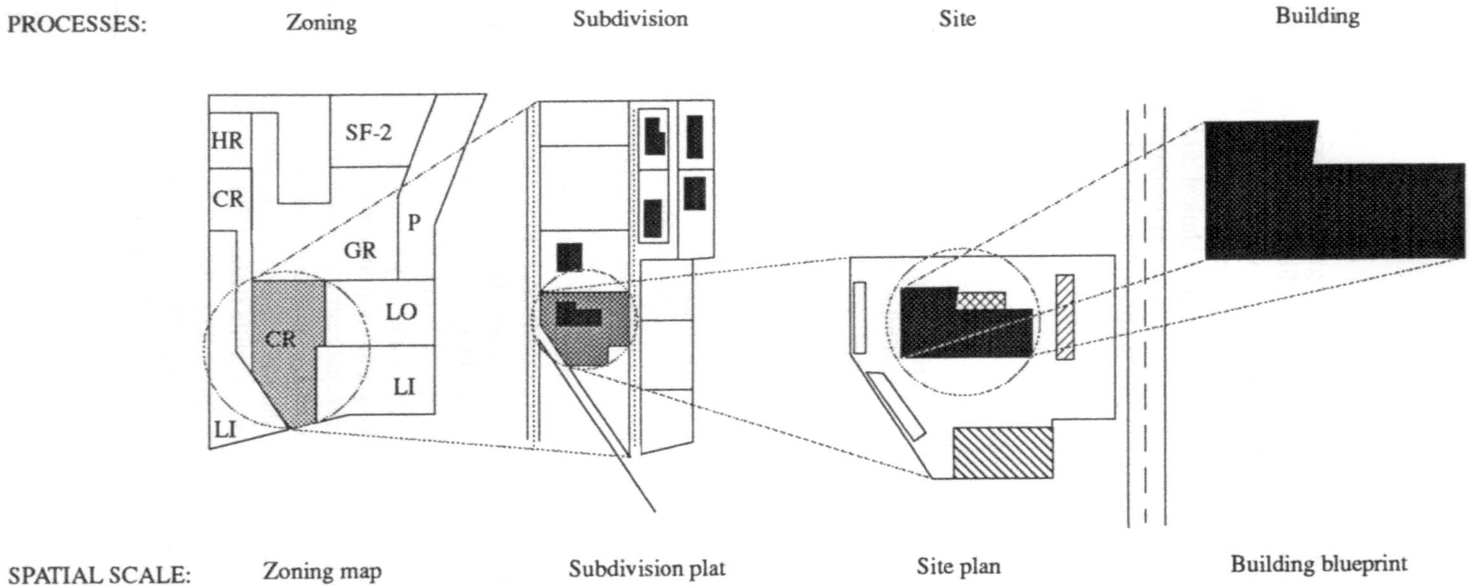

Figure 8.1. The permitting sequence for a municipal site development (from Shaw and Maidment, 1990).

The third step in permitting shown in Figure 8.1 is approval of the site plan for an individual lot. It is here that the detailed engineering design of the site facilities is performed and checked. Stormwater drainage, environmental design, parking and traffic access, and other technical issues must be addressed at this stage. As shown in the figure, the building is treated as a footprint on the site. Site planning focuses on the location of the building and the condition of the site and its surroundings, but not the internal configuration of the building. In spite of this limitation, site plan approval is the most technically complex of all the permitting steps and the most likely to benefit from automation. This area has been the focus of site code development.

The last step is approval of a building permit for each structure on the site. Issuance of a building permit accounts for the structural design, plumbing, electrical, mechanical, fire and safety provisions of the building, in accordance with one of the standard building codes, such as the uniform building code. Unlike the first three steps in the permitting sequence, building codes are relatively standardized throughout the United States and most communities adopt one of the standard codes with amendments as a requirement for city code compliance.

The names and requirements for permits in a city's permitting process vary from one city to another, but whatever the permitting sequence, each city's process ensures that the four steps shown in Figure 8.1 are accounted for.

Constraint tunnel

The telescoping action of progressively focussing in on a smaller and smaller area of land in increasing detail, implied by the sequence of dashed circles and lines from left to right in Figure 8.1, creates a tunnel of constraints which must be traversed to gain approval of the project, as shown in Figure 8.2. The dimensions of this tunnel are defined by the design requirements imposed on the site by each part of the regulatory code, the outer limits of which are set by the land use restrictions arising from the zoning category, and more restrictive limits occur as additional considerations are brought to bear on the design, such as transportation, environmental, drainage, and other factors. These factors are applied progressively as the project goes through the constraint tunnel, progress in which is measured by approval of the required sequence of permits. Of course, though the constraint tunnel is visualized here as a spatial object in three dimensions, in fact the constraints have many dimensions, some geometrical and others functional.

A development project may be conceived as a "brick" comprised of a tract of land and a concept of how it will be used. Like the tunnel, the brick has many dimensions, some geometrical, others functional. This brick moves through the constraint tunnel continually being refined as site design and approval proceed, until, hopefully, it emerges as an approved design at the end. Unfortunately, however, this is often not the case. It often occurs that the brick gets jammed part way through the tunnel because regulations require some site feature that is infeasible within the original design concept; this requires either that the shape of the brick be modified or that the constraints be relaxed to allow the brick to pass on.

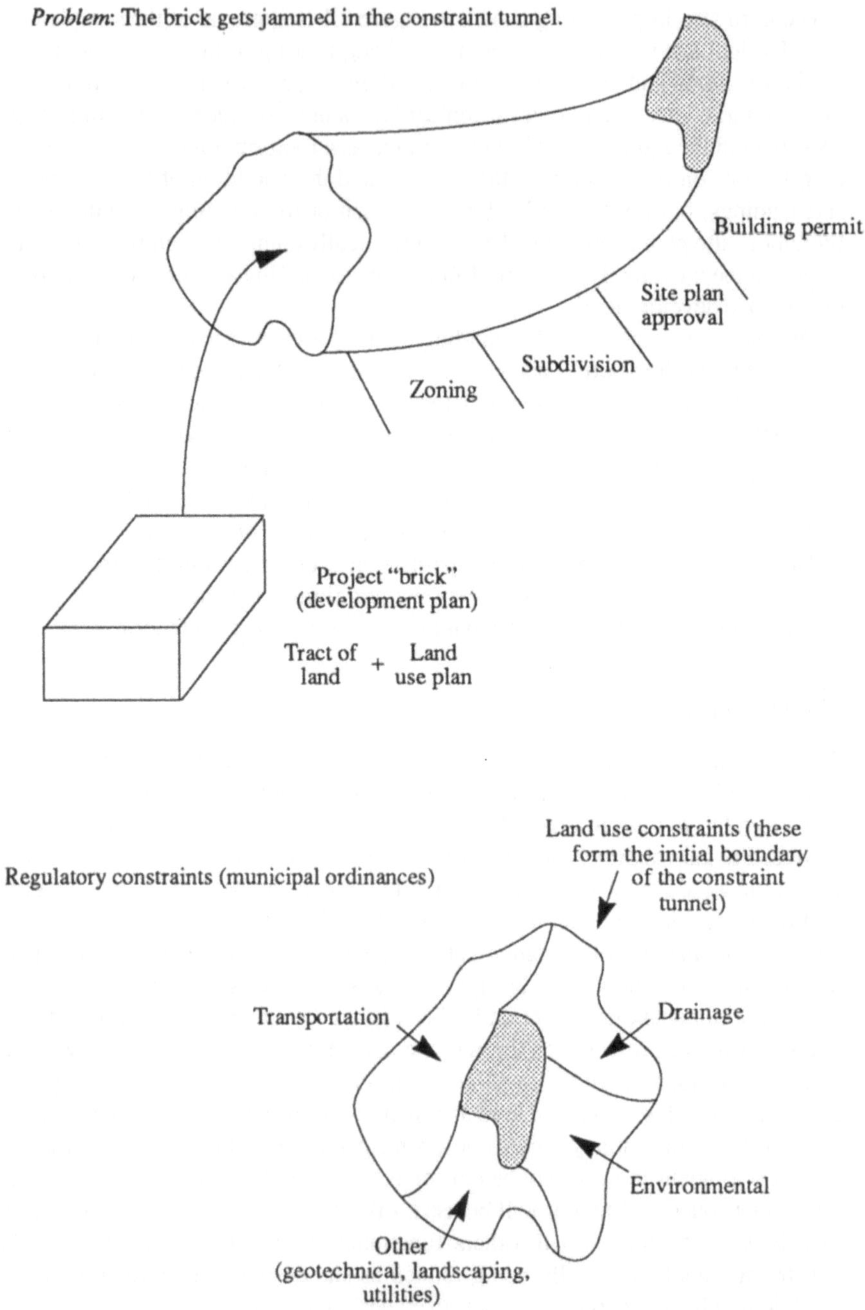

Figure 8.2. Approval of a proposed development plan (project "brick") may be measured by progress of the brick through a regulatory constraint tunnel.

For example, a common constraint for commercial building construction is that one parking space is required for each 250 square feet of floor space in the building. If the parking is laid out around the building, it can easily occur that insufficient space is left to accommodate the required number of parking spaces, and so at an advanced stage of design, the project becomes infeasible. If this could have been anticipated, a slightly smaller initial building design could have been adopted and the problem eliminated, but the financial pressures to get the largest site "yield" (amount of floor space or number of occupiable units) from the development frequently cause problems of this kind late in the design stage. The only alternative then is to seek a variance from the parking requirement, relaxing it to one space per 275 or 300 square feet of floor area, a process which can take weeks or months to accomplish.

At the outset of the project, the constraint tunnel twists and turns off into the future, a dimly-lit corridor whose dimensions are known only by the experience of having passed other bricks through the tunnel in the past. The more illumination that can be cast on the tunnel at the outset of the development project, the better the chance that the project "brick" can be shaped to fit the tunnel before the designs have solidified and changes become more and more expensive.

Regulation filters

The process of determining the dimensions of the constraint tunnel before detailed design and review requires that the regulations applicable to a particular project be isolated from the great mass of all regulations in the city code. It can be imagined that the process of isolating applicable regulations is accomplished by a series of filters (shown in Figure 8.3) into which one pours all the regulatory code and watches as the regulations fall through a sequence of screens each of which removes some regulations until the applicable set falls through the bottom, together with the set of constraints on design implied by those regulations.

The first filter relates to the type of permit which is being sought, that is, which of the four steps in Figure 8.1 is the subject of the current application. Land often changes hands after part of the development process is complete, so the applicant for site plan approval for a particular lot within a subdivision may not be the same person who accomplished the zoning and subdivision steps.

The second filter pertains to the time of application and determines which version of time-dependant parts of the regulatory code will be applicable, considering the "grandfather clauses" that exempt projects for which applications have already been filed from regulations passed subsequent to the application. The requirements to obtain a permit can be frozen in the form that existed at the time of application even though the regulations may be subsequently altered during the time which passes from first application to permit approval.

The third filter relates to the geographic location of the site in relation to the ordinance jurisdictions. The regulation is applicable if the site is located within its jurisdiction and not otherwise.

SITE CODE: Isolates from all municipal regulations applying to any kind of project anywhere in the city those particular regulations applicable to this particular project at this location. Extracts from these regulations the numerical constraints which will limit the size of the project.

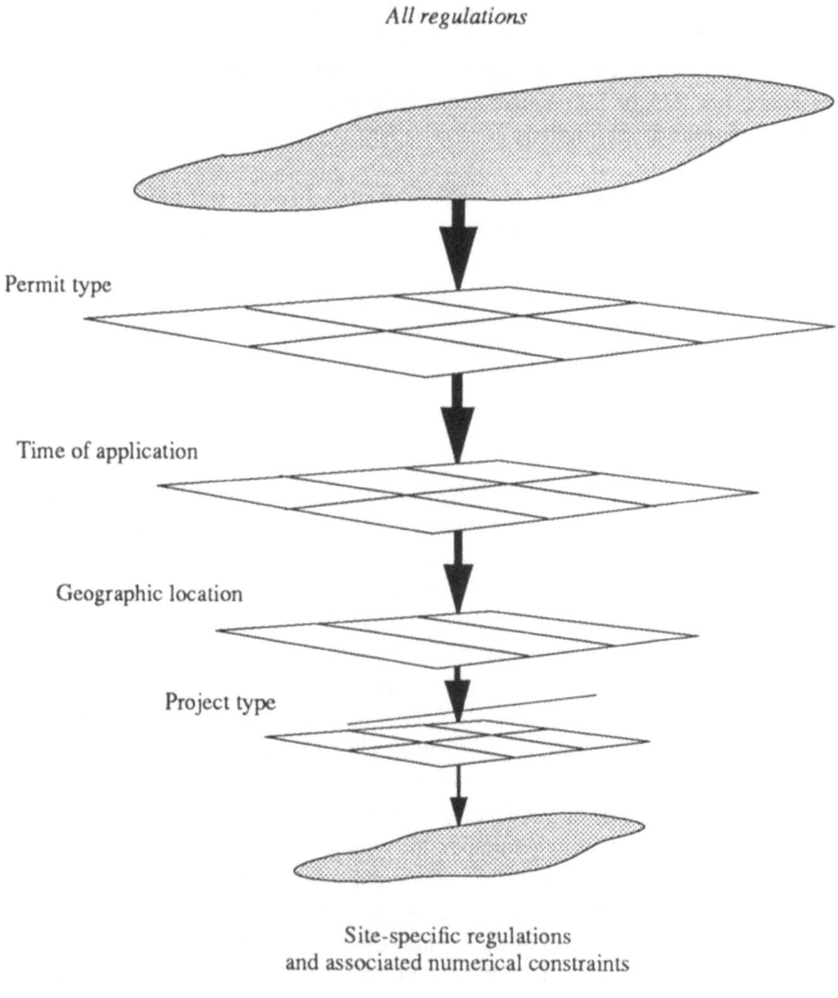

Figure 8.3. Site Code's regulatory filters.

The final filter is related to the project type, the substantive nature of what is proposed for construction on the site. The regulations for an automobile service station are different from those for a condominium regardless of where it is located in the city.

Spatial jurisdiction

All parts of the regulatory code have a spatial jurisdiction, or region of application, within the city (shown in Figure 8.4). These spatial jurisdictions can be classified into three categories: spatially uniform regulations which apply the same way everywhere in the city, spatially variable regulations which apply throughout the city with varying effect depending on location, and spatially-triggered regulations which apply in some areas but not in others. Regulations governing particular land uses, such as those for condominiums, are spatially uniform but those relating to zoning categories are spatially variable because they have particular values for each category on the zoning map. It may happen that these variable regulations override the uniform ones, restricting the density of condominium development within residential areas, for example. Principal and scenic roadway regulations are examples of spatially-triggered regulations since they apply only within a defined distance, say 1000 feet, of the designated roadways and not elsewhere.

Spatial jurisdictions can be imagined as a set of regulatory polygons floating over the city. If these polygons are drawn on a set of overlaid paper maps, a push-pin could be pressed through the maps at the site location to determine which regulatory polygons the site lies within. This push-pin function is readily performed by a GIS as a point in polygon test, or more strictly a polygon (the site boundary) within polygon (the jurisdictional boundary) test.

Types of regulations

The regulations within the text of the code are all numbered and some have labelled sections and subsections within them. If we define a regulation unit as a piece of text with a single regulatory intent, then these units can be classified into three categories: (1) administrative regulations, such as those defining the process by which one gets a permit, the time period of application or the spatial jurisdiction of particular parts of the code, (2) substantive regulations, which may impose design requirements on the site, and (3) explanatory regulations, which describe the meaning of particular terms within administrative and substantive regulations. The first three levels of the filters in Figure 8.3 are associated with screening out regulations on administrative grounds while the fourth level relates to substantive regulations. Administrative regulations determine what path will be followed through the code and substantive regulations determine whether the project will have to satisfy some requirements of the code along that path.

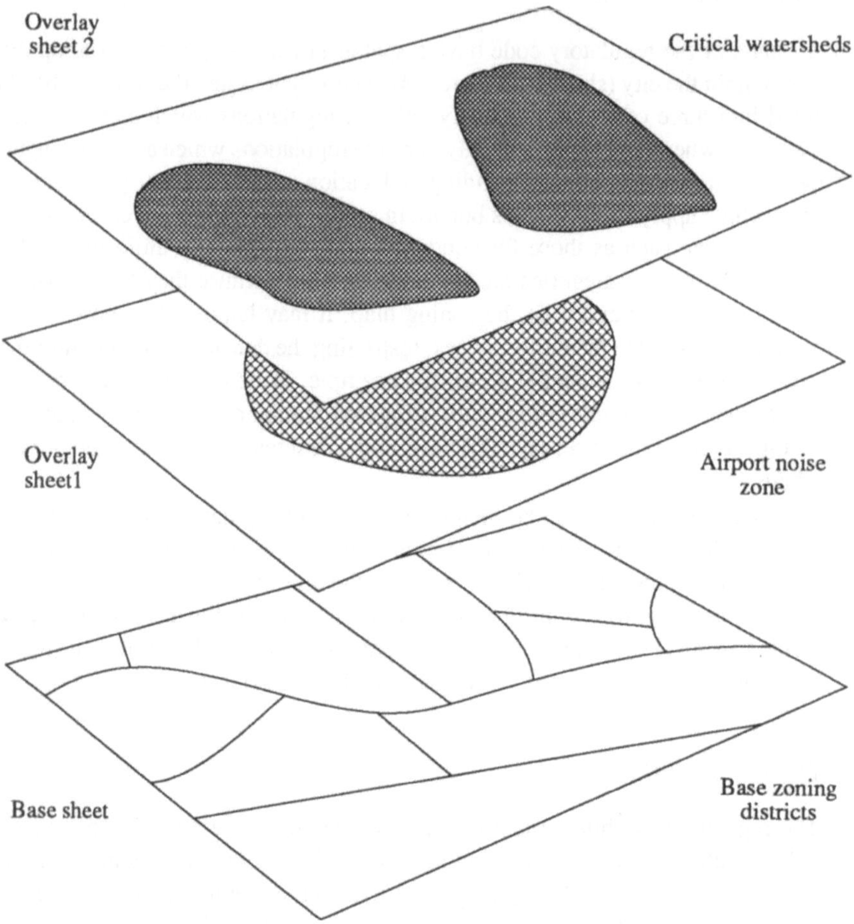

Figure 8.4. Overlaying regulatory polygons. Each polygon represents the spatial jurisdiction of a set of regulations.

Determining the project requirements

Determining project requirements is accomplished by a process of progressive refinement and interpretation of the regulations as illustrated in Figure 8.5. For this purpose, we define three sets: set P of potentially applicable regulations, based on the type of permit desired and the technical areas of compliance with that permit which are being addressed, set A of applicable regulations, which is a subset of P, and set R of project requirements, which is derived by interpretation of the set A.

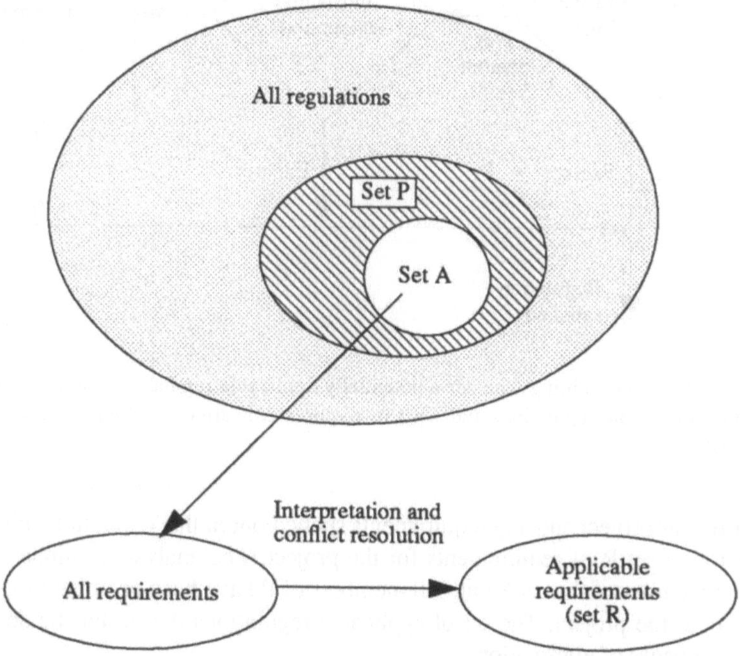

Figure 8.5. The process of identifying applicable regulations and their requirements.

The process by which the sets A and R are determined from P is illustrated in Figure 8.6. If we consider each regulation unit within the set P, the first question asked is "Is the project administratively exempt from this regulation?" The answer is determined by the first three filters of Figure 8.3: type of permit, time of application, and spatial jurisdiction of the regulation. If the project is administratively exempt, attention passes to the next member of set P. If the project is not administratively exempt, the next question is "Is the project substantively exempt from this regulation?" Answering this question requires specific knowledge of the project beyond its location in time and space. For example, a condominium project would be substantively exempt from regulations that apply only to single family homes. If the regulation is not substantively exempt then it becomes a member of the set A, its provisions become part of the "customized code" of this

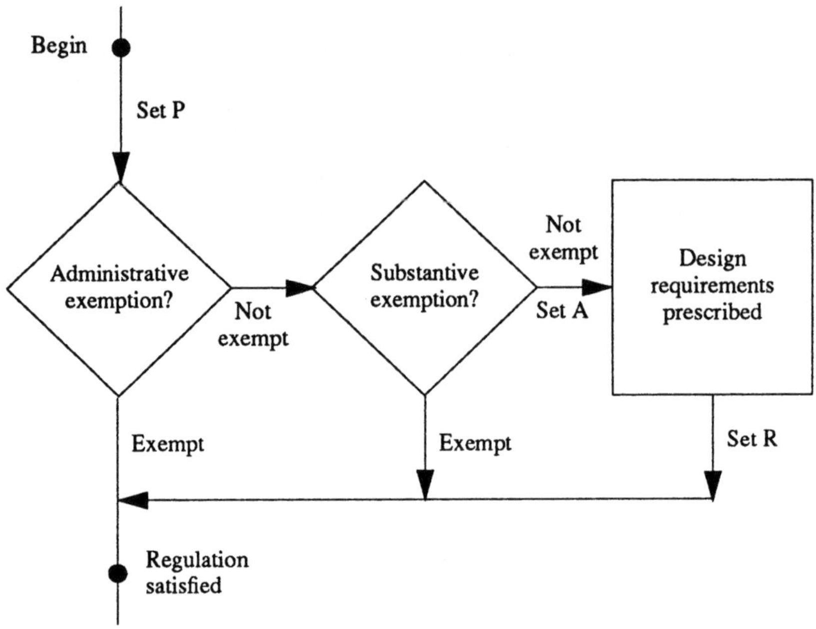

Figure 8.6. Determining whether a potentially applicable regulation is a member of the set of applicable regulations and what its requirements are (from Shaw and Maidment, 1990).

particular project and any requirements spelled out in this regulation unit are added to the set R of requirements for the project. The analysis continues with the next member of the set P until all members of P have been examined for applicability to the project. The set of applicable regulations A can thus be determined by a process of elimination.

After all regulations in set P have been examined, the set A is complete and the requirements R can be determined. Some requirements are easily evaluated and described, such as maximum impervious cover for a particular zoning category. Others are less clear because they are stated not as specific requirements but rather as performance goals, such as "the full effects of development on runoff shall be accounted for." Multiple requirements, from different sections of the code, may affect the same project parameters. Maximum impervious cover, for example, is specified both in the base zoning requirements and in water quality protection provisions. In most cases of multiple specification, the more restrictive requirement is applied to the project. As shown in Figure 8.5, a final stage of interpretation is thus required to determine how performance goals will be attained for this project and how conflicting requirements will be resolved to determine the requirements set R.

The process of reducing set P to set A and deriving set R is a shorthand version of the review procedure currently followed by city staff. They know from experience where potentially applicable regulations P are likely to be found; by checking off all the administrative criteria, they gradually focus on the applicable set A; by discussion among themselves and with the applicant they determine the requirements R. The problem from the perspective of city staff is that the code is complicated enough that some applicable regulations might be omitted by oversight, or by errors made in reducing the set P to the set A. Further errors may arise in extracting or interpreting the requirements R. Also, the searching required to determine A and R can consume time that would be better spent on interpretation which can be performed only by staff. The problem from the applicant's perspective is that the review and approval procedure can take a long time. Each city department is specialized in its own area of the review process and is not well-versed in the requirements of other departments, so an applicant with a complicated project may find himself caught between conflicting interpretations provided by different city departments. Some cities, such as Austin, have adopted a case-manager approach to minimize such conflicts by having all reviews pass through a single staff member.

Building a regulation processor

The goal in applying computers to regulatory review is to produce timely, accurate, and consistent information about city regulations. We suggest that a decision support tool can be developed to help city staff reduce the tedium of sorting through the regulations to determine their applicability.

A conceptual design for this decision support tool, which we call a regulation processor, is shown in Figure 8.7 and consists of the following seven components. The boxes shown in solid outline in Figure 8.7 already exist in Site Code II; the boxes and links shown with dashed lines have yet to be created.

1. *Regulation processing engine:* An expert system knowledge base that analyzes regulations in the manner depicted in Figure 8.6 and directs the flow of information between the user and the system's component programs.

2. *Project tracking system*: Many cities maintain such a system in order to record data and facts regarding ongoing projects.

3. *Permit index*: For a given type of permit and technical area to be examined, this index directs the search to the potentially applicable set P of substantive regulations.

4. *Regulation register*: Contains a list of all regulations and their titles, jurisdictions, effective dates, and a the text of each regulation.

5. *Regulation knowledge bases*: Express the logic of complex regulations and from which the user may call out to GIS, hypertext, databases, and other external routines to help the regulation processor determine the sets A and R from the set P.

6. *Requirements table*: Lists the requirements that particular regulations can impose on a project.
7. *Session report:* Presents a printed report of the results of the investigation.

The user of the regulation processor interacts primarily with the regulation processing engine, either directing its focus to the permit process of interest or responding to its requests for additional information about the project. When geographic information is needed, the regulation processing engine activates the GIS and the user interacts with it until the information is retrieved and session control is returned to the regulation processing engine. At any time the user may refer to the body of the code itself, available as a hypertext document on the computer.

Tasks to be performed

The tasks that a regulation processor might perform are arranged in an ascending fashion in Table 8.1. These tasks can be viewed in terms of the brick and constraint tunnel analogy drawn in Figure 8.2. The first task, process assessment, consists of determining whether the brick can enter the tunnel at all, that is, whether the land use desired on the property is permitted under its current zoning category, and if it is permitted, process assessment describes the process that will be required to get the required permit approvals. If the desired land use is not permitted, then the zoning category which is closest to the current zoning which does permit the required land use could be identified, or alternative land uses permitted under current zoning could be determined. In the City of Austin, process assessment is voluntary for all proposed projects and is done manually by the applicant and city staff by filling in the blanks on printed forms.

The second task, project assessment, consists of passing a conceptual brick through the constraint tunnel and determining its shape when it falls out the end, that is, for a project of given location and land use, the set of basic requirements R imposed by the main constraining regulations is identified, from which an approximate yield of the site for this land use can be calculated. The City of Austin currently supports a voluntary project assessment, which is also manually performed by city staff.

The third task in the hierarchy, compliance checking, applies when some preliminary design work has been done and the layout of the site determined: it consists in passing a brick of defined dimensions through the constraint tunnel, which differs from passing a conceptual brick through the tunnel because the brick's dimensions are known and are being compared against the dimensions of the tunnel to see if the project complies in its present form. One can imagine many degrees of complexity of compliance checking of this type, some automated and others manual.

The fourth task, site selection, arises when, instead of a single site, a number of potential sites are available for a given land use and the intent is to chose the best one. This task can be likened to passing a number of conceptual bricks down the

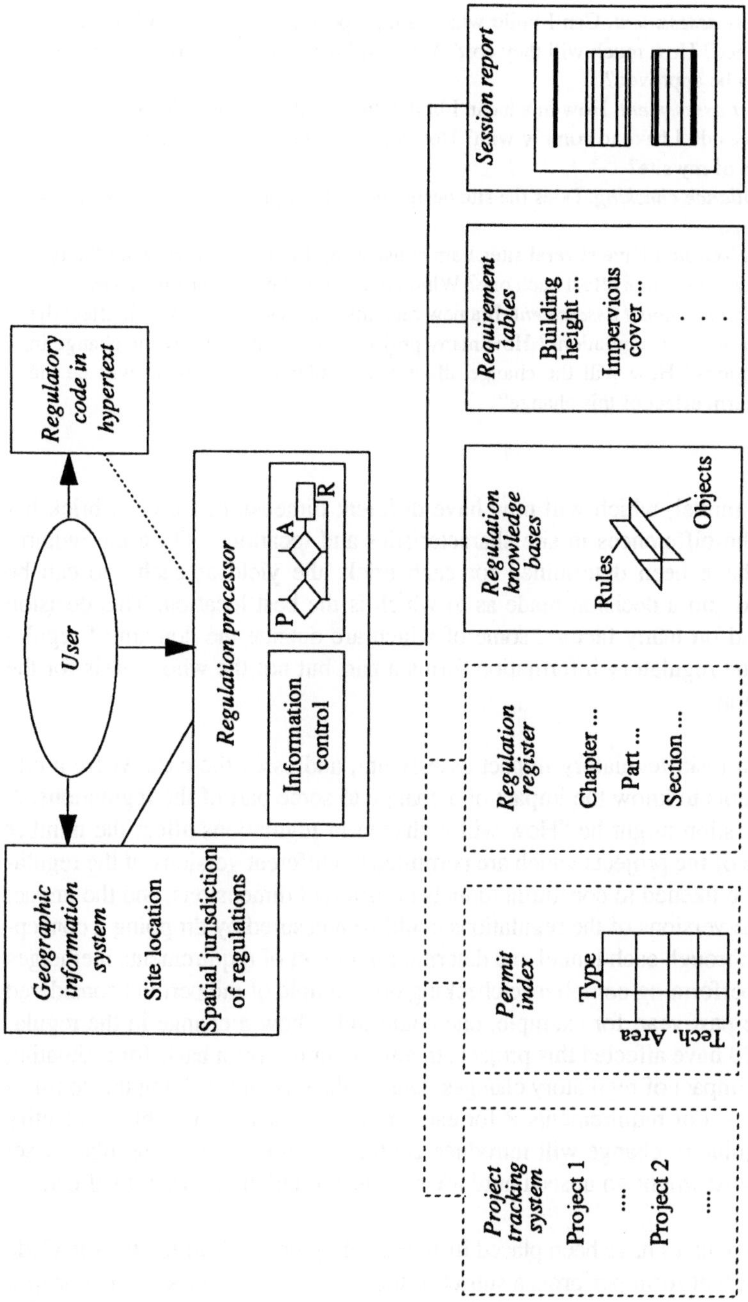

Figure 8.7. The conceptual design of a regulation processing system. Site Code II elements are shown with solid lines.

Table 8.1. Types of Enquiries a Site Code System Might Address

1. *Process assessment*. Can I build what I am proposing on this site? What permits do I need? How much will they cost? What do I have to submit? How long will it take to be approved?
2. *Project assessment*. How much can I build? What will my site yield be? What ordinances do I have to comply with? How will the regulatory constraints affect the design of my site?
3. *Compliance checking*. Does the site design that I have prepared comply with regulations?
4. *Site selection*. I have several sites I am considering for a project. How do the regulatory constraints affect each one? Which one would be best for my project?
5. *Regulatory impact assessment*. If a new regulation is passed, how will it affect the action of other regulations? How many projects will be affected by the change in regulations? How will the change affect a particular project? What will be the long-term effect of this change?

constraint tunnel, which will now have different dimensions for each brick because of the differences in site characteristics and location. Once the requirements R have been determined for each brick, the yield at each site can be determined and a decision made as to which is the best location. This decision will depend on many factors, some of which are outside the domain of regulations, so the regulatory information forms a part but not the whole basis for the site selection.

The fifth task, regulatory impact assessment, addresses the case where a city council wants to know the impact of a change to some part of the regulations. A typical question might be "How will a change in regulations affect the number and nature of the projects which are permitted?" Different versions of the regulations can be likened to constraint tunnels of different dimensions, and the impact of alternate versions of the regulations could be measured by dropping a conceptual brick through each tunnel and determining the set of requirements it emerges with. By performing compliance checking on a sample of the permits considered during the past year, for example, one could judge how a change in the regulations would have affected this project stream and thus have a basis for estimating the future impact of regulatory changes. During the step of resolving the conflicts among the set of requirements R for each project, it may be possible to identify how a regulatory change will introduce conflicts with the current regulatory set that otherwise might go unseen until after the new regulation goes into effect.

These five tasks have been placed in an ascending order of difficulty. Site Code II in its present form performs a subset of the first two tasks, process and project assessment. It will be some time before sufficiently mature systems are available to perform the other three tasks.

Implementation

Site Code II integrates three types of information: an expert system (Nexpert Object), a GIS (Arc/Info) and a database, NXPDB, which is a delimited ASCII file format native to Nexpert. The expert system handles interaction with the user and with the GIS and database. The GIS contains a parcel map (a map of the legal boundaries of all the lots) of a section of Austin overlaid by the zoning boundaries for that section of the city. The database contains data files listing the land use types permitted in each zoning category and a list of standard requirements for projects constructed in each zoning category.

Communication between Nexpert and Arc/Info is accomplished using the Arc-Nexpert interface illustrated in Figure 8.8. This interface is a command link which permits Nexpert to pass the name of a macro to Arc/Info, and to receive a single numerical or string variable, the result of the macro, in response. Messages are passed through two intermediate ASCII files: while one program is active, the other is dormant, waiting for a particular ASCII file to appear. When that occurs, the dormant program wakes up, opens and reads the file and processes it; meanwhile, the other program goes dormant while waiting for a different ASCII file to appear with a new message to process. In this way, control is transferred back and forth between the two systems and the full functionality of both systems is available.

Two limitations of the current Arc-Nexpert interface should be noted. First, while Nexpert and Arc/Info have compatible data structures, allowing data tables to be transferred back and forth between the two systems en masse, this can be currently accomplished only in the DOS environment where PC Arc/Info uses dBase III files to store its feature attribute tables and Nexpert reads them with its dBase data bridge. In workstation Arc/Info running under UNIX, the feature attribute tables are stored as Info files, and no data bridge currently exists between Nexpert and Info. At this time, the only way to transfer blocks of data between the two programs in the UNIX environment is through an intermediate database such as Oracle or Ingress to which both Arc/Info and Nexpert have access. Second, this interface was developed as a research tool at the University of Texas at Austin and it is not a supported software product in any sense.

An example enquiry

A sample session in Site Code II operates in the following manner. It is assumed that the user has a tract of land located and a permitted use in mind. Nexpert first prompts the user for the zoning category of the parcel. The user can enter the zoning category, if it is known, or select the "not known" option. If the zoning category is not known, Nexpert triggers the Arc-Nexpert interface and Arc/Info receives the message that the user wishes to determine the zoning category of a particular lot. An Arc macro written for this purpose then displays the parcel map of the area (a map of the legal boundaries of lots) and prompts the user to select the parcel of interest. Once the lot is identified, its zoning category is read from the feature attribute table for the parcel map and the address and zoning category are displayed for checking by the user. Once the user is satisfied, Arc/Info triggers the Arc-Nexpert interface and returns the zoning category to Nexpert as a text string such as 'SF-1' (single family residential, low density) or 'MF-4' (multifamily residential, high density), for example.

At this point, Nexpert knows the zoning category either from direct user input or from its query to Arc/Info, and it reads from a database the set of land use types permitted in this zoning category. Nexpert then displays the list of possible uses and asks the user to indicate which one is desired. Some of the titles of these land use types are fairly opaque, for example "lodging house residential" and "lodging house commercial," but it is assumed that the user either knows what these terms mean or has a city staff member or the hypertext version of the code available to define and clarify them. The user either selects one of the permitted uses or indicates a choice "not listed above" which terminates the session with a message that the land use required is not permitted under current zoning.

If a permitted land use is selected, Nexpert returns to the database and reads the list of basic requirements, such as building height, impervious cover, front and side setbacks that a design in this zoning category must conform to. In Austin, land uses may be fully or conditionally permitted, where conditionally permitted means that the proposed use must be reviewed by the planning commission. Currently, Site Code II indicates in its display message what the requirements are and if the use is conditionally permitted that fact is noted as well. This completes the investigation of this particular site and the user either quits or returns to the first question ready for examination of another site.

Assessment of Site Code II

As described above, the questions answered by Site Code II are fairly basic and could be answered manually by studying the zoning map and appropriate parts of the regulatory code. But even at this level, operating Site Code II is probably faster than doing the job by hand and the system has several positive attributes: it is exhaustive and complete in its search (all lots within the mapped area are identified with the appropriate zoning category) and all permitted land uses and base zoning requirements for these land uses are contained in the database. To extend this concept to the whole city would require a parcel map covering the whole city, a detailed mapping exercise that not many cities have yet completed. It would also require that all the zoning maps be digitized and kept updated in a digital form instead of in a paper form as is done currently. But it is technically possible that these mapping tasks could be accomplished so that the kind of enquiry just described could be automatically accomplished anywhere in the city.

Site Code should be extended to areas of the code beyond zoning and land use restrictions. We have studied the sections of Austin's Land Development Code dealing with water quality protection, for example. These require the establishment of a water quality buffer zone of 100 to 300 feet in width along any signifi-

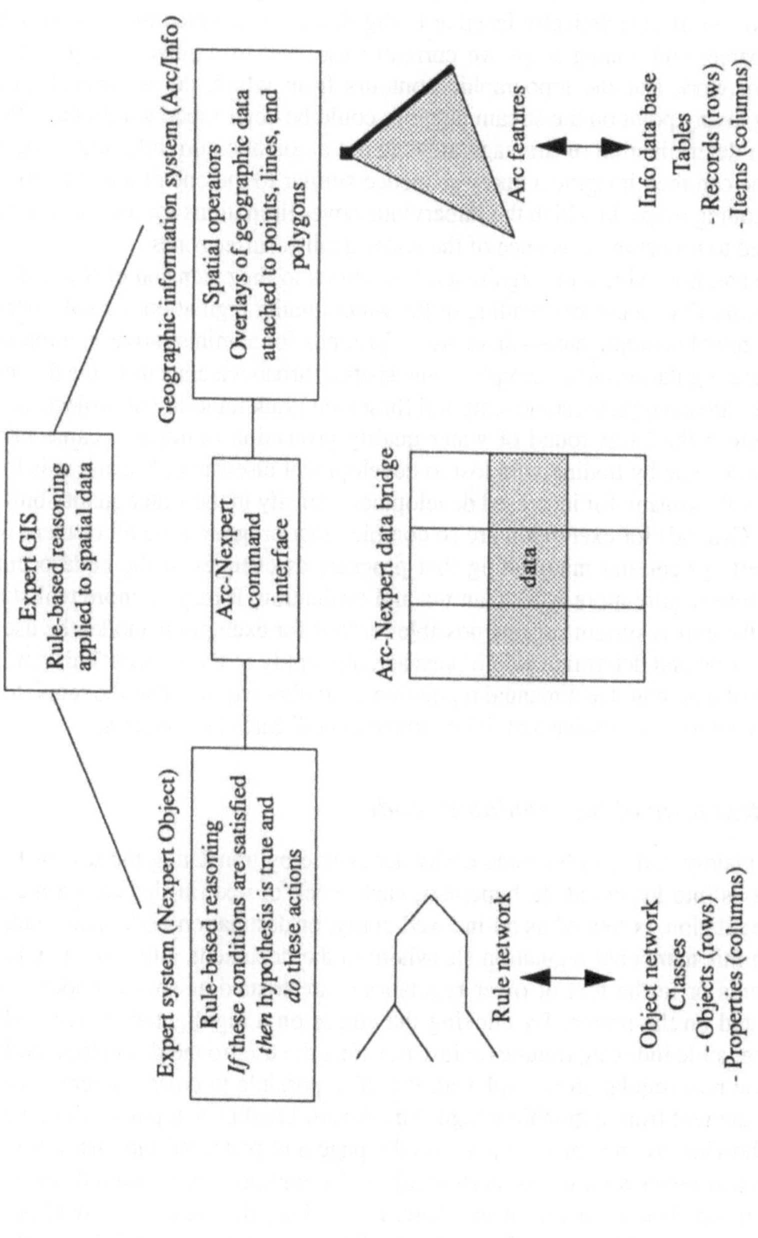

Figure 8.8. An expert geographic information system is formed by linking an expert system with a GIS.

cant waterway passing through the site. Within this buffer zone, the allowable impervious cover is reduced to 18% from the 30-40% allowed normally as a means of reducing pollutant washoff from concrete and asphalt surfaces. The width of the buffer zone depends on which of seven classes of watersheds the site lies within (these watersheds have varying degrees of sensitivity to pollution) and on the drainage area of the waterway above the site (larger waterways have larger buffer zones). It is technically feasible to digitize the watershed map, overlay it on the parcel and zoning maps we currently use, and to digitize a map of the stream network and the topographic contours from which the watershed area draining to any point on the stream network could be determined within the GIS, although determination of drainage areas is not automatic currently within Arc/Info. One can then imagine a query sequence similar to the one described above for the zoning maps, in which the impervious cover limitations for the site would be refined to reflect the existence of the water quality buffer zones.

There are, however, some significant limitations to the extension of Site Code in this form. Even a cursory reading of the water quality regulations reveals them to be extremely complicated—there are 13 grounds for administrative exemption from these regulations, for example, some spatial (urban watersheds in the downtown area are exempted), some temporal (based on grandfathering of projects initiated before the latest round of water quality protection ordinances came into force), and some by trading land lost to development elsewhere, because it is located in a floodplain, for increased development density in the water quality buffer zone. Grounds for exemption are so complex that programming all of them in the expert system and maintaining that program as changes in the code occur may simply require more work than manual evaluation. It may be more realistic to have the expert system suggest possible reasons for exemption and let the user read the code and determine which ones actually apply to the project. This combination of automated and manual regulation searching requires that the regulatory code be easily accessible and linked to the overall Site Code system.

Hypertext form of the regulatory code

The regulatory code can be made easily accessible by translating the text of the regulations into hypertext. In hypertext, each block of text, in this case a numbered regulation, is treated as an indexed entity, or destination node. Any references to this numbered regulation elsewhere in the document, either in the table of contents or in the text of other regulations, are treated as source nodes and highlighted on the screen. By clicking the cursor on a highlighted source node reference, a file indexing routine rapidly transfers the user to the destination node where the new regulation is displayed. It is also possible to cross-reference sections in the text from a graphic image. The Austin Land Development Code has tables showing zoning categories across the page and permitted land uses down the side and either a dash (not permitted), a C (conditionally permitted) or a P (fully permitted) in each cell of the table. By clicking the cursor on any C or P cell, access is provided immediately to the list of regulations pertaining to this particular combination of zoning category and land use type.

As an experiment, Volume 1 of Austin's Land Development Code, about 600 pages in length, has been converted to hypertext form using a process called Hypertrans, developed by Texas Instruments. The display environment within which the capabilities just described are available is called Discover which operates in DOS machines. The Hypertrans conversion process accomplishes the indexing and cross-linking of the text independently of the environment the text is ultimately displayed in, and the same text can be viewed in about a dozen other hypertext display environments, including one in UNIX called Knowledge Management Systems. KMS appears to have a significant potential for linking into Site Code because it operates as a UNIX process and has a command script language like Arc/Info's AML which means that it might be run as an external process linked to Nexpert, just as Arc/Info currently is operated.

Conclusions

The research described in this chapter is an attempt to build an information system called Site Code which will provide consistent, accurate information on a city's site development regulatory process to applicants for permits. It is shown that by using a combination of expert systems, geographic information systems, and databases, an applicant can determine for a section of the City of Austin, Texas, what the zoning of a particular lot is, what land uses are permitted under that zoning category, and what requirements the zoning regulations impose on the site plan. A unique interface between the Arc/Info GIS and the Nexpert Object expert system allows both programs to function interactively in a manner that has apparently not been previously accomplished between a GIS and an expert system, thus forming a new linked technology called EGIS. Before this interface was developed, Site Code was not feasible because the linkage between the regulatory logic and the spatial jurisdiction of the regulations in relation to the location of the site is a critical feature of the city regulatory process.

It also appears to be feasible to connect into Site Code a hypertext version of the regulatory code so that code text, maps, and data can all be accessed automatically and manipulated by the expert system. Together with manual interpretation of the code text, one could then determine the applicable regulations and the design requirements of a particular site by both automated and manual means, thus eliminating the need to code all the logic of the regulatory code within the expert system. While Site Code as described here applies to regulation in the municipal environment, the same concept seems applicable to more general environmental regulatory processes, such as those set out by EPA in the Code of the Federal Register.

While it is clear that linking GIS and expert systems does provide significant decision support capabilities for building site planning that have not previously existed, it should also be recognized that no set of computer routines will ultimately determine a city's regulatory process. This process will always be subjective to some degree, in part because it is so complex that some desirable goals cannot be reduced specific requirements which lend themselves to computer coding, and in part because the form and aspirations of a city reflect its unique culture and history which cannot be expressed in any quantitative way—they are simply matters of human perception.

Acknowledgment

The authors wish to acknowledge the contributions of Douglas Shaw and George Arimes to the development of Site Code I, Dean Djokic who developed the Arc-Nexpert interface, Steve Wilkinson and City of Austin staff who provided guidance for the development of Site Code II, and Steve Morris of Texas Instruments Hypermedia Systems and Services who provided the hypertext version of Austin's Land Development Code. This research was supported in part by the National Science Foundation under Grant MSM 8717452, the Environmental Systems Research Institute, Neuron Data, and the University Research Institute of the University of Texas at Austin.

References

Djokic, D., 1991. "An Expert Geographic Information System for Stormwater Network Analysis," Ph.D. dissertation, University of Texas, Austin, May.

Djokic, D. and D.R. Maidment, 1991. "Terrain analysis for urban stormwater modeling," *Hydrologic Processes 5*(1):115-124.

Garrett, J.H. and S.J. Fenves, 1987. "A Knowledge-Based Standards Processor for Structural Component Design," *Engineering with Computers*, Vol. 2, pp. 219-238.

Hosking, J.B., W.B. Mugridge, and M. Buis, 1987. "Fire Code: a case study in the application of expert-systems techniques to a design code," *Environment and Planning B: Planning and Design 14*:267-280.

Leary, M.E., 1988. "Knowledge and reasoning in development control and urban design: An expert systems approach," *Environment and Planning B: Planning and Design 15*: 383-398.

Rosenman, M.A. and J.S. Gero, 1985. "Design codes as expert systems," *Computer-Aided Design 17*(9):399-409.

Shaw, D.T. and D.R. Maidment, 1989. "Site Code: a Knowledge-Based Model of Municipal Site Development Regulations for Computer-Aided Design" in T.O. Barnwell, ed., *Computing in Civil Engineering*, Proc. 6th Conference on Computing in Civil Engineering, Am. Soc. Civ. Engr., pp. 707-714, New York.

Shaw, D.T. and D.R. Maidment, 1990. "Site Code: a Knowledge-Based Model of Municipal Site Development Regulations," Report CRWR-194, Center for Research in Water Resources, 230 pgs., University of Texas, Austin.

Wright, J.E. and K.A. Buehler, 1990. "A GIS Proving Ground for Water Resources Research," Indiana Water Resources Research Center, Purdue Univ., W. Lafayette, IN.

PART THREE
Environmental planning and management
Introductory remarks by Jeff R. Wright

Environmental planning and management problems are generally well suited to solutions using expert systems technology because: 1) the problem domain is generally well defined even though the physical system might be poorly understood, and 2) conventional wisdom regarding alternative solutions frequently resides with those individuals having a great deal of special experience and expertise within fairly narrow areas even though (or perhaps because) this knowledge is necessarily multidisciplinary. In this section, we present four examples of the use of this technology in addressing complex environmental management problems.

In Chapter 9, Stefano Schibuola and Phillip Byer describe one use of knowledge-based expert systems applied to the review of environmental impact assessments (EIA). They provide a solid argument that the quality of an environmental impact assessment and public participation can improve if the public can effectively review the EIA document. Public groups would benefit from having available specific knowledge needed to perform good reviews. A knowledge-based system (KBS) can potentially disseminate this knowledge in an inexpensive and readily usable form. Research presented in this chapter evaluates the usefulness of the KBS approach by developing a research prototype capable of critiquing a central component of EIA—the evaluation of alternatives. Case studies demonstrated the system's effectiveness and the quality of its interaction with non-experts. Schibuola and Byer provide a comparison of this approach with other, more traditional methodologies.

In Chapter 10, Elliott and Ciesielski describe the design and development of an expert system for managing river water quality. The State Electricity Commission of Victoria (Australia) draws large volumes of water from the Latrobe River for use in its electricity generation activities. The use of this water can result in large variations in river flow and water temperature and subsequent environmental damage. The Commission is concerned with keeping its use of water within the constraints imposed by the Environment Protection Authority while at the same time operating its generation plants to maximum efficiency. The flow of water can be controlled by valves in a number of water storages and a charge engineer is responsible for setting the valve openings to ensure that the river flows and water temperatures at a number of monitoring points are within limits. The water flow patterns are highly complex, involving a number of weirs, cooling towers, and pumps and not all necessary information is available. Decisions are made using a combination of available data and heuristic rules gleaned from previous experience with the system. The authors describe a rule-based expert system designed to capture the expertise needed to operate the valves in a way that ensures river flows and temperatures remain within limits. The prototype system, which was developed using the popular VP-EXPERT tool, is discussed in detail.

A problem of growing concern in the U.S. and around the world is the comprehensive evaluation of sites that either have been, or will be, used for the storage and disposal of hazardous materials. In Chapter 11, Fang, Mikroudis, and Pamukcu describe the development of a multi-domain knowledge-based expert system to be used to aid decision-makers in addressing these types of complex problems. After outlining the general approach to site analysis and investigation, the authors present a framework within which to conduct the assessment of technical, social, and economic factors, portraying the task as one that can benefit from the use of expert systems technology. A system for supporting this methodology has been developed using the GEOTOX shell tool. The details of system design and implementation, knowledge representation, and use are presented.

In Chapter 12, Chen and Sharpe describe the NOISEXPT knowledge-based system which was developed to help control the impact of noise resulting from high-speed trains. Because very few trains in the world operate at speeds above 300 kph, considerable expertise is required to assemble and extrapolate relevant sources of information to estimate the likely impact and to establish guidelines for noise control design. NOISEXPT is a knowledge-based system being developed on the basis of high-speed train experiences in Japan, France, and German as well as from other sources, including research expertise from four international workshops on railway noise and dozens of U.S. government reports. The NOISEXPT system is composed of five sections, including one in which nearly 400 paper and book references form an extensive knowledge base for the system. These also include standards from more than 10 countries and international organizations, noise source information, several prediction methods, and guidelines for noise control measures. NOISEXPT has been developed using the CRYSTAL shell and runs on PC-compatible computers.

9

Use of Knowledge-Based Systems for the Review of Environmental Impact Assessments[1]

Stefano Schibuola and Philip H. Byer

Introduction

Environmental impact assessments (EIA) have the potential for improving public decisions affecting environmental quality. Such assessments facilitate rationality, comprehensiveness, accountability, and public participation in the decision-making process. However, many reviews of the environmental assessment process show that EIA are often carried out in an adhoc and methodologically unsound way (Rosenberg, et al., 1981; Nichols and Hyman, 1982; Whitney and Maclaren, 1985). Too often, this results in questionable documents that merely attempt to rationalize and justify decisions not made with due concern for the environment.

One way to encourage meaningful and effective environmental assessment is to subject EIA documents to a review process. In Ontario, where EIA are called environmental assessments (EA), environmental assessments are routinely reviewed by government agencies and members of the affected public. However, the public review process is frequently inadequate, since public groups often lack the knowledge to adequately critique a document or the financial resources to retain an expert to do so.

Public groups would therefore benefit from having available specific knowledge needed to effectively review environmental assessments. The problem is to provide a source of such knowledge in a way that is effective and economical.

---

1. Reprinted by permission of the publisher from "Use of Knowledge-Based Systems for the Review of Environmental Impact Assessments", by S. Schibuola and P.H. Byer, *Environmental Impact Assessment Review*, Vol. 11(1):11-27, 1991, by Elsevier Science Publishing Co., Inc.

Several alternative knowledge sources exist, including:

1. Environmental assessment literature.
2. Environmental assessment review manuals or handbooks.
3. Paid specialists or consultants.
4. Volunteer specialists (e.g., university professors).
5. Knowledge-based systems.

The first four alternatives exist today. The fifth alternative source of knowledge is the use of a computerized knowledge-based system. A knowledge-based system is a computer program that uses human knowledge and expertise to solve problems ordinarily requiring human intelligence (Hayes-Roth, 1984). Such a system would contain some of the knowledge needed to review environmental assessments. By engaging an individual in a dialogue about the environmental assessment in question, the system could point out weaknesses and deficiencies in the document.

The knowledge-based systems field is relatively new, and its feasibility for addressing problems in environmental planning has received relatively little attention. Consequently, it is not certain whether it is possible or useful to represent planning knowledge in a machine-usable form. Further, it is unknown if such a system would perform effectively, both in an absolute sense and relative to traditional ways of carrying out environmental planning tasks.

This chapter presents research on the use of knowledge-based systems to aid in the review of environmental impact assessment documents. A prototype knowledge-based system was developed to investigate issues involved in the design of such systems and to test its usefulness and effectiveness. Full details are presented by Schibuola (1988).

Knowledge-based system design process

Knowledge-based system design is an evolutionary process. One begins building the knowledge base by gathering and coding the knowledge necessary to solve simple problems in the chosen domain. Once the system is capable of handling these simple problems, more knowledge is added. This is done through feedback with users, and through using the system on more complicated problems and comparing its output with a human's solution. This incremental process highlights weaknesses in the knowledge base, thereby focusing the knowledge acquisition process.

The simplest knowledge-based system, in terms of the size of its knowledge base and the complexity of problems it can address, is the demonstration prototype. It typically contains from 50 to 100 rules and requires about one to three months of development time. It usually involves only the system designer and one expert. The demonstration prototype is designed to handle a small subportion of the main problem and usually performs adequately for one or two test cases.

Its chief purpose is to gauge whether a knowledge-based system is an appropriate way of handling the problem. It also tests the designer's initial ideas of problem definition and scope. The knowledge-based system developed in this research is a demonstration prototype called REVIEW.

Waterman (1986) notes five major tasks in the knowledge-based system design process: identification, conceptualization, formalization, implementation, and testing.

Identification

Identification is the gathering of information needed to define the problem and its intended solution. This includes identifying the problem itself, the project participants, the necessary resources, and the goals and objectives that the system should meet. It is also here that the problem is scoped to a manageable size.

Whitney and Maclaren (1985) divide the ideal environmental assessment process into a series of eight interdependent steps based on the rational decision-making model. The steps are: (1) needs assessment, (2) alternatives generation, (3) scoping, (4) prediction, (5) significance assessment, (6) evaluation, (7) mitigation, and (8) monitoring. This is a useful framework, as a reviewer can systematically look at each of the eight steps, with the review of each step regarded as a subproblem in reviewing an environmental assessment document. Good design practice suggests building a system to deal with one subproblem when scoping the entire problem. For these reasons, it was decided to build a demonstration prototype to address one step, and the step chosen was step 6 (the evaluation of alternatives).

This step was chosen for two reasons. First, it is the one with which the authors are most familiar. Therefore, less time would be needed to research the necessary knowledge. Second, reviewing an evaluation is a good combination of "hard" technocratic questions regarding specific methods and "soft" questions regarding the general approach taken. Since most environmental planning tasks are also characterized by this presence of hard and soft components, the review of evaluations is a good choice as a representative planning task.

With the problem scoped to the reviewing of the evaluation alternatives within an environmental assessment, Table 9.1 outlines the project goal, main task, and subtasks for the scoped problem.

A final step in identification is to establish criteria along which the knowledge-based system will be judged during testing. These typically relate to ease of use, quality of results, and cost to develop and maintain. The prototype presented here was assessed using the following criteria.

1. Results produced by REVIEW:

   Validity, accuracy, correctness of critique produced

   Comprehensiveness (with respect to relevant issues of problem at hand and to applicability to different problems)

   Usefulness of critique

Table 9.1. Problem Definition

Category	Description
Project goal	Aid an environmental lawyer or affected individual in reviewing the evaluation portion of an environmental assessment.
Main problem	Compare the evaluation with the ideal evaluation process and the general requirements of evaluation.
Subproblems	- Selection of attributes or objectives - Attribute characteristics - Use of information
Problem characteristics	- Non-specialist users - Two levels of review - Two paradigms of review process
Information required	- Description of evaluation procedure as contained in environmental assessment document - Knowledge of what constitutes a good evaluation
Important concepts	- Goals and objectives of evaluation - General expectations from an evaluation - Who was involved at each step - Method used at each step - Key assumptions
Solution	A discussion of the major problems and/or questions to be resolved with the environmental assessment under review. The points may be used directly as criticisms, or they may spark the user into finding weaknesses with the environmental assessment that the knowledge-based system cannot find itself.

2. Usage characteristics:
   Ease of consultation
   Understandability of process
   Understandability of results
   Novice usability
3. Efficiency and performance relative to other knowledge sources:
   Time and money required
   Knowledge source development and maintenance
   Relative performance
   Availability of source
   Time or money required to consult source

Conceptualization

This is the thorough exploration of the problem to identify key concepts, relations, control mechanisms, subtasks, strategies, data, and constraints needed to solve the problem. Conceptualization for REVIEW consisted mostly of organizing and structuring knowledge from literature on evaluation and from the authors, who are experts in evaluation, so this knowledge would be amenable to machine representation. REVIEW addresses the following evaluation subtasks and related issues shown in Figure 9.1. The system reviews each evaluation on the basis of these items.

In terms of a paradigm, the knowledge-based system reviewing process can be regarded as being similar to what would occur in the following situation. A person is interested in reviewing the evaluation within an environmental assessment, has access to the document, but is not knowledgeable about evaluation. A professional reviewer having the required knowledge is available for consultation, but does not have direct access to the environmental assessment document and reachable only by telephone. The reviewer must ask the nonspecialist questions to get the information that is needed to review the document. Based on knowledge and the responses of the nonspecialist to the questions, the reviewer can ask further questions and finally produce a review. This is the process the computerized knowledge-based system attempts to emulate.

A knowledge-based system should be usable by people who are not specialists in environmental impact assessment. The system must be able to get the information it needs by asking questions to which a lay person can find answers directly in the document. It should not require the user to make value judgments or interpretations of questions, and it should avoid asking questions requiring subjective answers.

The review of an evaluation can be regarded as occurring at two distinct levels, with much interaction between the levels. The first level might be regarded as a direct, systematic, and rigorous review of each step taken in the evaluation. The methods and general approaches taken are critiqued both on technical grounds and on the grounds of what is expected of any evaluation. For example, a level one review might turn up the fact that an evaluation aggregates numerical scores measured on an ordinal scale. This is a meaningless mathematical operation that can distort an evaluation. A level one review is a straightforward process requiring knowledge of an "ideal" evaluation process and the specific details of the evaluation under review. This seems amenable to the knowledge-based systems approach.

The second level of review probes deeper into the evaluation. It does not rely on the knowledge used to prepare the level one review, but does use the results of that review. It also draws on deeper levels of knowledge, such as common sense, real-world general knowledge, and an understanding of the setting and context of the evaluation as it probes for the motives or reasons behind the problems uncovered by the level one review. For example, a reviewer may notice the preferred alternative in an evaluation is the same one recommended in an older study con-

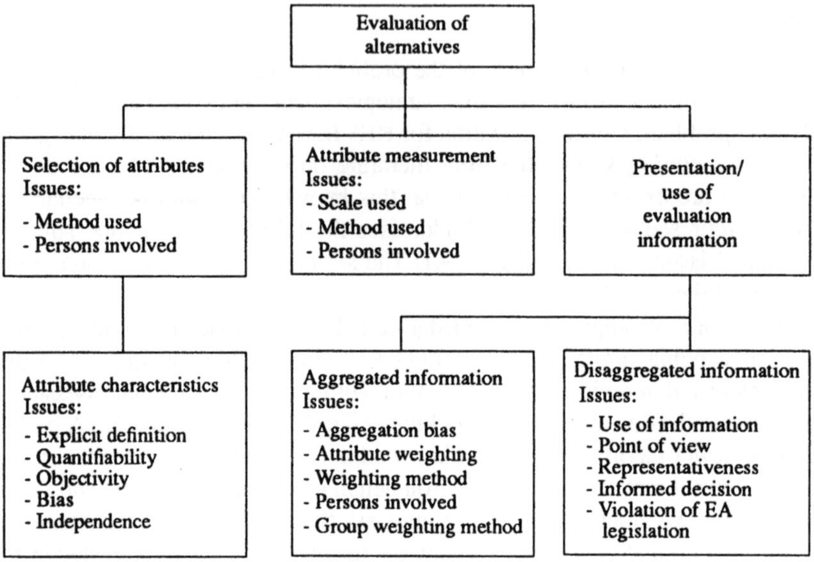

Figure 9.1. Conceptual framework for REVIEW.

ducted by the proponent without considering environmental impacts. Coupling this fact with the possibility of a distorted evaluation as noticed in the level one review, the reviewer may argue that the evaluation is biased and simply rationalizes the previous decision. Due to the present limitations of knowledge-based systems in handling common sense and real-world knowledge, a knowledge-based system cannot adequately prepare a level two review.

The two-level nature of the review process is mirrored in two possible paradigms: forward chaining or backward chaining (Waterman, 1986). In forward chaining, a system or person starts with the available facts and tries to match a known hypotheses to them. In backward chaining, the system or person tries to prove each known hypothesis by searching for supporting facts. The review process can proceed in either a backward chaining, hypothesis-driven mode or a forward chaining, data-driven one.

In the hypothesis-driven review mode, the reviewer starts with knowledge of the ideal evaluation process. With this knowledge in mind, the reviewer systematically searches the evaluation and notes how it diverges from the idea.

In the data-driven mode, the reviewer reads sequentially through the evaluation document and makes notes about what is wrong with it. Since the reviewer is not specifically looking for a certain fault, the reviewer's mind is freer to draw on knowledge other than that related specifically to evaluation. During this process, the reviewer may draw on such diverse and fuzzy knowledge as the context of the project for which the evaluation is being done, common sense, and general real-world knowledge.

The distinction between the two paradigms is not simple or clear cut. Most human beings switch from one mode to another as they review a document. However, it was useful to make this distinction when designing the knowledge-based system. The hypothesis-driven paradigm is relatively straightforward to implement in a knowledge-based system. It represents a systematic, rigorously logical approach to review. It is ideal for turning up problems associated with a level one review as it leaves "no stone unturned" in looking for possible problems. However, it cannot produce a level two review since it does not explicitly bring other knowledge into the problem.

The data-driven approach is less structured and more natural. Because of its free-ranging style and its facility for bringing in more knowledge, this approach suits a level two review. Although it can be used to produce a level one review, its lack of logical rigor makes it easy to miss potential level one problems.

The REVIEW knowledge-based system was therefore designed for a level one review using the backward chaining, hypothesis-driven approach. As this system is used, the user gains insights into the evaluation through the comments produced by the system, as well as by questioning the system via an explanatory interface. Then, using these level one results and the user's own common-sense and real-world and project knowledge, the user can prepare the level two review. In this way, synergy is developed between the knowledge-based system and the human user, with the strong points and knowledge of each being combined to strengthen the review. Figure 9.2 illustrates the interplay and synergy between a human and a knowledge-based system in preparing a review.

Formalization and implementation

Formalization is the expression of the concepts acquired in conceptualization in a formal knowledge representation scheme. The REVIEW formalization scheme was based on the standard paradigm of diagnosis, since this most closely matched the process of review.

Implementation is the transformation of the formalized knowledge into a functional computer program. The knowledge for REVIEW was implemented in Prolog based on an IF-THEN knowledge representation scheme.

In knowledge-based system design, the designer faces two decisions in choosing the software with which to implement the system. First, a choice must be made between using an expert system shall or an artificial intelligence (AI) programming language to create the system.

For this project, the designer chose a language over a shell for two reasons. First, since the application area was previously unexplored, there was great uncertainty regarding system requirements. The designer considered it undesirable to be locked into a specific knowledge representation scheme in a shell. Second, the designer felt he could learn more about knowledge-based systems and their development techniques using an AI programming language rather than a shell.

Given this choice, the second decision was to choose a programming language. In AI today, the two dominant languages are Prolog and LISP (Waterman, 1986;

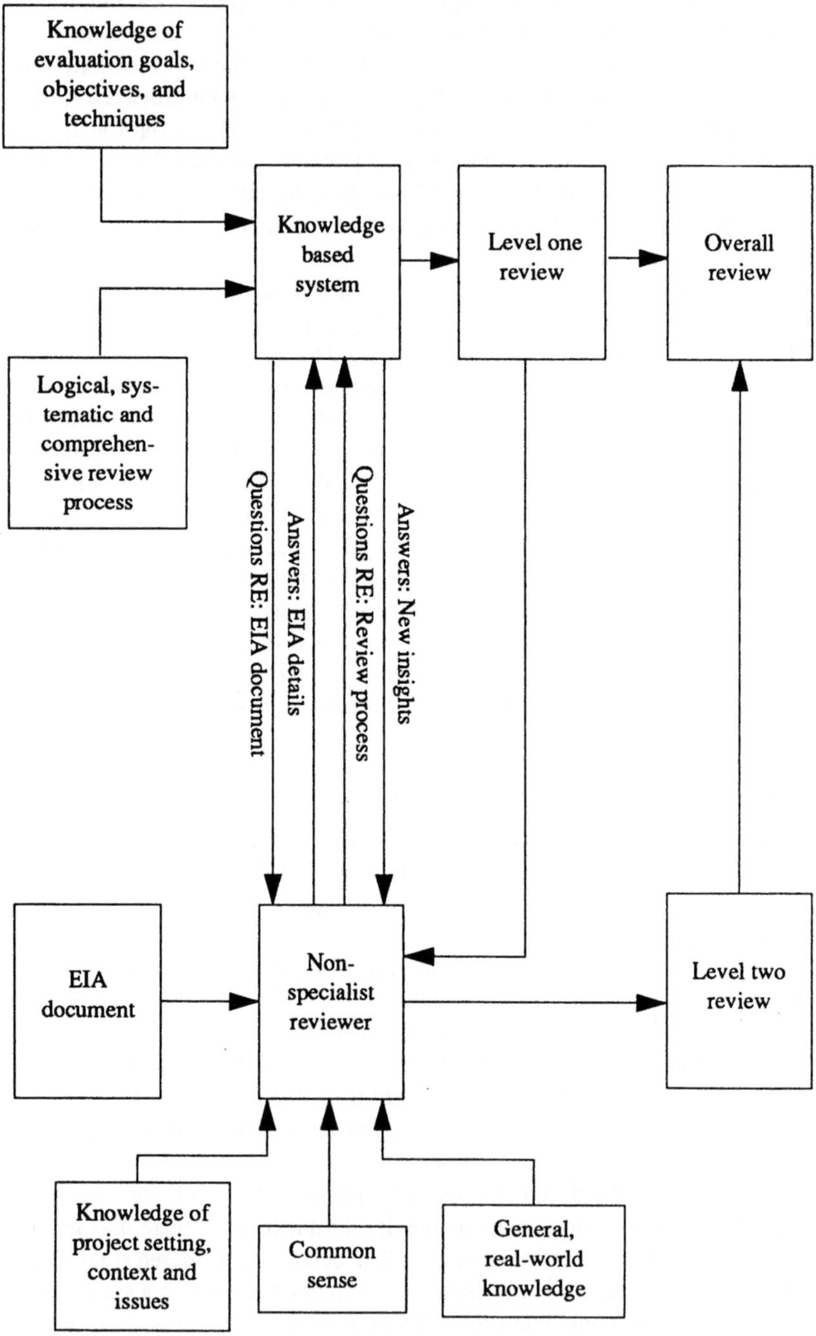

Figure 9.2. Human - knowledge-based system interaction in REVIEW.

Townsend, 1987a, 1987b). Prolog is an easier to learn high-level language, while LISP is a more powerful low-level one. The designer chose Prolog because its ease of learning outweighed the more powerful features of LISP in this case. The designer felt there was a good chance that he could build a knowledge-based system in this application area using Prolog. The possible extra insights or power gained from using LISP would be outweighed by the increased time taken to learn the language and develop the system. Overall, the Prolog language seemed a good compromise between using a shell and using LISP.

The following is an example of how knowledge was conceptualized, formalized, and implemented in REVIEW. In reviewing the example, the reader may be struck by the level of domain-specific jargon in the system given its nonspecialist audience. The issue of jargon, and ways of eliminating or reducing its effects, is discussed in the "Testing and assessment" section.

A subtask in evaluation is attribute measurement. In an evaluation, once the distinguishing characteristics or important impacts of the alternatives have been selected, each alternative must be measured against each chosen attribute. For review purposes, the relevant issues here are the types of scales and methods used to measure each attribute, and the persons involved in the measurement exercise.

Thus, the REVIEW system must contain knowledge of the kinds of scales used in evaluation, the various impact measurement methods, and the possible groups of people who may be involved. It also must know the kinds of attributes that can and cannot be measured on each scale, the kinds that can and cannot meaningfully be measured under each measurement method, and the groups of people who should be involved in impact measurement.

In the following example, a user has input "air quality" as an attribute in an evaluation under review. The system has already asked if this attribute is objective and quantifiable and has received a "yes" to both questions. It has also asked if the scale used to measure the attribute was explicitly stated in the document and also received a "yes" answer. The system is now processing this attribute, and the scale used to measure it, via the following Prolog clause:

```
process_attribute2([Attribute|Tail]):-
  issue(Attribute,scale),
  problem(Attribute,scale),
  problem(Attribute,measure_who),
  problem(Attribute,measure_how),
  check_scale(Attribute),
process_attribute2(Tail).
```

The "issue(Attribute,scale)" clause causes the following question to be asked of the user:

```
What scale was used?
1.  Ordinal (ranking)
2.  Cardinal (interval)
3.  Ratio - no units
4.  Ratio - with units
5.  Qualitative, relative
```

```
6.  Qualitative, absolute
7.  Standards (meet/does not meet)
```

Note that an on-line definition, with examples, for each of these terms was available at the user's request.

Suppose a qualitative, relative scale was used to measure the air quality impacts of each alternative. The user therefore picks item five from the menu.

The "problem(Attribute, scale)" clause causes the system to review its list of problems related to attributes and scales. One such problem is:

```
problem(Attribute,scale):-
  xpositive(Attribute,objective,0),
  xpositive(Attribute,quantifiable,0),
  xpositive(Attribute,scale,5),
  conclude(Attribute,qual_scale).
```

In English, this rule states that there is a potential problem with using a qualitative scale if the attribute is both objective and quantifiable. Since all conditions for this problem are met, the "conclude(Attribute,qual_scale)" clause causes REVIEW to output the following comment:

```
The attribute is objective and quantifiable, yet a
qualitative scale was used to measure it. Is there some
reason why a more reliable and objective numerical
measurement was not performed?
```

Testing and assessment

Testing is the systematic assessment of the results produced by the system. In the first phase of testing, the authors rigorously studied the system's rules to spot mistakes and make generalizations from overly specific rules, and by judging the system's solutions to a series of test problems against the criteria established in the identification phase. The designer then used the errors or omissions in the system's review as guides in either modifying the existing knowledge base or adding new rules.

During the development and testing of the REVIEW system, a major issue was comprehensiveness. A knowledge-based system is only as good as the level of knowledge available in building it. As with any human knowledge, it is sometimes difficult to tell when the level of knowledge is complete or high enough to do a task well.

In knowledge-based system design, the only workable test for comprehensiveness is to apply the system to more and more test cases, and see what changes are required. Design is considered to be completed when a satisfactory performance level is reached and most new cases do not require additional changes.

Once REVIEW was developed, it was assessed with respect to the three general criteria established and discussed in the identification stage. Three case studies of evaluations from environmental assessment documents were used. All three case studies are projects that underwent an environmental assessment in Ontario, Canada. Two are proposed roadway extensions in urban areas (Regional Municipality of Hamilton-Wentworth, 1982; Municipality of Metropolitan Toronto,

1985); the other is a proposed liquefied natural gas (LNG) facility (Consumer's Gas Co. Ltd., 1985). All three were subjected to public hearings (Ontario Municipal Board and Ontario Environmental Assessment Board, 1985; Ontario Energy Board, 1986; Ontario Environmental Assessment Board, 1988). In each case, either the authors or other people knowledgeable about environmental assessment and evaluation used REVIEW to create a review of the evaluation. An assessment of REVIEW with respect to the three general criteria follows.

Results produced by REVIEW

Table 9.2 gives a sample output from one case study and lists what REVIEW judged to be faults in the evaluation. Each knowledge-based system review was compared to a human review of the evaluation, which was compiled from a variety of sources including review board decisions and formal reviews by experts.

In general, the comments produced by REVIEW appeared to be valid and accurate. With only a few exceptions, all of the comments produced by the REVIEW system either matched those of the human review or lent support to them. In addition, the knowledge-based system reviews were generally comprehensive; few faults not addressed by the REVIEW system could be found.

The exceptions are interesting as they occur for different reasons. Some represent a lack of knowledge in the knowledge base. A lack of knowledge is a significant but not fundamental problem to the use of a knowledge-based system. It simply requires an expansion of the knowledge base to allow the system to deal with the errors it missed. It should be noted that a lack of knowledge could be seen in some of the human reviews as well. Sometimes, the REVIEW system was able to spot errors missed by the human reviewer (e.g., point 8 in Table 9.2).

Other exceptions are examples of errors noticed not through detailed knowledge of evaluation but by common sense. For example, in the LNG project a human reviewer was surprised that the factors of land availability and price were not included in the evaluation, though it is obvious these would play some role in the siting study. This is an example of the kind of common sense that is extremely difficult to include in a knowledge-based system. Though it is easy to include this specific bit of common sense in the system, what is difficult, if not impossible, is to include every bit of knowledge potentially relevant to all evaluations. Further, it is probably unnecessary, since the knowledge-based system is designed to produce a review in cooperation with the non-specialist user. Since the user has this common sense, it is pointless to try to replicate it in the system.

A final type of exception occurred when an item of critique that is usually applicable is not important in a particular situation. For example, since aggregating evaluation information into a single score may hide or distort impacts, REVIEW notes this when it is informed that there was such aggregation (see point 6 in Table 9.2). However, this potential problem would not be important if there are no significant differences in impacts between alternatives.

In judging the usefulness of results from REVIEW for public hearings, mixed conclusions were reached. In some cases, the results would have been very useful

Table 9.2. REVIEW Critique of Proposed Finch Avenue Extension

General problems

1. Only experts chose attributes, and there was no guarantee that experts took into account any but their own points of view while doing so. The evaluation might yield a different preferred alternative if some other group chose the attributes.
2. Attributes may not be relevant, meaningful, or comprehensive since the rationale behind their choice was not discussed.
3. Choice of attributes should have been related to concerns and issues raised by the scoping exercise. Since they were not, two problems may arise. First, attributes may not cover all concerns so they might obscure some impacts. Second, the attributes themselves may have little meaning and it may be difficult to assess importance since they aren't related to specific values, concerns, or goals.
4. Impacts were not distinguished according to incidence group. This may result in "tyranny over the minority" where the considerable impacts on a small group might be ignored or drowned out by the smaller impacts on a large group.
5. The values assigned to the alternatives are questionable since the meaning of each point on the scoring scale is not clear. There is great potential for bias toward one alternative here.
6. The evaluation information was summed to yield an overall score for each alternative. This may hide information. For example, if one alternative is slightly acceptable for most attributes but highly unacceptable for only one, then adding all impacts together will make this obscure.
7. Only experts weighted attributes. There was no guarantee that all relevant points of view in their assignments of weights were taken into account. If a different group did the weighting, a different alternative might result.
8. Equal weight was given to each attribute. Setting all attribute weights equal is still a judgement and may bias the evaluation.
9. Due to previous errors or ambiguities in attribute choice, measurement, or weighting, final numbers could be misleading and should be questioned.

Problems with specific attributes

For various subjective attributes, including "land use and development," "social impact," and "natural environment":

10. Only experts were used to judge a subjective attribute. There was no guarantee that the experts took any but their own points of view into account while scoring the alternatives. The evaluation might yield a different preferred alternative if some other group did the scoring.

to groups opposed to a project. In other cases, the most contentious issues surrounding a project did not revolve around evaluation; consequently a critique of only the evaluation would have been of limited use. However, the performance of REVIEW shows that the knowledge-based system approach is valid, and a system could be developed to review the entire environmental assessment process.

System usage characteristics

In assessing REVIEW's usage characteristics, a variety of opinions were sought from people of different backgrounds, including an environmental planner and an ecologist familiar with environmental assessment but unfamiliar with evaluation. It should be noted, however, that since all users were at least somewhat familiar with environmental assessment and the use of computers, none of them could give the perspective of a true non-specialist member of the public.

These users found REVIEW to be quite easy to use. They compared it favorably to many other PC-based programs, and felt the menu structure and help systems were fairly intuitive and easy to learn and use. The problems that were encountered can all be remedied by additional work on the user interface or some enhancement to the knowledge base. For example, all users commented on the need for a "back-up" facility which would allow the user to undo a mistake or revise an answer based on subsequent questions.

Most other problems appeared to be due to the particular wording of questions or explanations. Such problems can be solved by changing the wording or adding examples to the explanations to further clarify a point. A related problem is the use of jargon in the system. There are two basic types of jargon: words unfamiliar to the non-specialist and familiar words that take on different meanings in the knowledge domain. In the evaluation field, the jargon tends to be of the latter type. Words such as "evaluation," "objectives," and "alternatives" have fairly precise meanings in evaluation literature. However, they are also very common words in daily life used in a variety of ways by non-specialists. The words will connote different concepts to people of different backgrounds.

The problem of ambiguous jargon interpretation in evaluation probably cannot be eliminated. However, it can be minimized in three ways. First, jargon should be avoided wherever possible. Second, if jargon must be used, it should be standardized so that the same term is used for the same concept throughout the knowledge base. Third, an on-line or paper glossary should be provided for the jargon used, giving precise definitions and examples. REVIEW did contain a rudimentary on-line glossary in the form of help screens tied to specific questions. These screens were found helpful by the users.

However, a fundamental problem may exist. If users have little or no knowledge of evaluation, they may have difficulty finding answers to questions posed by the system. This problem may be partially solved by adding to the glossary.

The users generally felt that the process by which REVIEW produced its critique is understandable. The availability of an explanatory interface was particularly important since it allowed the user to query the system about its reasons for asking a particular question. Users generally felt that the comments produced by REVIEW were clear, unambiguous and understandable. No fundamental problems were encountered, and problems in understanding the results can be addressed by re-wording the comments in the knowledge base.

All users felt that REVIEW is at least potentially useful. A fundamental problem, however, is that the system is inherently limited in its ability to assess the significance of problems it identifies. This problem cannot be solved solely by enhancing the knowledge base. It is impossible to gauge the significance of an issue without knowledge of the context of the project under review. Since it is difficult for the system to gather this knowledge, it relies on the user to bring to bear project-specific knowledge and common sense to assess if the problems are significant. This problem might be lessened by equipping the system with the ability to give some indication about the potential significance of a problem.

Efficiency and performance relative to other knowledge sources

As indicated in the introduction, a knowledge-based system is only one of several alternative sources of knowledge available to help a non-specialist review an environmental assessment. No source is always best since the choice involves trade-offs among cost, time, availability, and quality of the results.

While retaining a human specialist would be the preferred approach, it is usually the most expensive. A less expensive alternative would be to retain the services of a volunteer specialist such as a university professor. However, it will often be the case that no such volunteer is available. Consulting the literature for self-learning is often impractical since lay reviewers may not be able to devote the time and effort needed to educate themselves in environmental assessment methodologies.

A paper manual or a knowledge-based system should be able to provide reasonably good results for a relatively small investment of time and effort by the reviewer. However, the availability of either the paper manual or the knowledge-based system depends on their development by a sponsoring agency.

Based on the case study findings, the knowledge-based system did fairly well in system usage and produced critiques of good quality. A knowledge-based system can also be readily available since it can be run on widely available microcomputers. Further, the question-answer format makes it fast to use, and the financial and time investment to the user would be comparable to that of a handbook provided that a computer is already available. Development requirements would also be comparable to that for a manual. The most time-consuming aspect of creating a knowledge-based system is knowledge acquisition, something that would also have to be done for a manual. Developing that knowledge into a knowledge-based system should take no more time than would be required in designing a handbook that presents the same knowledge in a useful format.

These two knowledge sources are therefore quite similar since they are simply two ways of conveying the same knowledge. However, the knowledge-based system's interactive nature gives it a potential edge in ease of use, understandability of process, usage time, and understandability of results.

Conclusions: The usefulness of the knowledge-based systems approach

The kind of knowledge amenable to knowledge-based system representation is "systematic" knowledge. In the application under study, "systematic" knowledge is knowledge of the objectives of evaluation, the subtasks involved in meeting those objectives, and the best methods for carrying out those subtasks. In evaluation, much of this knowledge exists. We know the objectives of an evaluation and what subtasks must be carried out in preparing an evaluation. Though we do not know the best way to evaluate alternatives in all cases, we do have knowledge of different proposed evaluation methods and their strengths and weaknesses.

However, this knowledge alone is not enough to adequately critique an evaluation. One needs not only evaluation knowledge but also real-world knowledge, common sense, and knowledge of the project setting. It is infeasible, and perhaps impossible, to equip a knowledge-based system with the latter three kinds of knowledge. Further, humans need not be experts in evaluation to have these.

The real strength of the knowledge-based systems approach is it makes the most use of the best knowledge of both the knowledge-based system and the human user. It thus creates synergy between them. The rigorously logical approach and detailed process knowledge of the system combine with the human's common sense, real-world knowledge and knowledge of project setting to produce a review of a quality neither can achieve alone.

An environmental assessment usually includes not only an evaluation but also an assessment of need, a scoping exercise, an identification of alternatives, a prediction of impacts, an assessment of impact significance, and a mitigation and monitoring plan (Whitney and Maclaren, 1985). A knowledge-based system may also be useful in reviewing these other aspects of environmental assessment, as well as other planning tasks.

As noted above, the usefulness of the knowledge-based system depends on the degree to which the task it is reviewing is systematic. Some parts of environmental assessment, such as prediction, are well-suited to the use of a knowledge-based system. Although one could argue that our track record as prognosticators is dismal, methods or at least guidelines for predicting most impacts do exist and are accepted as the best available planning tools. As such, the knowledge needed to critique predictions and prediction methods might be readily and usefully translated into a knowledge-based system.

Other environmental assessment tasks, however, are less systematic. In scoping or significance assessment, we are less precise in knowing the objectives of these tasks, the subtasks involved, and the best methods to use. Still, some general guidelines exist, and to this limited extent a knowledge-based system would be useful. The usefulness of such a system would grow as we gain more experience with environmental assessment and develop better guidelines and methods in these areas.

In summary, the knowledge-based systems approach is valid for planning problems that have a certain systematic component. Knowledge-based systems in areas of planning such as the evaluation of alternatives and the prediction of impacts are feasible and useful even with today's limited technology and knowledge. In the near future, developments in both planning methods and in knowledge-based systems will increase their natural synergy leading to useful knowledge-based systems in a variety of planning tasks.

References

Consumer's Gas Company Limited, 1985. *LNG Storage Project, Submission to the Ontario Energy Board, Site Selection and Environmental Assessment, Schedule 1: Initial Site Screening and Schedule 2: Preferred Site Selection*, Toronto, Ontario.

Hayes-Roth, F., 1984. "The knowledge-based expert system: A tutorial," *Computer IEEE*, *17*(3):11-28, September.

Municipality of Metropolitan Toronto, Department of Roads and Traffic, 1985. *Finch Avenue West Environmental and Needs Assessment*, Toronto, Ontario, April.

Nichols, R. and Hyman, E., 1982. "Evaluation of environmental assessment methods," *ASCE Journal of Water Resources Planning and Management 108*:87-105.

Ontario Energy Board, 1986. *Report of the Board: The Consumer's Gas Co., Ltd. Proposed Liquefied Natural Gas Plant*, Report No. E.B.R.L.G. 29, Toronto, Ontario.

Ontario Environmental Assessment Board, 1988. *Reasons for Decision and Decision of the Board: Finch Avenue West Extension*, Toronto, Ontario, February.

Ontario Municipal Board and Ontario Environmental Assessment Board, 1985. *Joint Board Decision: Mountain East-West and North-South Transportation Corridor Study*, Toronto, Ontario, October.

Regional Municipality of Hamilton-Wentworth, 1982. *Mountain East-West and North-South Transportation Corridor Environmental Assessment Submission*, Volumes I and II, Hamilton, Ontario, December.

Rosenberg, D.M., et al., 1981. "Recent trends in EIA," *Canadian Journal of Fisheries and Aquatic Sciences 38*:591-624.

Schibuola, S., 1988. *A Knowledge-Based System for the Review of Environmental Assessment*, Master of Applied Science Thesis, Department of Civil Engineering, University of Toronto, Canada.

Townsend, C., 1987a. *Introduction to Turbo Prolog*, Sybex, Berkeley, CA.

Townsend, C., 1987b. *Mastering Expert Systems with Turbo Prolog*, Howard W. Sams and Company, Indianapolis, IN.

Waterman, D.A., 1986. *A Guide to Expert Systems*, Addison-Wesley, Reading, MA.

Whitney, J.B.R. and Maclaren, V.W., 1985. "A Framework for the Assessment of EIA Methodologies," in J.B.R. Whitney and V.W. Maclaren, eds., *Environmental Impact Assessment: The Canadian Experience*, Inst. for Environmental Studies, Univ. of Toronto.

10

An Expert System for the Management of the Water Quality in the Latrobe River

Glenn Elliott and Victor B. Ciesielski

Introduction

Within the Latrobe Valley region of Victoria (Australia), the State Electricity Commission of Victoria (SECV) currently operates five brown coal-fired power stations. These stations have a number of features in common, one being that each requires large volumes of water for operation. Presently, three power stations, the Yallourn Power Station (YPS), the Yallourn "W" Power Station (YWPS), and the Loy Yang "A" Power Station (LYAPS), draw a significant percentage of water from the Latrobe River. This massive consumption of water has created significant environmental problems in the Latrobe River. In conjunction with the Environment Protection Authority (EPA), the SECV is attempting to minimize the environmental impact of its operations.

The volume of water drawn from the Latrobe River has increased as the number and size of the power stations has increased and to compensate, a number of water storage facilities have been constructed. These weirs are located on the Latrobe River in the vicinity of the YPS and YWPS. The first of these water storage facilities to be constructed was the Power Station weir in the 1930s. This weir provided the SECV with minimal control over the flow of the Latrobe River. A second weir, the Lake Narracan weir, was constructed in the early 1960s upstream of the Power Station Weir and provided additional control of the flow in the river and increased water reserves. Most recently, the construction of the Blue Rock Dam in the early 1980s has provided further reserves. The Latrobe River management system can therefore be considered to incorporate four major components: the Blue Rock Dam, the Lake Narracan weir, the Power Station weir, and the Latrobe River itself. A map of the system is shown in Figure 10.1.

Detailed problem description

A recent study by the SECV (Longmore, 1987) analyzes the environmental problems associated with the SECV's use of the Latrobe river water. The key results of this study are summarized below.

Operating constraints

The water requirements of the power stations drawing on the Latrobe River are very high. Due to the scale and the "around the clock" nature of the SECV's operations, control of the discharge rates from the Lake Narracan weir and the Power Station weir, which are closest to the power stations, are the responsibility of the SECV. As well as dramatically affecting the flow in the Latrobe River, the operations of the SECV also affect the temperature of water in the river. A substantial percentage of the water drawn for the YPS is used for cooling; this can result in heated water being discharged back into the river and increasing the temperature of river water. However, the SECV is not the only user of the Latrobe River, and to ensure that users down river of the SECV have adequate water and to protect the ecological balance in the river, the EPA has set legally binding river flow and temperature constraints. The SECV must maintain the Latrobe River within these constraints or face court action and public criticism. Unfortunately, a number of breaches of the constraints have been recorded in the past few years and the SECV is concerned with ensuring no further breaches take place. The flow and temperature constraints currently in effect are:

- The minimum flow of the Latrobe River allowable at the Thoms Bridge Monitoring Station is the minimum of 350 megalitres per day, or 125% of the natural inflow to the river upstream of Lake Narracan. (The additional 25% constitutes the water released from the Blue Rock Dam.)
- The temperature rise between the Lake Narracan weir outlet and the Brown Coal Mine Bridge must not exceed a maximum figure depending on the month of the year, as shown in Table 10.1.
- The rate of change of temperature at the Brown Coal Mine Bridge must not exceed 2°C per half hour.
- The river water temperature at the Brown Coal Mine Bridge must not exceed 26°C at any time.

Unfortunately, these constraints affect power generation efficiency of the YPS and balancing environmental considerations with economical power generation is a major problem, particularly in the summer months.

Options for controlling the Latrobe River

The responsibility of maintaining the Latrobe River within the constraints is one

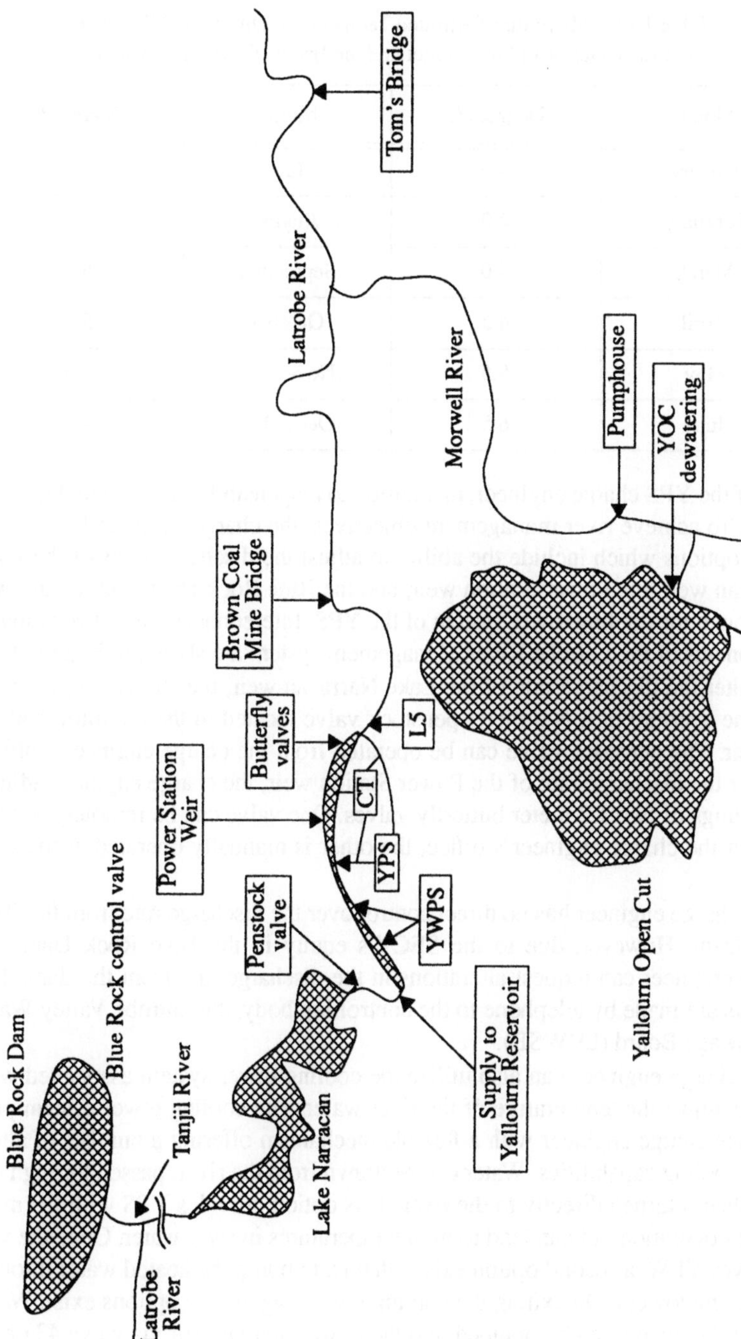

Figure 10.1. The Latrobe River management system.

Table 10.1. Maximum Permitted Temperature Differential Between the Lake Narracan Weir Outlet and the Brown Coal Mine Bridge

Month	Degrees C	Month	Degrees C
January	4.0	July	6.5
February	4.0	August	6.5
March	4.0	September	6.5
April	4.5	October	5.5
May	5.5	November	4.5
June	6.5	December	4.0

duty of the YPS charge engineer, in addition to responsibility for day to day operations. To achieve river management objectives, the charge engineer has a number of options which include the ability to adjust the discharge rates of the Lake Narracan weir, the Power Station weir, and the Blue Rock Dam, and to pass water through the cooling tower system of the YPS. Interconnections between major components of the Latrobe River Management system are shown in Figure 10.2.

To alter the discharge rate of the Lake Narracan weir, the charge engineer adjusts the aperture of a two meter penstock valve located in the retaining wall of the weir. The penstock valve can be operated from the charge engineer's office. To alter the discharge rate of the Power Station weir, the charge engineer adjusts the settings of two 1.2 meter butterfly valves. One valve can be remotely operated from the charge engineer's office, the other is manually operated at the weir wall.

The charge engineer has no direct control over the discharge rate from the Blue Rock Dam. However, due to the SECV's equity in the Blue Rock Dam, the charge engineer can request alterations in the discharge rate from the dam. The requests are made by telephone to the controlling body, the Latrobe Valley Water and Sewage Board (LVWSB).

The charge engineer can also utilize the cooling tower system associated with YPS to adjust the temperature of the river water. The cooling tower system provides the charge engineer with a flexible mechanism offering a number of alternative cooling capabilities. Water can be drawn from the river, passed through the YPS, then returned directly to the river. This option provides YPS with the most efficient operation but can lead to high temperatures in river water. Once the water leaves YPW, a second option exists, that is, to pump the heated water through the cooling towers. On exiting the cooling towers, again two options exist. Water leaving the towers can be pumped into the Power Station weir via valve 43 or recycled back through the YPS. By using various pumps and valves in the cooling tower system, many combinations of paths are possible.

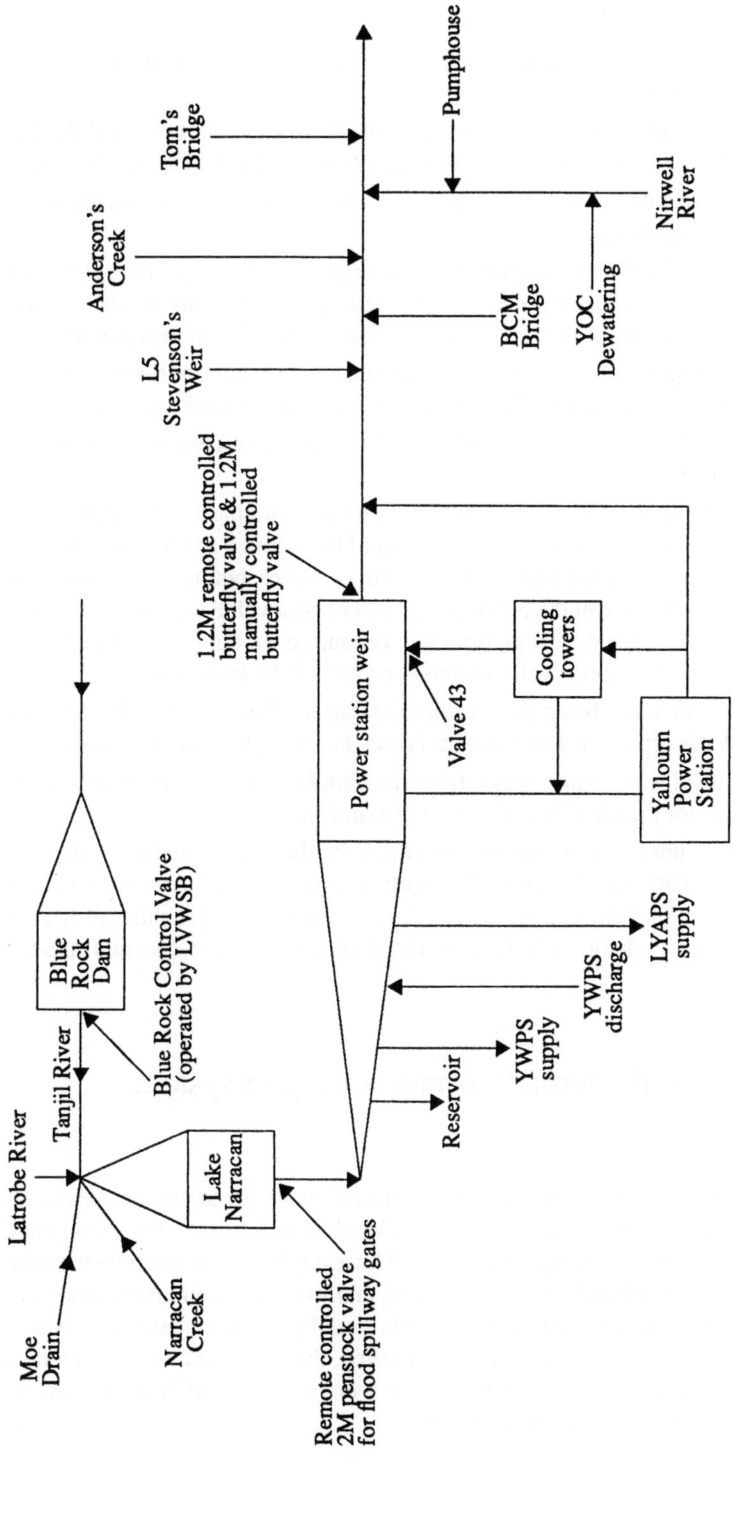

Figure 10.2. Interconnections between major components of the Latrobe River management system.

The decision process

To assist in the decision making process, the charge engineer is presented with data in various forms.

- Monitored parameters of Lake Narracan, the Power Station weir, and the Latrobe River are displayed on a series of gauges in the charge engineer's office.
- Trends of some of the monitored attributes are recorded on chart recorders located in the charge engineer's office.
- A collection of tables and graphs give discharge rates from Lake Narracan for a range of water levels and penstock valve settings, the rise rate of Lake Narracan, and an estimate of the Latrobe River flow at the L5 monitoring point.
- Data, including water demand, river flow, and Blue Rock Dam discharge rates, are obtained from the LVWSB via telephone on a daily (weekday) basis.

A major problem in decision making is that not all the necessary information is available. This includes:

- Data relating to the Morwell River flow. EPA measurements are taken at Thoms Bridge, downstream from the junction of the Latrobe and Morwell Rivers. The charge engineer has control of the Latrobe River, but no information and no control of the flow in the Morwell River. The Morwell River flow is dramatically affected by the discharge from the Yallourn open-cut dewatering pumps. These pumps are automatically switched on and off by level sensors.
- The volume of water being drawn from the Latrobe River by the YWPS at any time. Water is drawn on a demand basis and therefore the flow is irregular.
- The volume of water being drawn from the Latrobe River by the LYAPS at any time. Water for the LYAPS is also on a demand basis.

This lack of information creates difficulties for the charge engineer. Through experience in controlling the river, the charge engineers have learned to compensate for the lack of data. This has led to the development of a number of heuristics that are used, along with the monitored variables, in the decision making process.

The traditional approach versus an expert system approach

The traditional approach to a problem of this kind would be to use network models to compute the flows at various critical points and to use these computed flows to determine the appropriate actions. However, before an accurate network model could be developed and implemented, many of the quantities presently not monitored would need to be monitored. This would prove to be extremely costly. Even if all the necessary points were monitored, the sensors can only provide a picture of the instantaneous condition of the weirs and river. This alone may not be adequate to properly control the river.

Monitoring stations, that is, stations where it is necessary to compare observed values and conditions with expected or planned values and sound some kind of alarm if a critical value is out of range, have previously been identified as suitable for expert system development. A number of expert systems for monitoring problems have already been built (Fagen, et al., 1984; Thompson, et al., 1987). On this basis, it seemed reasonable to consider an expert system for monitoring the Latrobe River.

Further analysis of the problem and the methods employed by the charge engineers revealed that the problem was a "heuristic classification" problem. In a heuristic classification problem, the basic idea is to fit the data pattern to one of a set of mutually exclusive categories or classifications, for example, "Open_Lake_Narracan_Valve," "Close_Lake_Narracan_Valve," etc. Characteristic patterns of data are associated with each of the classifications. If the necessary data are available, statistical pattern recognition methods give optimal solutions. However, in the more usual case when not all data are available, heuristic rules are used. Heuristic rules reflect "rules of thumb" or "educated guesses". More detail on the recognition of classification problems is given in Clancy (1985).

Well established methods for building heuristic classification systems have resulted from research in expert systems (Weiss and Kulikowski, 1983). The use of heuristic classification, particularly in monitoring situations, requires interpretation of quantitative data in qualitative terms. For example, a water temperature of less than 20°C might be "normal," while a temperature between 24°C-26°C might be "dangerously high." Other examples are "within normal operating limits," "within limits but rising," etc. These qualitative states are then used in the classification rules. Some examples of rules using these concepts are shown in Figure 10.3.

Heuristic classification provides a method of decision making based on a combination of hard data and experiential, heuristic knowledge. Since this is the case in the Latrobe River management problem, it was decided to develop an expert system on the hypothesis that a detailed flow model is not necessary, but that a qualitative model of the river system could be embodied in the rules of an expert system, and that such an expert system would be adequate for the purpose of determining the necessary management and control actions.

The prototype expert system

A prototype system of the Latrobe River management system has been designed and built. The system has been implemented in the VP-EXPERT shell, and is presently being evaluated by the charge engineers involved in the management of the Latrobe River. The evaluation procedure being employed is to compare the actions recommended by the expert system with those of the charge engineer for a given set of data. Any discrepancies observed are being noted for correction.

```
IF      current_month = Jan or Feb or Mar or Dec
THEN    latrobe_river_safe_temp_diff = 3.5
        latrobe_river_maximum_temp_diff = 4.0

IF      latrobe_river_current_temp_diff >
           labrobe_river-maximum_temp_diff
THEN    latrobe_river_temp_state = above_limit

IF      power_stn_weir_remote_valve <> fully_open
        latrobe_river_flow_state_ = below_limit
or      latrobe_river_temp_state = above_limit
THEN    new_lake_narracan_penstock_valve =
           ((current_lake_narracan_penstock_valve)+ 0.05)
        new_power_stn_weir_remote_valve =
           ((current_power_stn_weir_remote_valve) + 0.125)
        new_power_stn_weir_manual_valve =
           (current_power_stn_weir_manual_valve)
        new_valve_43 = (current_valve_43)
        required_action = open_power_stn_weir_remote_valve

IF      latrobe_river_flow_status = low_and_falling
        power_stn_weir_remote_valve <> fully open
THEN    new_lake_narracan_penstock_valve =
           ((current_lake_narracan_penstock_valve) + 0.05)
        new_power_stn_weir_remote_valve =
           ((currrent_power_stn_weir_remote_valve) + 0.125)
        new_power_stn_weir_manual_valve =
           (current_power_stn_weir_maual_valve)
        new_valve_43 = (current_valve_43)
        required_action = open_power_stn_weir_remove_valve
```

Figure 10.3. Some rules from the system.

The version of the prototype that is under evaluation by the charge engineers has approximately 80 "IF condition THEN action" rules in its rule set. The system is best described as a "hybrid" system, as both forward and backward chaining is employed. The rule set is partitioned into a number of groups and each group of rules is organized around a sub-goal of the problem. For example, there are five or six rules used to determine the current qualitative state of the Power Station weir level. Once this is achieved, this intermediate state is used in other rules to determine the required actions, as illustrated in Figure 10.3.

In order to tie together the sub-goals established during a consultation a dummy variable, "required_action," was required to focus the inference engine of the VP-EXPERT shell on the action determining rules. Presently, there are 25 rules that utilize the dummy variable in the prototype rule set. Additional rules may be required in this area, and the evaluation procedure should identify these rules.

The rule set also includes approximately ten rules which perform a form of data validation by concentrating on the calculated rate of change in attribute values. For the rules to provide data validation, it is necessary to make an assumption about the manner in which the system would be used, i.e., the system would be run at hourly intervals. Thus the variation in the rate of change for a given attribute should fall within some upper and lower limit. If the rate of change is not within the set limits, then the data is said to be inconsistent and ignored. This should detect obviously erroneous data entered into the prototype; clearly an expert system's conclusions are only as good as the quality of the input data.

During the early stages of development, the following categories for water levels were used: below the minimum level, within the minimum and maximum levels, and above the maximum level. These three categories later were proven to be inadequate. In order to operate the YPS at an efficient level, it was necessary to maintain the river attributes at a level close to the maximum allowable. This provides little margin for error. Because altering discharge rates of various weirs does not have an instantaneous effect (for example, the effect of opening the Lake Narracan valve is not reflected at the Power Station weir until an hour later), it was necessary to add two additional categories. These categories were "within limits but approaching a limit" and "within limits and falling from a limit". The addition of these two categories also provided a mechanism capable of providing an early warning of a potential violation of a constraint.

The inclusion of two new categories created additional problems that had to be overcome and the number of rules had to be increased. For the new categories to be useful, it was necessary to calculate the rate of change in a number of numerical variables. For the prototype, a crude technique was adopted in order to save the value for the required attribute in a file. On the next consultation with the system, a new value was entered for each attribute. The current value of the attribute, minus the saved value, divided by the time differential between the consultations, gave the average rate of change for each variable. This calculated rate was then used in the expert system.

The calculation of the rate of change of each attribute is performed by a C program invoked from the prototype. The C program also performs a series of calculations based on the graphical and tabular data that the charge engineer normally performs manually. The resulting values from the calculations are imported into the prototype expert system for use within the decision making rules set.

A second consequence of the addition of classifications based on the rate of change is the need for information from previous consultations to be stored for referencing during later consultations. As stated above, a very crude method was employed in the prototype system. It has been recognized that in the final implementation of such a system, a database of measurements will be required. By maintaining a database of measurements, it will be possible to more accurately assess current trends in each monitored quantity and validate input data.

The knowledge acquisition phase was a slow and painstaking process. This was partially due to inexperience with knowledge engineering techniques and partially due to difficulty in extracting the real criteria used in control decisions from the great mass of information available. The identification of the limits for converting quantitative data into qualitative states used in the decision rules was also difficult. Another frustration in determining the limits was that different charge engineers had different opinions concerning the values that should apply to each of the limits.

The development time for the system, so far, has been approximately 60 man days over an elapsed period of seven months. This development period includes time to: (1) determine the scope of the problem, (2) assess the suitability of an expert system for use in managing water quality in the Latrobe River, (3) acquire knowledge for use in the expert system, (4) develop the rule set, including time to learn the VP-EXPERT package, and (5) write the "C" programs, including time for familiarization with the "C" language.

In the future, we hope to extend the system to continuously monitor sensors sampling the necessary data quantities and to prompt the charge engineer only when an adjustment is required, thus alleviating the charge engineer of the burden of entering required data manually. A natural progression from this is for the system to control the values directly. However, extensions may not be possible because of costs involved in providing necessary equipment (for the monitoring and subsequent collection of data), and the inherent dangers associated with a basically unmanned system managing the river.

Conclusions

Overall, the development of the prototype system has been regarded as a highly successful venture. The prototype has demonstrated that:

- An expert system based on the concept of heuristic classification and utilizing a set of "IF condition THEN action" rules can represent the expertise needed to manage the flows and temperatures in the Latrobe River. However, there is evidence to suggest that a frame representation may result in a simplified rule set.
- The simplified approach of observing the changes in qualitative states rather than absolute values within the major components of the river system is adequate for high quality decision making.
- A system which obtains data from a series of sensors could be developed to monitor the river system and only alert the charge engineer when a change of valve settings is required.
- There is a need for data validation within the system to ensure the suggested actions are correct. This is especially true for a monitoring system.
- There is a need to maintain a database to assist in the validation of data by observing trends in the historical values of the variables.
- It is possible to develop useful prototypes of expert systems in inexpensive, commercially available expert system shells.

References

Ciesielski, V.B., 1988. "Expert Systems: In Search of the Right Applications," *Professional Computing*, Sydney, Australia, March.

Clancy, W., 1985. "Heuristic classification," *Artificial Intelligence* 27:289-250.

Fagan, L.M., J.C. Kunz, E.A. Feigenbaum, and J.J. Osborn, 1984. "Extensions to the Rule-Based Formalism for a Monitoring Task," in B.G. Buchanan and E.H. Shortliffe, eds., *Rule-Based Expert Systems: The MYCIN Experiments of the Stanford Heuristic Programming Project*, Addison-Wesley, pgs. 397-423, Reading, MA.

Longmore, G., 1987. "Latrobe River Management at Yallourn," Technical Services Division, State Electricity Commission of Victoria, Technical Report Number ES87 008, July.

Thompson, J., J. Delaney, B. Marksjo, R. Sharpe, A. Grant, D. Raimondi, and M. Prior, 1987. "An Expert System to Give Advice to Operators of a Metropolitan Water Supply, Drainage, and Sewerage Network," *Proceedings of the Australian Joint Artificial Intelligence Conferences*, pgs. 86-99, Sydney, Australia.

Waterman, D.A., 1986. *A Guide to Expert Systems*, Addison-Wesley, Reading, MA.

Weiss, S.J. and C.A. Kulikowski, 1983. *A Practical Guide to Designing Expert Systems*, Chapman and Hall, London.

11

Multi-Domain Expert Systems for Hazardous Waste Site Investigations[1]

Hsai-Yang Fang, George K. Mikroudis, and Sibel Pamukcu

Introduction

Because of population growth, a progressive living standard, and industrial progress in recent years, much of our air, water, and land resources have become polluted. Open dumps and chemical and industrial wastes contribute to pollution problems such as those listed in Table 11.1. Improper disposal and management of hazardous wastes and toxic chemicals in numerous unidentified locations is one of our most pressing environmental problems. In 1978, the U.S. Environmental Protection Agency (U.S. EPA) estimated that approximately 60 million metric tons of hazardous wastes are generated annually in the United States at more than 750,000 sites. At the present time, these figures have dramatically increased. Figure 11.1 illustrates various types of generated wastes among several developed countries and Table 11.2 shows the typical hazardous substances in a selection of industrial wastes.

The objectives of this chapter are: (1) to introduce a newly developed multi-domain knowledge-based expert system as an aid for hazardous waste site investigations, and (2) to develop a better knowledge-based database and data management system.

---

1. Reprinted with permission from Fang, H.Y., Mikroudis, G.M., and Pamukcu, S., in *Expert Systems for Environmental Applications*, Hushon, J.M., ed., ACS Symposium Series 431, American Chemical Society: Washington, DC, 1990, pp. 146-159. Copyright 1990 American Chemical Society.

Table 11.1. Some Causes of Soil/Water Pollution in the United States

Southwest	South Central	Northeast	Northwest
Acid rain	Acid rain	Acid rain	Abandoned oil wells
Animal wastes	Animal wastes	Acid mine drainage	Acid drainage and mine tailing
Disposal of oil field brines	Disposal well	Buried pipelines and storage tanks	Acid rain
Evapotranspiration from vegetation	Evapotranspiration from vegetation	Hazardous chemical wastes	Brine injection
Hazardous chemical wastes	Hazardous chemical wastes	Highway de-icing salts	Disposal wells
Injection wells for waste disposal	Irrigation return flow	Landfills	Dry land farming
Irrigation return flow	Landfills	Mine fire	Highway de-icing salts
Leaching	Nuclear wastes	Nuclear wastes	Hazardous chemical wastes
Nuclear wastes	Oil field brines	Petroleum exploration and development	Irrigation return flow
Saltwater intrusion	Solid wastes	Radon (gas)	Landfills
Solid wastes	Waste lagoons	River infiltration	Mine fire
Spills of hazardous materials		Saltwater intrusion	Nuclear wastes
Water from fault zones and volcanic origin		Septic tanks	Septic tanks
		Surface impoundments	Sewage treatment plant discharges
			Surface impoundments

Basic information required for hazardous waste site investigations

The description and quantification of actual and potential hazards associated with a waste disposal site incorporates elements of site evaluation, chemical fate and transport evaluation, basic toxicology, exposure, and risk assessment. Many factors must be considered before a site characterization is possible. Assessment of the hazard level is a complex engineering problem which requires interdisciplinary knowledge. Understanding of the interaction between toxic wastes, water, soil, and rock is based on concepts of environmental geotechnology, geology, hydrology, climatology, chemistry, and toxicology. Basic information required for hazardous site investigations (Desmarais and Exner, 1984; Fang, et al., 1987) is discussed as follows.

Sources

The major sources of soil/water pollution are listed in Table 11.1. Regardless of source, there are three basic mechanisms by which ground soil/water can become contaminated.

1. Contamination may occur from rainfall such as acid rain or rain falling into a sanitary landfill, or oil or chemical wastes spilled onto the soil/water systems.
2. Contamination may result when pollutants are introduced as leakage from disposal wells or constructed waste disposal facilities such as landfills, septic tanks, laterals, and lagoons.
3. Contamination may also result from hydraulic, chemical, or physico-chemical alterations which allow polluting substances to move within or between soil layers. In this category, contamination involves chemical, physico-chemical, and microbiological factors.

Receptors

In order to identify potential receptors and the contaminant levels to which they are exposed, the following points where exposure to contaminants may occur are commonly evaluated.

- Groundwater (used as drinking water)
- Surface water (used for drinking or involving human contact)
- Air (airborne contaminants upwind from the site)
- Direct contact (with contaminated soils or exposed wastes)
- Ingestion (of contaminated plants, or animals/birds/fish)

Table 11.2. Typical Hazardous Substances in Industrial Waste Streams (reprinted with permission from Yong, 1989, *Proc. 2nd Int'l. Symp. on Environ. Applications*, Envo Publishing).

Industry	Arsenic	Heavy metals	Chlorinated hydrocarbons	Mercury	Cyanides	Selenium	Miscellaneous organics
Chemical		X	X	X			X
Electrical & electronic		X	X	X	X	X	
Electroplating & metal industry	X	X		X	X		X
Leather		X					X
Mining & metallurgy	X	X		X	X	X	
Paint & dye		X		X	X	X	X
Pesticide	X		X	X	X		X
Pharmaceutical	X			X			X
Pulp & paper				X			X
Textile		X					X
MSW	X	X	X	X	X	X	X

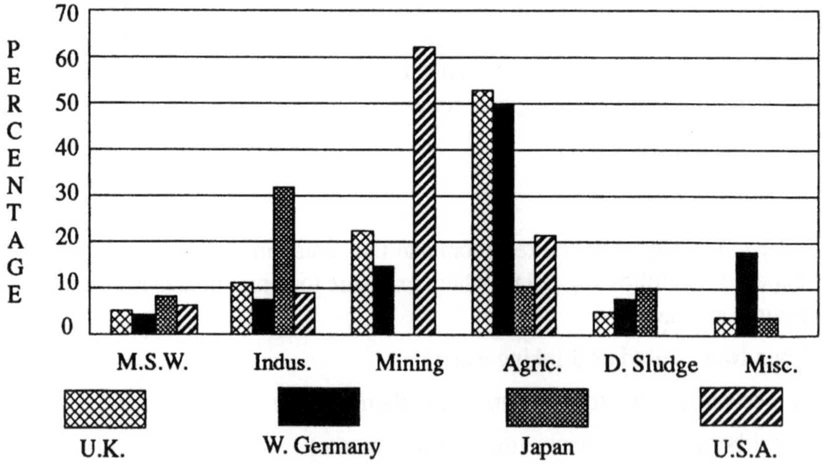

Figure 11.1. Various types of generated wastes (reprinted with permission from Yong, 1989, *Proc. 2nd Int'l. Symp. on Environ. Applications*, Envo Publishing).

Pathways

Populations and/or environments around the hazardous waste site may be affected via several pathways of exposure. Consideration should be given to the following factors.

- Geology and soils (rock joints, porosity/voids, profile of soil layer, hydraulic conductivity)
- Hydrology (groundwater flow direction, gradient, discharge)
- Climate (rainfall, evaporation, wind direction)
- Topography (land slopes, wetlands, rivers, lakes)
- Demography (population, distance from site, land use)
- Ecology (endangered species, plant and animal habitats

Once the sources, receptors, and pathways of contamination have been identified and evaluated, various methodologies can be applied to give a quantitative or qualitative measure of potential threats to human health and welfare or the ecosystem. In all cases, a series of investigations and data collection is required before one can proceed to the site evaluation.

A complete assessment of hazardous waste sites involves consideration of not only technical aspects, but political/social/economic aspects which are also important. In many cases, non-technical concerns may be even more important than technical ones. Let's examine the whole assessment system which covers three major phases: the transverse, diagonal, and longitudinal interactions (as illustrated in Figure 11.2). On the surface, these three phases are not directly interrelated, however, without considering each phase, the assessment of site investigation cannot be effectively undertaken. For example, without political and social involvement, technical aspects are only a small part of an overall system. In order to produce efficient, economical results, a well-planned system is necessary. Three phases involved in a hazardous site investigation are discussed as follows.

- *Transverse interaction* covers most of the non-technical factors such as politics (legislation, zoning, public opinion, or the news media) and social traditions. These factors often influence decision-making, for example, in determining whether or not a given hazardous waste site will be designated as an EPA Superfund clean-up site. Therefore, in some cases non-technical factors may be more important than technical one.
- *Diagonal interaction* basically involves economic factors such as total investment, interest rates, time constraints, availability of labor, construction equipment, and annual maintenance costs.
- *Longitudinal interaction* involves technical aspects of hazardous waste site investigation and will be further discussed in the following sections.

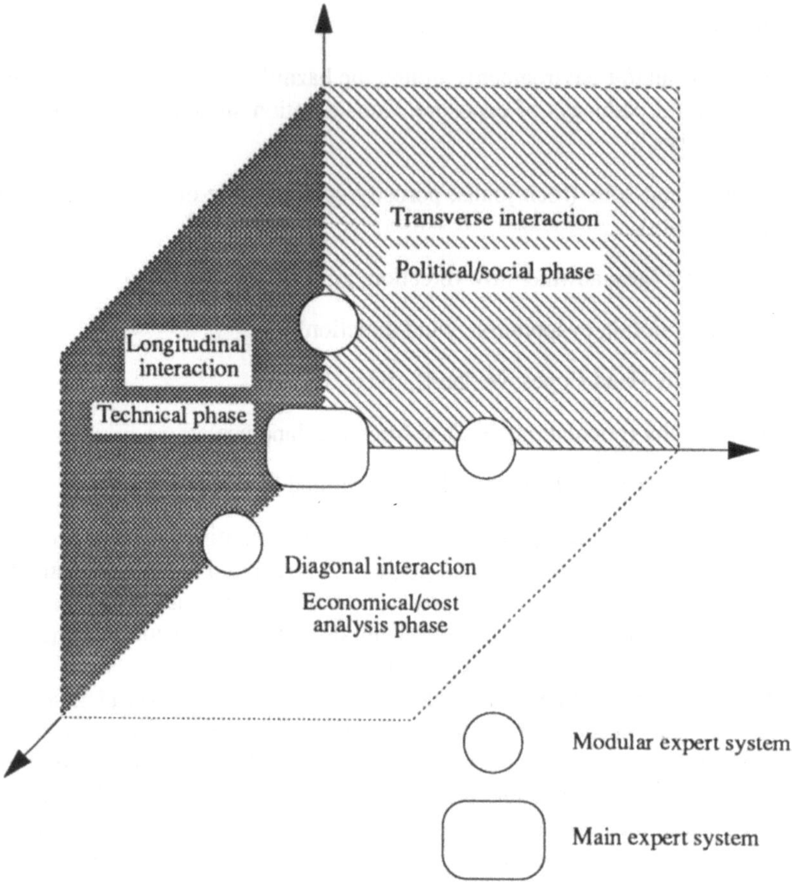

Figure 11.2. Three phases of interaction in environmental assessments.

Nature of knowledge-based expert systems

Knowledge-based systems (KBS) are computer programs that contain expert knowledge about a specific domain and are able to apply this knowledge to make useful inferences and provide expert-level advice to the user of the system. In addition, expert systems are capable, on demand, to 'justify' their own line of reasoning in a manner directly intelligible to the inquirer.

The knowledge includes facts and rules, both of which are interchangeable. Facts represent declarative knowledge and provide an actual database. Rules represent procedural knowledge, as well as rules of judgment and plausible reasoning. Inference includes the techniques used to generate new information from old and is applied according to a control strategy, e.g., choosing which rules to apply and trying different alternatives. The advantages and limitations of KBS and comparisons between conventional programming and KBS are discussed by Fang, et al., 1987.

The effectiveness of KBS

The effectiveness of KBS depends on how one selects the system and shell to be used for a project. If the project is straightforward, a KBS is probably not necessary. If the project is too complicated, an expert system cannot provide effective results. At present, most available expert systems are of a single-domain type (Figures 11.3 and 11.4). A single-domain system consists of two parts: human-computer interaction as shown in Figure 11.3 and the main portion of the expert system known as the shell (shown in Figure 11.3B) which generally consists of artificial intelligence (AI) techniques, graphics, and risk analysis. Each shell is designed for a particular application and some shells are more sophisticated than others. There are several commercially available shells on the market. GEOTOX, developed by the Envirotronics Corporation, is one such shell and is useful for analysis, design, construction, and management (Mikroudis and Fang, 1987; Pamukcu, et al., 1987). The GEOTOX shell was originally designed for a main-frame computer system, but has since been updated and converted into a personal computer system.

Structure of the GEOTOX shell

The theoretical background of GEOTOX has been discussed by Fang, et al. (1987) in a separate paper. The features of the GEOTOX shell, together with recent improvements on the shell and other related features, is discussed here.

GEOTOX is a knowledge-based expert system primarily designed for hazardous waste site evaluations. More specifically, it is intended to assist in preliminary investigations, although its knowledge base can be expanded to accommodate detailed investigations and field work. It can be used for multiple site comparisons, prioritizing, and ranking. Besides evaluation of existing sites, GEOTOX can be applied to assess potential hazardous waste sites and assist in the site selection process for new facilities. The various processes supported by GEOTOX are summarized as follows.

- *Interpretation*: Assessment of existing hazardous waste sites and evaluation of potential waste disposal sites.
- *Classification*: Ranking existing sites and screening potential sites.
- *Diagnosis*: Contamination problems at hazardous waste sites and selection of remedial alternatives.

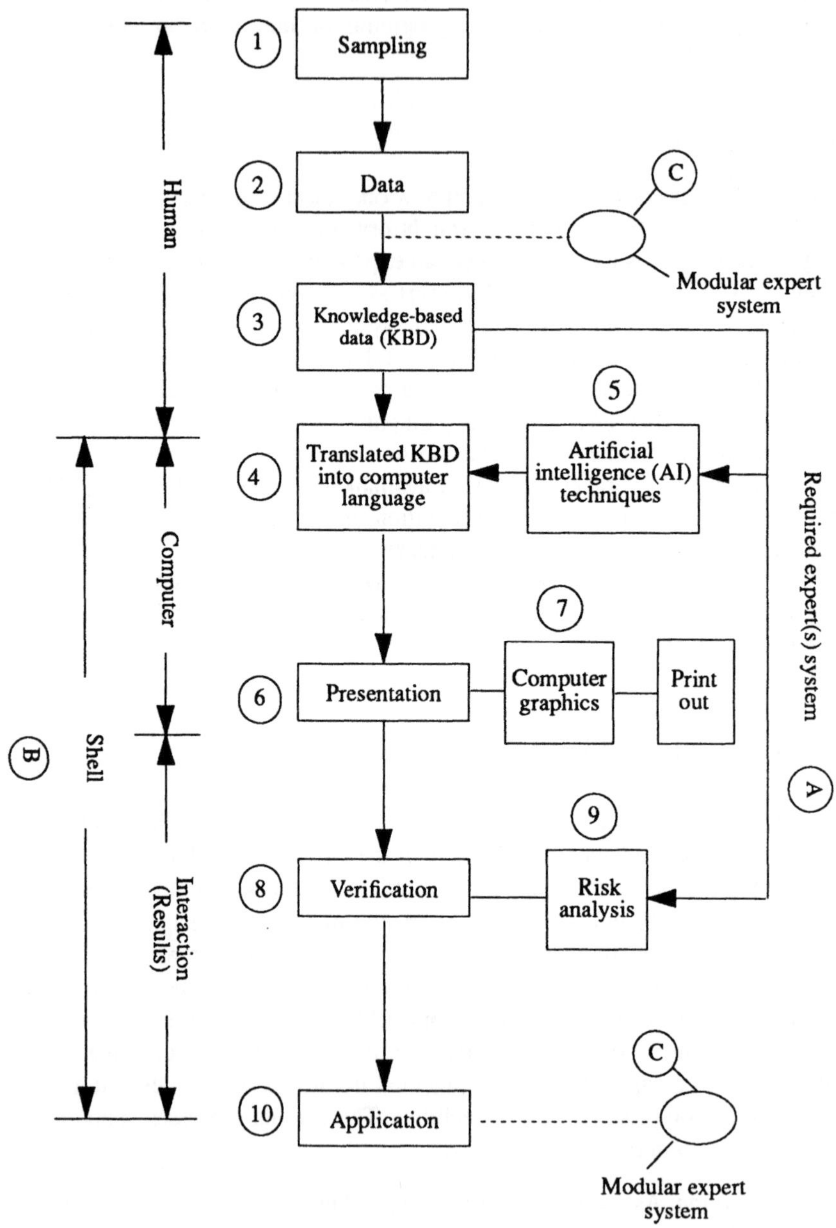

Figure 11.3. Human-computer interaction in knowledge-based expert systems.

General design requirements for 'expert' systems have been identified by researchers (Buchanan and Shortliffe, 1984) in artificial intelligence (AI). More recently, human behavior in engineering or psycho-engineering concepts have been introduced into the 'expert opinion' (Fang and Mikroudis, 1987).

Figure 11.4 shows human-computer interaction using the GEOTOX shell. As illustrated, a user (U) communicates with the subsystem of the database (KB), graphics (G), and analysis design (IM) through a human-machine interaction (IF) which is shown by the shaded surface. This interface can support 'customized' communications that allow a user to selectively choose from a wide variety of options throughout these subsystems. The linkages for data transfer among these systems are indicated by the lines with arrows. At any stage, the user can move backward and forward to see what has been done.

The main portion of GEOTOX covers the knowledge base, associative network, production rules, frames, and inference rules. A simplified view of the knowledge representation scheme is shown in Figure 11.5A. The hybrid system of an associative network and production rules is implemented using logic programming via the language Prolog. The associative network provides structure to the overall knowledge representation scheme. This network defines associations between data and site parameters as conceived by the domain expert(s). Production rules are used at conjunctive nodes (Figure 11.5A) of the network to determine what value should be inherited by the network according to existing conditions at the node.

Frames in Figure 11.5A are a collection of associative net nodes and slots. They are used to represent the final conclusions of the expert and to describe different situations or possible scenarios of contamination at the site. The application of the inference rule in GEOTOX is shown in Figure 11.5B. GEOTOX uses the hazard value h and confidence level c provided by the production rules and propagates them to all the associated nodes following the links in the associative network. Figure 11.5B(a) and 11.5B(b) show the h-c pair at typical nodes i and j of the GEOTOX associative network. Figure 11.5B(c) illustrates the effect of applying the inference rule for a given h-c pair. It shows that the updated h in Figure 11.5B(c) is closer to the value provided with more confidence. For detailed theoretical background on GEOTOX, see Mikroudis and Fang (1987). Main features and advantages of GEOTOX are summarized in Figure 11.6 and discussed by Fang, et al. (1987).

Multi-domain knowledge-based expert systems

If a project consists of multi-characteristics such as large structural systems, i.e., tall buildings, offshore structures, or complex projects such as a large construction operation or hazardous waste control facility, a single-domain system may not be used effectively. Therefore, a new multi-domain system (MDS) has been developed (Fang and Mikroudis, 1987) as shown in Figure 11.7.

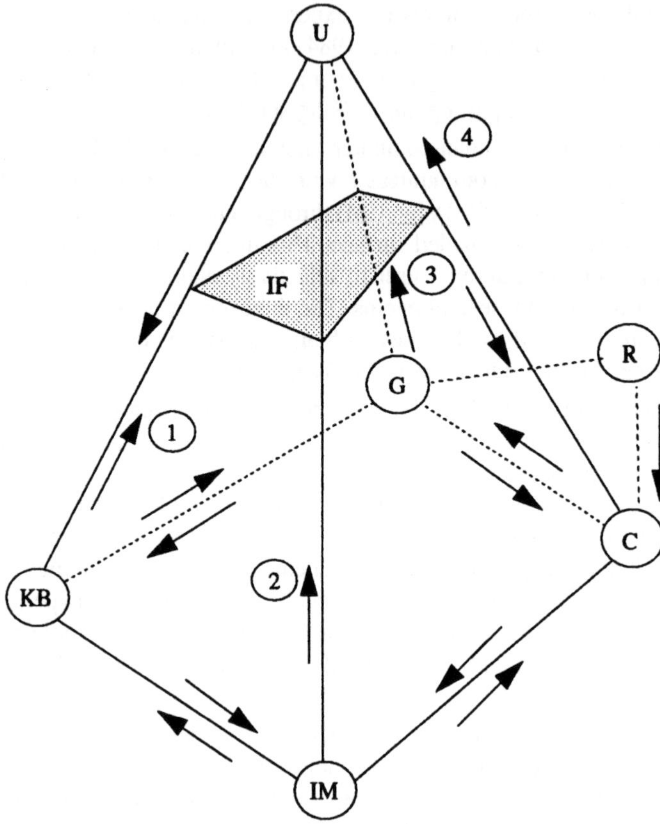

C	=	Context	G	=	Computer graphic
KB	=	Knowledge base (database)	IM	=	Inference mechanism (data analysis)
IF	=	Explanation interface	R	=	Remote-sensing devices
U	=	User			

Figure 11.4. GEOTOX model for data management, analysis, and presentation.

MDS is an extension of GEOTOX and contains additional features for addressing widened applications. The central component of MDS is a user-system and system-system interface which provides the link among the diverse software subsystems and establishes a seamless transition from application to application. This interface provides facilities for process-to-process communication in addition to standard facilities for communicating and reporting the processes of the system and the results to the user (as in the GEOTOX model). This interface establishes an open architecture where more satellite subsystems can tie-in and communicate with all the other direct link programs using the interface facilities.

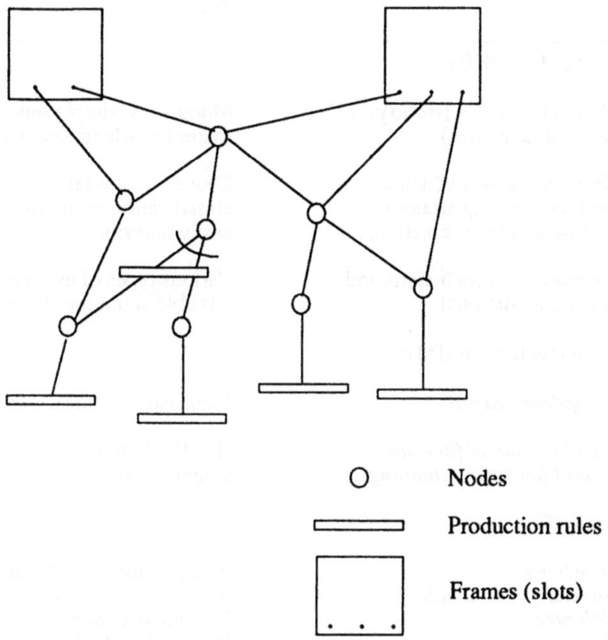

A. Simplified view of the knowledge representation scheme.

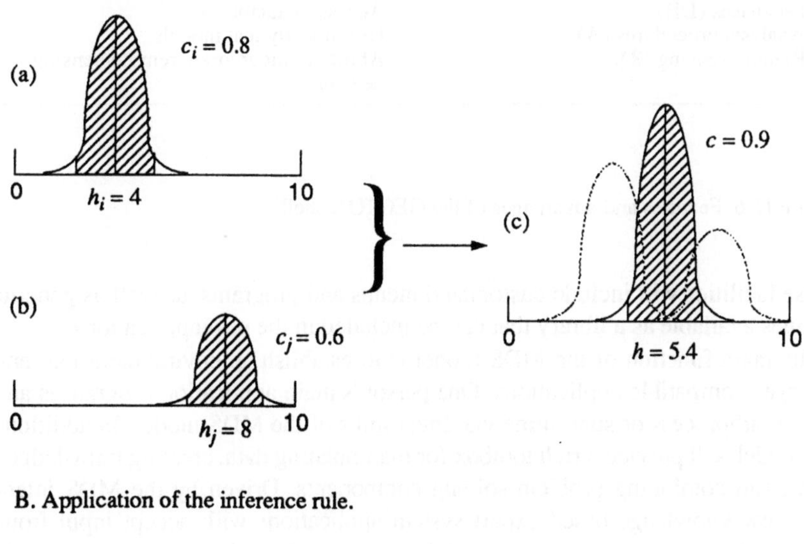

B. Application of the inference rule.

Figure 11.5. Structures of GEOTOX shell.

Features	Benefits
Knowledge base (KB):	
Production rules (for expert derived heuristics)	Modularity, simple semantics, easier knowledge acquisition.
Semantic network (defines problem solving strategy and parameter interactions)	Cause-effect relationships, classification properties easily described
Frames (for conclusions and recommendations)	Various types of evaluations and possible situations defined
Inference mechanism (IM):	
Confidence factors	Expresses confidence in data
Combination of forward and backward chaining	Handles both interpretive and diagnostic problems
Interface (IF):	
How/why	Examine the line of reasoning
Summarize/conclude	Review the state of knowledge
Volunteer	Flexible data entry
Change	Data update, "what if" questions
Revise	Modify the knowledge base
User specified options include:	
Computer graphics (G):	Visualization of conditions
Databases (DB):	Access to databases
Analysis procedures (A):	Use of analytical models
Remote sensing (R):	Ability to incorporate remote-sensing devices

Figure 11.6. Features and advantages of the GEOTOX shell.

These facilities will include customized menus and programs, as well as generic routines available as a library that can be included in the new applications.

The main function of the MDS model is to establish an environment that encourages compatible applications. One person's main application program is another's subprocess or subroutine via direct links of the MDS model. In addition, this model will provide a rich toolbox for manipulating data, creating knowledge-bases, and combining problem-solving components. Driven by the MDS interface, new knowledge-based expert system applications will: accept input from users, real-time sources, sensors, and databases; make inferences or retrieve all available information; and request additional information when needed. Because of the complexity of such applications, the interface should provide facilities for generating the code, testing the consistency of new KBS with old, and ensuring the concurrency of databases and knowledge-bases.

Thus, a real-world system based on an architecture that will provide all the KBS tools, databases, analytical models, and interfaces to the outside world for building an MDS application is needed. As long as a general protocol for transferring information among applications is established, the whole MDS model does not need to be present for developing practical applications. A partial implementation of subsystems of the model can be combined when needed with other compatible applications at a later stage. In this way, real-world knowledge contained in existing databases and analytical models can be made accessible and can be manipulated by MDS application programs. The expertise of such real-world programs relies on communication with other machines and applications that provide the data and knowledge necessary to solve real-world problems.

The main features of the MDS include:

- IF/IPC A generalized user-system and system-system interface (IF) with built-in interprocess communication channels.
- KR Several knowledge representation languages and techniques.
- IE Several inference engines that can be combined with the various KR techniques.
- DB A database manager that provides access to various databases.
- AM An analytical/algorithmic model library that allows execution of different algorithmic programs.
- C/R Communication/real-time access of other computer systems (e.g., via a network), or instruments, sensors, and real-time devices.
- G/CAD Graphics and computer-aided design packages.

Modular expert system (MES)

The modular expert system is an integral part of multi-domain expert systems. In multi-characteristic projects requiring interdisciplinary knowledge, such as hazardous waste site investigations, it is difficult to put all the necessary knowledge-based data into one system when using a personal computer. Therefore, a MES is needed to integrate the separate systems (Barnett, 1973).

The structure of MES is similar to the GEOTOX model shown in Figures 11.3 and 11.4. The major difference between these two systems is the knowledge-based data bank (see Figure 11.3, steps 2 to 3, and Figure 11.4).

A modular expert system is just like a list of menus. A menu is connected with the main multi-domain expert system as needed and disconnected when not needed. In Figure 11.2, the interactions among technical and non-technical aspects require additional expert systems to incorporate the unrelated systems into a related unit. Such a temporary expert system unit is referred to as a modular expert system. When organizing a large amount of experimental data into a knowledge-based data system, expert opinion and an expert system is also required. In such

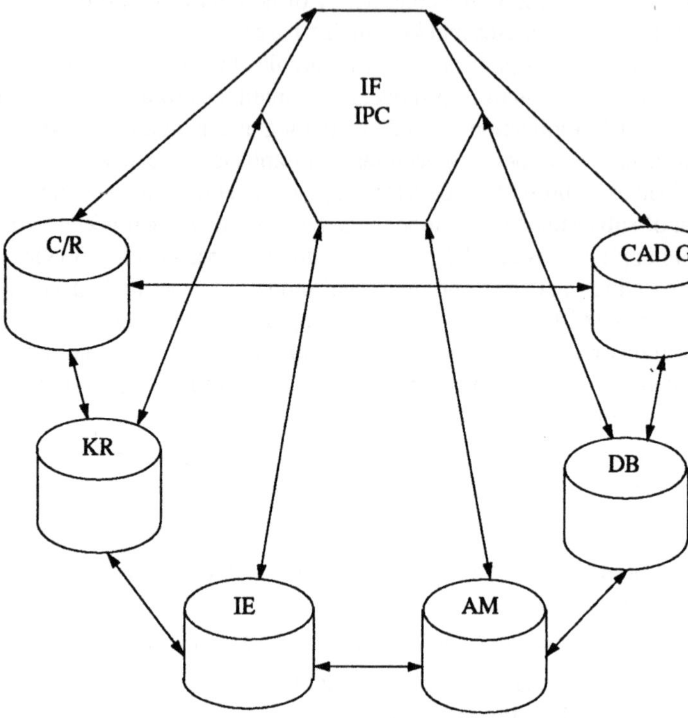

IF - Explanation interface
IPC - Inter-process communication
C/R - Communication/real time
KR - Knowledge representation
IE - Interface engines
AM - Algorithmic model
DB - Database manager
CAD - Computer aided design
G - Computer graphics

Figure 11.7. Fang-Mikroudis model of a multi-domain knowledge-based expert system.

cases, a modular expert system is needed to integrate the data. Similarly, when the GEOTOX model is used for various applications, each application will be different. Therefore, certain modifications are needed for each case and MES can integrate these modifications into a reliable output.

The major function of MES is to aid in solving defined problems under specified conditions. It is a temporary system as far as the main expert system is concerned. In other words, the MES unit is used to customize individual or particular projects as needed. Several basic MES units have been developed and are dis-

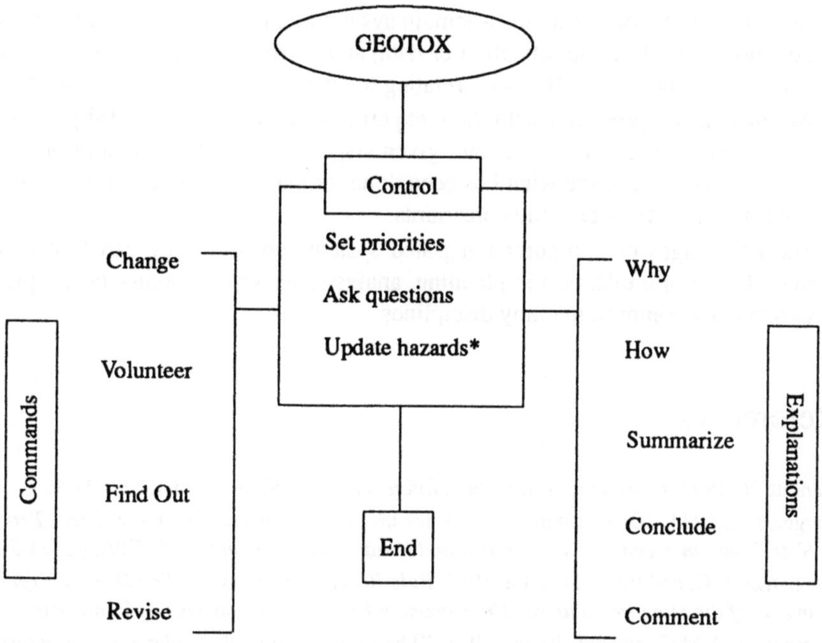

Figure 11.8. A flow chart illustrates how GEOTOX or MES operates during a typical consultation session.

cussed in a separate report (Mikroudis and Fang, 1987). Figure 11.8 illustrates how GEOTOX or MES operates during a typical consultation session.

Summary and conclusions

Assessment of site investigations for hazardous wastes requires interdisciplinary knowledge. Due to complex interactions among many variables, the GEOTOX knowledge-based expert system is introduced.

In many cases, it is difficult to put all the necessary knowledge-based data into one system when using a personal computer; therefore, it is necessary to have modular expert systems to integrate these into the main system. The structure of MES is similar to the GEOTOX model but with different knowledge-based databases.

The multi-domain expert system is an extension of GEOTOX which covers more features and widens the applications of GEOTOX. In addition to hazardous site investigation, this model has been used successfully in various environmental and geotechnical problems, such as radon assessment, landslide control, and large structural systems.

Additional functions of a multi-domain system includes a database for storing large amounts of data, classification of data, and correlation studies such as that for identifying the degree of hazard relating to geotechnical/hydrological data.

All information produced with these expert systems includes colorful pictorial displays and/or tabular results at any given stage of interaction. In addition, the user can move back to see what has been done, or may interactively alter technical and/or financial criteria and constraints.

The advantages of computer integrated systems are that they can lead to a greater degree of unification in planning, analysis, design, and management processes that are common in many disciplines.

References

Barnett, V., 1973. *Comparative Statistical Inference*, John Wiley & Sons, New York.

Berger, I.S., 1984. "Determination of risk for uncontrolled hazardous waste sites," *Proc. Natl. Conf. on Mgmt. of Uncontrolled and Hazardous Waste Sites*, U.S. EPA, pp. 23-26.

Buchanan, B.C. and E.H. Shortliffe, 1984. *Rule-Based Expert System, The MYCIN Experiments of the Stanford Heuristic Programming Project*, Addison-Wesley Publishers.

Desmarais, A.M.C. and P.J. Exner, 1984. "The importance of the endangerment assessment in Superfund feasibility studies," *Proc. National Conf. on Management of Uncontrolled and Hazardous Waste Sites*, U.S. EPA, pp. 226-229.

Fang, H.Y., 1989. "Expert systems for assessment of radon gas," *ASCE Environmental Engineering National Conference*, J.F. Molina, ed., ASCE, New York, pp. 97-104.

Fang, H.Y. and G.K. Mikroudis, 1989b. "Modular expert system: An expert system between expert systems," report prepared for Envirotronics Corporation, International.

Fang, H.Y., G.K. Mikroudis, and S. Pamukcu, 1987. "A unified approach to the assessment of waste disposal sites," in G.N. Pande, ed., *Computers and Geotechnics*, Vol. 3, pp. 129-156, Elsevier Applied Science Publishers, Ltd., England.

Fang, H.Y. and G.K. Mikroudis, 1987. "Multi-domains and multi-experts in knowledge-based expert systems," *Proc. 1st. International Symposium on Environmental Geotechnology*, Vol. 2, pp. 355-361.

Fang, H.Y., 1987. "Soil-pollutant interaction effects on the soil behavior and the stability of foundation structures," in A.S. Balasubramaniam, et al. eds., *Environmental Geotechnics and Problematic Soils and Rocks*, A.A. Balkema Publishers, pp. 155-163.

Feigenbaum, E.A., 1977. "The art of artificial intelligence: themes and case studies of knowledge engineering," *IJCAI 5*, pp. 1014-1029.

Mikroudis, G.K. and H.Y. Fang, 1987. "Classification of waste disposal sites using GEOTOX," *ASCE Geotechnical Practice for Waste Disposal '87*, R.D. Woods, ed., ASCE Geotechnical Special Publication No. 13, New York, pp. 105-120.

Pamukcu, S., G.K. Mikroudis, and H.Y. Fang, 1987. "GEOTOX: A new knowledge/data base management system for controlling wastes," *Proc. Environment '87 Intern. Sym. on Environmental Management*, Istanbul, Turkey, pp. 78-98.

Yong, R.N., 1989. "Waste generation and disposal," *Proc. 2nd. International Symp. on Environmental Geotechnology*, Vol. 1, Envo Publishing Co., Bethlehem, PA, pp. 1-24.

12

NOISEXPT: An Expert System for Noise Control in Very Fast Train Design

Qian Chen and Ronald Sharpe

Introduction

Noise is an important environmental factor which affects construction and engineering, especially in the design of high-speed train systems. For example, noise reduction measures in Japan have been undertaken for an area containing about 32,000 houses and institutions which are affected by Shinkansen system at a cost of about 76 billion yen (Kikuchi, 1988).

Several acoustic projects have explored the use of expert systems. Harris, et al. (1985) and Cohn, et al. (1986) developed a system called CHINA (Computerized HIghway Noise Analyst) which addresses the problem of designing highway noise barriers. It is a rule-based system written in FRANZ LISP to run on a VAX 11/780 computer using an inference engine called GENIE. Ng (1988) in and Yanagida, et al. (1988) have investigated knowledge-based systems for architectural acoustics. The latter involves a conceptual design for an acoustic field simulator and an expert system for room acoustics (ARDEX), and has been developed for a mini computer (DG MV/7800 XP) using Prolog and FORTRAN. Ng reports that a demonstration prototype also written in Prolog has been developed but no technical details are given. An expert system called SKUA has been developed by Stanford University (SEAI, 1986) to interpret vibration data for acoustical analysis of machinery.

As a part of an extensive feasibility study in progress for the proposed Melbourne-Sydney very fast train, a rule-based system called NOISEXPT has been developed at the Commonwealth Scientific and Industrial Research Organization (CSIRO), Division of Building, Construction, and Engineering to:

- Synthesize the existing relevant regulations and standards from various countries and international organizations to derive acceptable train noise tolerance levels for the community.

- Analyze the component noise sources from trains and their characteristics to determine the effects of various parameters on noise emission and possible noise reduction measures.
- Find a suitable prediction method for noise impacts including corrections for ground effect, screen effect, view angle correction, track condition effects, and so on.
- Provide advice for noise control engineering design to those who may have little knowledge of acoustics and lack easy access to the small number of local and international train noise experts.
- Provide ready access to the extensive literature on railway noise for further information.

The extent and complexity of expertise on railway noise makes knowledge acquisition and representation a major problem to be resolved, in addition to the selection of a suitable expert system shell. There are now many commercial expert system shells to choose from, so many in fact that it is too time consuming to evaluate them all before making a choice. Since extensive experience had been obtained previously using a shell called CRYSTAL in another project (Sharpe and Marksjo, 1989), this shell was utilized in this project because it had most of the desired features. These features included the ability to handle very large knowledge bases quickly on a personal computer, a good development environment, user-friendly displays including menus and character graphics, an ability to handle complex mathematics, and an interface to a database system. The main shortcoming is that CRYSTAL is limited to production rule knowledge bases and does not permit frames or object-oriented programming. It is also limited to low resolution graphics which are unsuitable to many applications.

Since proposed noise standards vary greatly from country to country, the effort to establish an acceptable noise tolerance value has been very difficult. A technique for treating this as an uncertainty problem has been developed and reviews the available standards and recommends a standard for very fast train (VFT) design. This technique, of course, may be modified by the user.

The prediction of noise involves a complex set of equations and calculations. The prediction method adopted in NOISEXPT for VFT design combines a British formula for wheel/rail interaction noise (Peters, 1974), a German formula for high-speed aerodynamic noise from the turbulent boundary layer (King and Bechert, 1979), an American formula for the relation between peak noise level L_{max} and the equivalent noise level L_{eq} (Kurzweil, et al., 1979), plus a Nordic formulae for various corrections (Ringheim, 1984). Each noise component submodel may also be run separately or in combination with other submodels. Noise contour maps may be produced via an interface to another CSIRO package, POLDIF (Taylor and Anderson, 1988).

An interface has been also used to enable communication between the reference knowledge base (REF) and a dBase file so that nearly 400 book and paper reference details can be easily found by a matching technique. Knowledge about the noise sources for railway traffic is organized to be viewed both analytically and synthetically with options to have quantitative details for components such as engines, air inlets, and exhaust systems displayed by means of empirical equations.

The development effort for NOISEXPT has been approximately one person-year. Since the system includes a lot of knowledge about conventional trains, it can be easily extended to handle all kinds of train noise. NOISEXPT may also be useful in developing other expert systems for noise control engineering.

Due to a lack of medium and high resolution graphics facilities in CRYSTAL, it is unlikely that a system with detailed acoustic diagnosis and design refinement including detailed drawing of rail, wheel, and other vehicle parts will be achieved in the near future. An interface to enable communication between CRYSTAL and a commercially available CAD system might be necessary.

The noise control problem

Although several hundred papers on railway noise have been published, the number of experts in this area is unlikely to number more than ten in developed countries. Most countries do not have experts on railway noise or VFT design.

Acoustics is a difficult science involving complex theories. The computing and modeling of sound fields requires powerful computers and is time consuming. In practice, wide ranging engineering experience is the only method that can be used for noise control reduction design. This usually means a lot of knowledge from different specialities has to be combined to solve a single problem. To establish a standard for environmental noise from railway traffic disciplines, such as law, psychology, statistics, medical science, and ergonomics as well as engineering, need to be combined with acoustics so that a subjective assessment of railway noise can be developed. To control or reduce noise from railway traffic requires not only knowledge about sound emission and propagation, but also expertise on mechanics, material science, and so on.

High-speed trains

Although railways have been in use for over a century, until recently they have been losing passengers to other modes of transportation such as air and road travel. While only about 12% of travelers use railways, in recent years high-speed trains have been viewed as a viable competitor to air travel. At present, commercially available high-speed trains can be found only in France, Germany, and Japan, thus limiting the availability of noise data. In addition, noise data for the French TGV railway lines have never been published. The InterCity Experimental (ICE) train in Germany has made great progress. 430 km of railway track is being built for it, but the ICE will not be operating commercially until sometime in the early 1990s. More than $160 billion of investment in fast trains is now being planned in Europe over the next decade, in addition to extensions to the Shinkansen system in Japan and the Los Angeles/Las Vegas train in the United States.

Despite the aforementioned problem with a lack of noise data, expertise on very fast trains and their design can be acquired through an international exchange with overseas experts in addition to the literature.

In Australia, because of a lack of expertise on VFT design, an expert system containing a detailed knowledge base of international experience can play a major role as an adviser for planners and design engineers making environmental impact assessments of very fast trains. The system also has potential for use in overseas projects where experts are not readily accessible.

CSIRO, Division of Building, Construction and Engineering has been well qualified to develop NOISEXPT because of experience in research on construction and acoustics as well as in building expert systems (Sharpe and Marksjo, 1989). To quote one keynote speaker at the Second Australian Conference on expert systems (P.H. Winston, 1987):

> *"Often the best exploiters of AI technology are people who are not expert in AI but who need AI technology to solve a problem of importance to their work and hence to their careers.*
>
> *Also, the basic principles of AI can be learned much faster than domain expertise in most cases. A person with some programming experience and skill with computers can do useful AI work in a year, and occasionally less. To become good at anything for which experience counts can take many years.*
>
> *Consequently, the best people for tackling AI problems are those who cope well with things that lack precise definition."*

NOISEXPT is, therefore, a trial system developed by an acoustical expert with assistance from experts on expert systems and based on an expertise which is not readily available and understandable.

NOISEXPT structure

The modular structure of NOISEXPT is shown in Figure 12.1.

Synthesis of standards

Unanimous agreement on international standards for acceptable noise levels for general traffic has not been established, and in the case of high-speed trains is unlikely to be reached for at least another decade. The railway noise problems which regulation makers face are numerous and very complex.

Railway noise criteria are rather uncertain. Although L_{eq}, the average acoustic energy noise level, is considered a good method to predict general annoyance to the community, some prefer to use L_{max}, the peak noise level. Researchers in Japan have developed a standard for the Shinkansen VFT system using L_{max}, which is an average of ten peak noise levels out of twenty trains in sequence passing the location under consideration. Many European countries prefer L_{eq}.

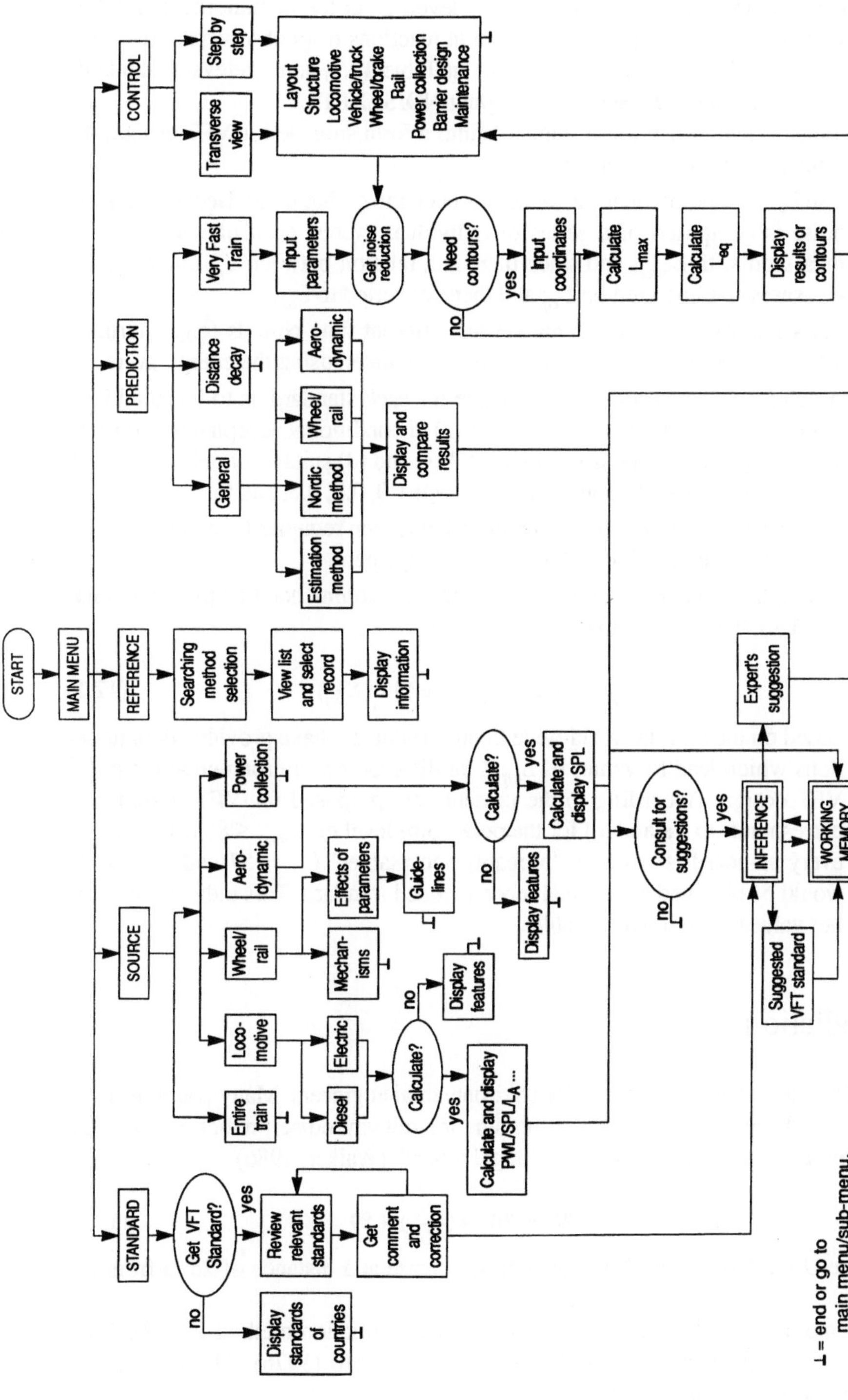

Figure 12.1. Modular structure of NOISEXPT.

The variation in proposed tolerant noise levels by different authorities in different countries is quite large. The variation in reactions from observers at the same noise level is also very large. As shown in Table 12.1, some relevant standards have been collected and used in developing NOISEXPT.

In order to synthesize these standards into a form suitable for VFT design, the following method is introduced:

1. Usc L_{eq} as a rating quantity since this been the global trend. Because railway noise is composed of a series of individual events, each of which can be measured with L_{max}, and since there is a relation between L_{eq} and L_{max}, all measures can be based on L_{eq} and then converted to L_{max}.
2. For standard values which are set for different time periods (e.g., morning, afternoon, or evening) a single value is calculated using the energy average.
3. A subjective assessment is then made for each standard as to its suitability. There are four options to be selected: (a) unconditional acceptance; (b) probable acceptance; (c) possible applicability; and (d) suitability unknown. Each option corresponds to a weight factor (w_i): 1.0, 0.8, 0.6, and 0.
4. A suggested correction for standard value is then requested. As shown in Table 12.2, these vary for each subjective judgement.
5. A weighted average of corrected values is recommended as the VFT noise standard. The equation is:

$$L_{eq,vft} = [\Sigma w_i (L_{eq,i} + \delta_i)] / (\Sigma w_i) \qquad (12.1)$$

6. Based on the current world-wide trend, the authors have provided default options which lead to values of $L_{eq} \leq 58$ dBA as the environmental target for VFT design. According to the present data proposed for VFT design, this corresponds to a standard for the peak noise level of: $L_{max} \leq 85$ dBA. In case every standard is rated as "probably acceptable," $L_{eq} = 57$ and $L_{max} = 84$ would be obtained as the final recommended standard. This indicates the earlier values are relatively stable.

Prediction

Prediction is an important task for environmental engineers. Many countries have developed their own computer programs for railway noise prediction, with the simplest method being called the "klogU model" (Walker, 1986):

$$OASPL = 30\log\left(\frac{U}{100}\right) + 89 \qquad (12.2)$$

where *OASPL* is the predicted overall train noise at a distance of 25 m from the track.

In more detailed models, a widely accepted prediction method for wheel/rail noise has been developed by Peters (1974) and Cato (1976). The A-weighted noise level at a distance *d* can be expressed as:

Table 12.1. Some Relevant Standards

Source		Standard (dBA)	
NSW, Australia, SPCC Environmental Noise Control Manual 1985		Planning levels	Max. levels
	$L_{eq,24h}$	55	60
	L_{max}	80	85
VIC, Australia Guidelines followed in recent environmental impact statements	Acceptable level ($L_{eq,24h}$)		65
	Desirable level ($L_{eq,24h}$)		55
Japan Shinkansen	Residential area	(L_{max})	70
	Other areas	(L_{max})	75
Netherlands Decree on railway noise 1987	L_{eq} (0700 - 1900h)		60
	L_{eq} (1900 - 2400h)		55
	L_{eq} (2400 - 0700h)		50
France TGV lines to Atlantic and Belgium	L_{eq} (0800 - 2000h)		65
Denmark	$L_{eq,24h}$		60
United Kingdom Greater London Council 1976	$L_{eq,24h}$		65
UK, Midland Joint Advisory Council 1983	$L_{eq,16h}$		55
	$L_{eq,night}$		50
UK Bolton MB	$L_{eq,24h}$		60
	$L_{max,ave}$		80
UK, Docklands Light Railway London 1982	L_{eq} (0700 - 1900)		60
	L_{eq} (1900 - 2300)		55
	L_{eq} (2300 - 0700)		50
Germany Guidelines (applicable to railway noise)	Area	$L_{eq,day}$	$L_{eq,night}$
	Hospitals, schools	60	50
	Residential	62	52
	Commercial/residen.	67	57
	Light industry	72	62

Table 12.2. Corrections for Standards

Subjective judgement of standards	Comments on the standard	Corresponding correction (dBA)
a	It is just acceptable	0
b	It is just acceptable	0
	It seems a bit high	-3
	It seems a bit low	+3
c	It is just acceptable	0
	It may be too high	-5
	It seems a bit high	-3
	It seems a bit low	+3
	It may be too low	+5

$$SPL_{wr} = L_G + 10\log\left(\frac{a}{(4N)}\right) - 20\log\left(\frac{l_c}{26.4}\right) + C_s\log\left(\frac{U}{120}\right) \tag{12.3}$$

with

$$L_G = 10\log\ \left\{\left(\frac{L}{d}\right)\right\} arctg\left(\frac{L}{2d}\right) + \left(\frac{L}{2d}\right)\sin\ \left\{2arctg\left(\frac{L}{2d}\right)\right\} + 86 \tag{12.4}$$

where

a = Mean number of axles per coach
N = Total number of individual vehicles for one train including power cars
l_c = Coach length, m
L = Total length of the train, m
U = Train speed, km/h
C_s = Speed coefficient (ten times the power of the train speed to which sound intensity is proportional). C_s is taken as 35 ± 5 for most Western European trains, and
d = Distance to the near rail, m.

For aerodynamic noise prediction, King (1979) developed a semi-empirical equation:

$$SPL_{ae} = 57\log\left(\frac{U}{160}\right) + 10\log\left\{\left(\frac{H}{d}\right)[arctgD + \sin(2arctgD)]\right\} + 80 \tag{12.5}$$

where

SPL_{ae} = aerodynamic sound pressure level, dBA
D = $L/2d$, and
H = height of the sound radiation area of the turbulent boundary layer, m.

Combining these two methods and taking the effect of noise control into account, a recommended prediction method for VFT design is obtained with correction for screens, ground absorption, and track conditions. The corrections are obtained through a Nordic method in which the following equations are used:

$$\delta L_{view-angle} = 10\log\left(\frac{a}{180}\right) \tag{12.6}$$

$$\delta L_{ground} = -12\log\left\{\frac{d}{1+\frac{d}{10}}\right\} + 3\log(h_{ave}) + 7.76 \tag{12.7}$$

$$\delta L_{screen} = -10\log d_s - 10\log\left\{\frac{e+\frac{1}{4}(d_s+1)}{1+\frac{e}{3}}\right\} - 7.54 \tag{12.8}$$

where

a = View angle of the train from receiver in degrees
d_s = Distance from screen to track centerline, m
d = Distance from receiver to track centerline, m
h_{ave} = Average height of the connection line between the receiver and the track (0.5m above the track surface) above ground.

To convert L_{eq} to L_{max} and vice versa, a simplified equation is deduced:

$$L_{eq,24h} = L_{max} + 10\log\left[\left(\frac{L_{24}}{U}\right)\cdot(1.2D+1)\right] - 43.8 \tag{12.9}$$

where

L_{24} = total length of trains passing in 24 hours, m.

NOISEXPT is designed so that users can invoke every prediction method in the knowledge base and make comparisons of different methods or different parameters within one prediction method. For example, the results from several methods using typical parameters are shown in Table 12.3. The train parameters such as train length, coach length, total train length passing by the place under consideration in 24 hours, and so on are chosen according to VFT design data. Graphical contours of noise levels may be displayed via POLDIF as shown in Figure 12.2.

After noise control treatments are made through consultation with the sub-system CONTROL, the effects of the reduction measures are fed back to get new

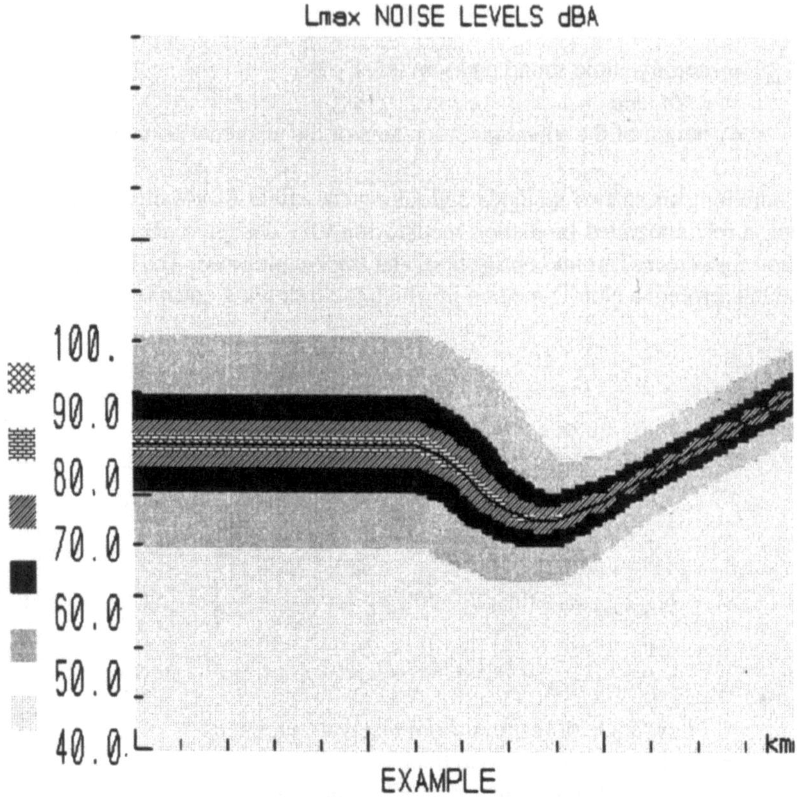

Figure 12.2. Typical noise contour plot generated by NOISEXPT using POLDIF contour plotting package. This shows a plan view of noise emanating from the train moving along the black line from left to right. The train slows from 350 kph to 240 kph as it enters the

prediction results which can show whether or not the VFT design standard can be met.

Conclusions

Noise control engineering, as part of environmental planning, is a suitable area in which to apply knowledge-based systems because of the complexity, uncertainty, and lack of experts in this field and to the need to synthesize globally-acceptable noise levels.

The CRYSTAL shell provides a suitable development environment which includes a blackboard to enable forward chaining inference engine and interfaces with other software packages. However, the shell has limitations, especially the lack of medium and high resolution graphics for this application.

Table 12.3. Typical Results from Several Methods [dBA, at 25m]

$U_{(km/h)}$	SPL_{wr}	SPL_{ae}	L_{maxa}	L_{maxb}	L_{maxc}	L_{eqa}	L_{eqb}	L_{eqc}	$OASPL$
40	71.4	38.4	78.7	71.4	56.4	61.1	54.3	39.3	77.1
80	81.9	55.6	87.9	81.9	67.2	68.1	61.8	47.1	86.1
120	88.1	65.6	93.3	88.1	73.8	72.3	66.2	51.9	91.4
160	92.4	72.7	97.1	92.5	78.7	75.2	69.4	55.6	95.1
200	95.8	78.3	100.0	95.9	82.7	77.5	71.8	58.6	98.0
240	98.6	82.8	102.5	98.7	86.2	79.3	73.8	61.3	100.4
280	100.9	86.6	104.5	101.1	89.3	80.9	75.5	63.7	102.4
320	103.0	89.9	106.3	103.2	92.0	82.3	77.0	65.9	104.2
360	104.8	92.8	107.8	105.0	94.6	83.5	78.4	67.9	105.7
400	106.4	95.4	109.2	106.7	96.9	84.6	79.6	69.7	107.1

where:

SPL_{wr} = Wheel/rail noise predicted by Peters' (1974) method
SPL_{ae} = Aerodynamic noise predicted by King and Bechert's (1979) method
L_{maxa} = Peak noise predicted by Ringheim's (1984) method
L_{maxb} = Peak noise predicted by $SPL_{wr} + SPL_{ae}$
L_{maxc} = Peak noise predicted by (SPL_{wr} - reduction) + SPL_{ae}
L_{eqa} = Equivalent noise predicted by Ringheim's (1984) method
L_{eqb} = Equivalent noise by wheel/rail + aerodynamic
L_{eqc} = Equivalent noise by (wheel/rail-reduction) + aerodynamic
$OASPL$ = Peak noise predicted by Walker's (1986) klogU method

There is a marked lack of measured noise data for VFT transportation systems that are currently operating in France, Germany, and Japan and hence much knowledge must be derived from conventional railway traffic. This situation is expected to be corrected in the future as more data becomes available.

NOISEXPT might also prove useful as an advisor for conventional railway noise control and could possibly be extended into a larger system for other noise sources such as road traffic.

Acknowledgments

The VFT design project is acknowledged for indirectly contributing to the funding support for this project. M. Anderson kindly offered the POLDIF system and modified it to meet the requirement of NOISEXPT. B. Barsikow of DFVLR (Germany), B. Ehling of DE-Consult (Germany), J. Kragh of DAI (Denmark) and M. Ringheim of KILDE (Norway) provided useful additional material as well as details on research in progress.

References

Cato, D.H., 1976. "Prediction of Environment Noise from Fast Electric Trains," *J Sound Vib. 46*:483-500.

Cohn, L., R. Harris, and W. Bowlby, 1986. "Using Expert Systems for Transportation Noise Design Decision Making," *Transportation Policy and Decision Making*, Martinum Nijhoff, Dordrecht, the Netherlands.

Ng, F.F., 1988. "Knowledge Base System for Architectural Acoustics," *3rd Westpac Acoustical Conference*, pp. 427-430, November 2-4, Shanghai, China.

Harris, R., L. Cohn, and W. Bowlby, 1985. "An Application of AI in Highway Noise Analysis," *Transportation Res. Record 1033*, TRB, National Academy of Science, Washington, D.C.

Kikuchi, I., 1988. "Shinkansen Noise: Research and Achievements in Countermeasures for Shinkansen Noise", *J. Sound Vib. 120(2)*:381-389.

King III, W.F. and D. Bechert, D., 1979. "Aerodynamic Noise Generated by High-Speed Trains," *Noise Control Engineering 13*:13-23.

King III, W.F. and B. Barsikow, B., 1987. "Aerodynamic Fluctuations as the Dominant Source of Radiated Noise Generated by the ICE at High Speeds," Fortschritte der Akustik - DAGA '87, pp. 213-216.

King III, W.F. and H.J. Lettmann, 1988. "On Locating and Identifying Sound Sources Generated by the German ICE Train at Speeds up to 300 km/h," *Inter-Noise 88 Proceedings*, pp. 1391-1396, Avignon, France.

Kurzweil, L.G., W.N. Cobb, and R.P. Kendig, 1979. "Propagation of Noise from Rail Lines," *J. Sound Vib. 66(3)*:389-405.

Peters, S., 1974. "The Prediction of Railway Noise Profiles," *J. Sound Vib. 32*: 87-99.

Ringheim, M., 1984. "Noise from Railway Traffic, A Nordic Prediction Method for Noise Emitted by Rail Traffic," Prepared for the Nordic Council of Ministers' Noise Group.

Ringheim, M., 1984. "Background Material for the Nordic Rail Traffic Noise Prediction Method," KILDE Report 130, December.

SEAI, 1986. "Expert Systems 1986," SEAI Technical Publications, PO Box 590, Madison, GA 30650.

Sharpe, R. and B.S. Marksjo, 1989. "Expert Systems for Engineering Codes," *Proc. IABSE Conference on Expert Systems*, Bergamo, Italy.

Taylor, M.A.P. and M. Anderson, M., 1988. "Modeling the Environmental Impacts of Urban Road Traffic with MULATM-POLDIF: a PC-Based System with Interactive Graphics," *Environment and Planning B: Planning and Design 15*:413-431.

Walker, J.G., 1986. "The Quiet Route to Higher Speeds," *Railway Gazette International*, pp. 640-642, September.

Winston, P.H., 1987. *Application of Expert Systems* in J.R. Quinlan, ed., Proceedings of the 2nd Australian Conference.

Yanagida, M., T. Hidaka, and T. Yamaguchi, 1988. "A System Concept for Knowledge Based Architectural Acoustics," the 2nd Joint Meeting of ASA & ASJ, November 14-18, Hawaii.

PART FOUR

Knowledge acquisition, evaluation, and validation

Introductory remarks by Tschangho John Kim

Knowledge acquisition, evaluation, and validation are often regarded as major bottlenecks in the development of expert systems. The knowledge acquisition process is a difficult and time-consuming task with current technology. At the same time, any expert system can provide advice only as good as the knowledge it has in its knowledge base. Moreover, expert systems have two intrinsic problems: who determines the level of acceptance and how it is determined. These problems result because expert systems are designed to provide acceptable solutions most of the time, unlike algorithmic programming systems which are designed to derive the optimal values all the time. In this book, we define evaluation as that which measures the system's accuracy and usefulness, while validation determines whether the correct problem was solved.

Since environmental planning and engineering is a knowledge domain that is not only large but difficult to confine to boundaries, it becomes important to devise an effective way of acquiring knowledge in these fields. Some of the methods which have been suggested to facilitate knowledge acquisition include the use of machine learning algorithms. In fact, machine learning has been applied to many areas, such as physics and mathematics.

In Chapter 13, Choi, Wisseman, Fittipaldi, and Kim provide an example of how a machine learning program, ID3, can be used in environmental fields to facilitate the tasks of knowledge acquisition and to reduce the number of rules in the knowledge base without losing the quality of the results. They begin with a detailed description of an Expert System for Archaeological Research and Analysis (ESARA) and explain how both conventional knowledge acquisition methods and induction methods are used to extract knowledge from the domain experts.

In Chapter 14, Tracz and Wawrzynkiewicz describe how they obtained knowledge from thirty experts in evaluating transportation options available for a community in Poland. The way they obtained experts' knowledge is conventional; what is innovative, however, is their estimation of the weights of the experts' opinions. Experts were asked to answer questions and complete pairwise matrices in order to apply combinatorial rules.

In Chapter 15, Perman and Ortolano test an expert system called SLUDGECADET in order to evaluate the system's ability to represent the participating experts' diagnostic techniques. They also describe appraisal of the system's performance using the standards of an expert who was not involved with system development.

In Chapter 16, Wiggins, Kim, and Jain examine how expert systems are actually being used in environmental planning by interviewing field users. A system called Cost of Remedial Action (CORA) has been chosen since it had been used by several hundred users evaluating over 1,000 sites. The use of CORA has been evaluated in regard to: (1) adoption characteristics, (2) manner of use of the system, (3) pattern of use of the system, (4) overall acceptance, and (5) productivity increases due to system use.

13

Knowledge Acquisition and Representation in Building an Expert System for Archaeological Research and Analysis: ESARA

Keechoo Choi, Sarah Wisseman, John Fittipaldi, and T. John Kim

Introduction

ESARA (Expert System for Archaeological Research and Analysis) has been developed to help archaeologists and urban field planners faced with the problems of artifact analysis both before and after excavation in land development and planning processes. The planning profession has, in general, failed to reduce the destruction of artifacts which archaeologists study, interpret, and archive even though both the intrinsic and economic benefits of preservation are widely recognized. Profit-driven motives for land-use allocation decisions for historic preservation planning have led to frequent neglect of archaeological treasures due to complicated jurisdictional issues[1].

Due to the wide scope of its concerns and activities, urban planning must take into account specific land development considerations that arise when an archaeological site is discovered at locations intended for public purpose. Implementation of ESARA as a decision support tool was motivated not only to facilitate a systematic planning process but to provide basic archaeological knowledge.

---

1. In the past two decades, urban renewal projects involving the Department of Housing and Urban Development (HUD), Department of Commerce (DOC), and Federal Highway Administration's (FHWA) highway projects were the most detrimental for archaeological sites.

This chapter focuses on the knowledge acquisition and representation of ESARA and its connection features with a hypertext system which supports the expert system. Instead of the conventional interview-style knowledge acquisition method, the induction method was used to obtain a set of rules for each sub-domain. Then, the induced rules were modified and rewritten in the form of a Personal Consultant Plus expert system shell which was used for the development of ESARA. For detailed and further discussions on applications of expert systems in urban planning, and descriptions of available expert system development shells, see Kim, Wiggins, and Wright (1990).

ESARA overview

Goal

ESARA addresses issues in building and applying a prototype expert system for archaeological research. The system is designed in such a way that it covers various instrumental techniques according to the type of archaeological problems addressed in the system (Wisseman, et. al, 1989).

Contemporary analysis techniques, when coupled with an expert system which can incorporate the judgement and experience of experts, should help the researchers in archaeological planning. Therefore, the goal of ESARA includes suggesting the most appropriate analysis techniques for each archaeological material (especially ceramics, metal, and stone) through consultation with the system, as well as assisting archaeologists in planning excavations.

Functional structure

ESARA is organized into two parts. First is the problem-oriented part which contains the knowledge base of four sub-domains and suggests the most appropriate techniques through consultation. The second is a hypertext information system which enables the user to obtain more in-depth information on the technique selected from previous consultation.

ESARA deals with four archaeological sub-domains.

1. *Dating* of artifacts is a crucial step toward understanding cultural and behavioral changes through time. Dating helps a client choose a technique for determining the age of artifacts.
2. *Composition* of artifacts is defined as physical and chemical characterization in terms of microstructure, mineral compounds, or individual elements. This sub-domain helps a client select a technique for determining the composition of artifacts.
3. *Provenience* is defined as the source of raw material or the place where an artifact was made, and helps a client choose a technique for determining the place of manufacture of ceramic artifacts or the source of stone.

4. *Technology* is a sub-domain which helps a client choose a technique for determining the technology used to manufacture or create artifacts.

The system is arranged so as to provide easy movement through its sub-domains, and has the option of moving back only one step or initiating a new start. Figure 13.1 displays the schematic structure of the ESARA and flow of the consultation. The same structure is true for all materials, i.e., ceramic, stone, and metal.

A typical consultation session proceeds as follows:

Step 1 Choose either CONSULT for consultation or INFO session for detailed descriptions of techniques.

Step 2 If INFO is chosen, go to step 7, otherwise go to step 3.

Step 3 Choose a particular sub-domain (for example, dating).

Step 4 Consultation ending in a specific technique.

Step 5 Consult again with changes, if necessary, or go back to step 2 or INFO hypertext system[2].

Step 6 Activate INFO.

Step 7 Consult INFO.

Step 8 Review information on the selected technique in step 4.

Step 9 Go back to CONSULT, review the technique with different questions, or EXIT to change the sub-domain.

Knowledge acquisition

Two types of knowledge acquisition methods were used to construct the ESARA knowledge base. One was the interview-style method and the other was induction. The dating sub-domain and the ceramic parts of composition, provenience, and technology knowledge bases were developed using the first method, while the latter method was used for the development of the knowledge base for other materials.

Conventional interview method

Waterman (1986) examined seven techniques for extracting knowledge from a domain expert. Among these, the *problem description* method was most frequently used for the early stage development of ESARA. This method requires the expert to describe a prototypical problem for each category of answer in the domain. Three other methods (system refinement, examination, and validation) were partly used to prepare the feedback input in testing ESARA.

---

2. Hypertext information system was constructed and is included in ESARA to explain archaeometric technique(s) in detail, once suggested as a conclusion by ESARA.

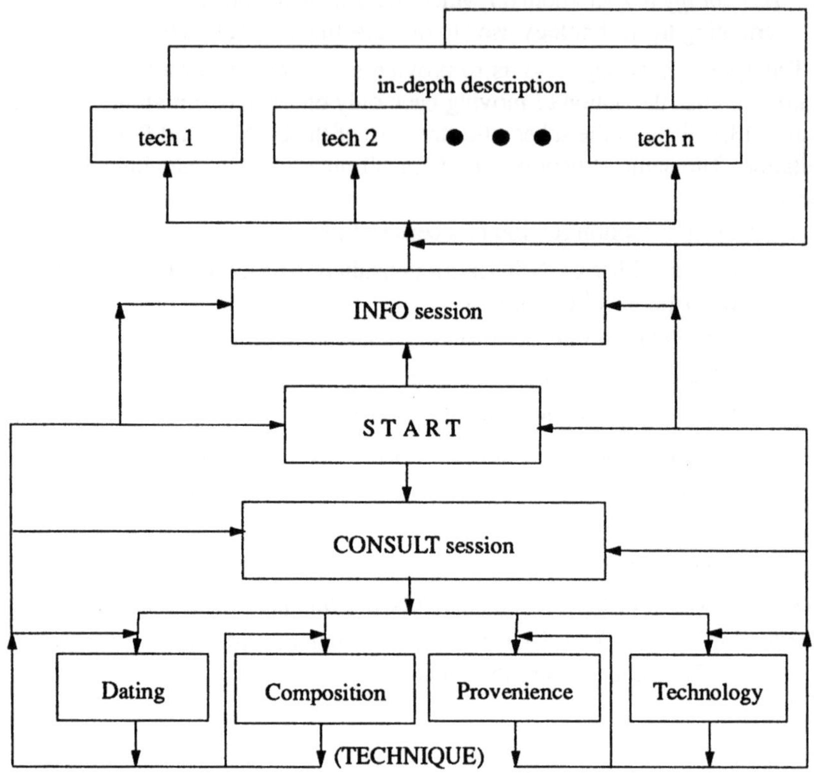

Figure 13.1. Schematic structure of ESARA.

During the interview, basic factors which play an important role in selecting an appropriate technique for each sub-domain were chosen. The size of these basic factors (attributes or parameters) depends on individual sub-domain. The process of obtaining knowledge from experts is described as follows:

Step 1 Obtain the initial sub-domain knowledge.

Step 2 Prototype knowledge data and put it in the knowledge base according to the ARL (abbreviated rule language) format of the Personal Consultant Plus.

Step 3 Try a sample session with the prototype knowledge base.

Step 4 Let the experts comment on the prototype knowledge base.

Step 5 If the knowledge base is reasonably complete and covers the diverse consultation combinations, then exit.

Step 6 Let the experts comment on what is missing or incorrect in the current knowledge base.

Step 7 Acquire the missing knowledge and add it to the current knowledge base.

Step 8 Go to step 3.

Rule induction

The induction method has been studied in the machine learning[3] area as a supplement of the cumbersome interview technique.

A knowledge acquisition bottleneck can be overcome to some degree since the induction method can produce more compact and direct rules compared with verbose and chunk rules, and it may expedite the progress of the system development. While knowledge acquisition in a completely automated manner seems to be an ambitious goal, the potentials of the induction method is a valuable tool to refine a partial and inconsistent knowledge base derived from a human expert. To date, few induction methods have been successfully applied to the knowledge acquisition problems (Michalski, 1985; Shaw, 1988; Liang, 1989).

Tool used to induce rules

In developing ESARA, the analog concept learning system (ACLS) was used to induce rules from pre-classified examples having *cases* as rows and *attributes* as columns.

ACLS is a system-building aid that induces rules from examples of decisions provided by a domain expert based on Quinlan's ID3[4] algorithm which is rooted in information theory on entropy (Jackson, 1988). In ACLS run, the user specifies the problem in terms of the set of attributes relevant to the decision and possible decision classes. The user then gives ACLS examples of situation-decision pairs, from which it creates general rules that classify the examples in terms of their attribute. It outputs a rule either as a decision tree or as a PASCAL code. Developed at the University of Edinburgh, ACLS was implemented upon UCSD-PASCAL first and became commercially available through Intelligent Terminal Limited (see Waterman, 1986).

According to Shapiro (1987), historical development of ACLS is as follows:

- Hunt, et al.'s CLS (concept learning system) was the first to generate rules automatically from an example set, and it is implemented based on the entropy concept in information theory (also see Jackson).

---

3. According to Forsyth and Rada (1986), the learning algorithm attempts to achieve four goals: provide more *accurate* solutions, cover a *wider* range of problems, obtain answers more *economically*, and *simplify* codified knowledge. Even though the first two criteria usually have the highest priority, the last goal is valuable enough that a system can rearrange its knowledge base to be more intelligible to human readers. The last goal in a sense supports the action of using ACLS in the development of ESARA.
4. Interactive Dichotomizer 3.

- Michalski (1980) proved that an induced expert system outperformed an expert system generated by conventional means of rule acquisition from plant pathology[5].
- Quinlan (1979) redesigned CLS on a more rational and efficient basis using a refutation filter and then adapted it to huge example files.
- Quinlan (1983) produced machine-executable descriptions with a run-time efficiency five times greater than the best human-produced description.

Procedures and benefits in using ACLS

A generalized procedure for developing a knowledge base using ACLS can be described as follows:

1. Identify crucial attributes based on previous interview sessions.
2. Construct the knowledge matrix (pre-classified example set).
3. Break down the *decision vector*[6] which has multiple outcomes with the same attributes. (This was done due to the unique conclusion production feature of ACLS, but does not mean that Personal Consultant Plus can't have multiple actions in the conclusion.)
4. Codify the matrix input (see Appendix C).
5. Prepare attribute file (see Appendix B).
6. Run ACLS (see Appendix D).
7. Obtain either a decision tree or it's equivalent PASCAL code version. Interpret each conclusion and implant it into the knowledge base using the Personal Consultant Plus format (see the section "Knowledge representation").

As can be seen in the Table 13.1, the number of rules and the size of the knowledge base were reduced. This reduction was of great help in developing the system for use on IBM XT compatible personal computers, where only 640 K-bytes are available for the total application area[7].

A significant reduction in rule size without losing generality is possible, since a small group of rules can be integrated into a single rule by using the relational operators in Personal Consultant Plus' ARL language as shown in the following example.

```
<<before induction>>
   IF:: ART=2 AND MUT=1 AND DES=2 AND BUD=3 AND NUM=1 AND TIM=4
THEN:: LASS=3
   IF:: ART=2 AND MUT=2 AND DES=3 AND BUD=4 AND NUM=2 AND TIM=3
THEN:: CLASS=3
```

---

5. AQ family type algorithm was used.
6. Sometimes called the *feature vector*.
7. Realistically, however, 640 K-bytes are not usually available since DOS occupies about 80 K-bytes, and PC Scheme (Personal Consultant Plus *Lisp* language) needs 380 K-bytes.

```
<<after induction>>
  IF::  ART=2 AND MUT<3 AND DES>=2 AND BUD>=3 AND NUM<=2 AND TIM>=3
THEN::  CLASS=3.
```

Note that the two rules before induction can be written as a single rule.

Knowledge representation

Personal Consultant Plus adopts the *rule and frame-based* feature to represent its domain knowledge.

The typical rule used in ESARA development is as follows:

```
  IF [condition 1] AND/OR [condition 2] AND/OR....
THEN [action or inference 1] AND/OR [action or inference 2]....
```

Though both Lisp and ARL are available in the Personal Consultant Plus environment, ARL seems to be a more compact form of English-like rules. ARL allows the knowledge engineer to combine names and values of parameters with words and symbols called functions that Personal Consultant Plus can understand. For example, ‘+’, ‘–’, ‘=’, ‘!=’ are easy to understand functions in knowledge bases.

Table 13.1. Comparison Between the Original and ACLS Induced System (Original rules contains a set of rules derived from experts)

Technology sub-domain		
Dimension	Original rules	ACLS induced rules
Number of rules	384	212
Size (kilobytes)	169.7	87.2

The term *abbreviated* indicates the following simplification in coding the knowledge base. If the English-like rules are written as below:

```
  IF    the temperature is higher than 80,
THEN    the weather is hot.
  IF    the temperature is 65 or higher and less than 80,
THEN    the weather is warm.
```

For Personal Consultant Plus to understand these rules, these should be written in ARL as follows:

```
  IF statement:        TEMPERATURE > 80
THEN statement:        WEATHER = hot
  IF statement:        TEMPERATURE BT 65 81
THEN statement:        WEATHER = warm
```

In the above two rules, `TEMPERATURE` and `WEATHER` are parameters in the knowledge base, and `hot` and `warm` are possible values of `WEATHER`. Similarly, the numbers `80`, `65`, and `81` are possible values of `TEMPERATURE`, and `BT` and > are ARL system functions.

System evaluation

A *summary* evaluation of ESARA, to determine whether the system accomplished its objectives, was performed by an expert team composed of a knowledge engineer and a domain expert (as usual in evaluating prototype systems). Testing was an important stage in developing ESARA: uncovering problems and redirecting the construction of the system, reformulating concepts, and revising the flow.

During the test-trial, rules were revised and added for accuracy and completeness. At this stage, many default answers were added to meet varieties of combinations which could happen during the consultation. System refinement and examination were completed to a certain degree, but system validation by outside experts still remains to be completed.

A *formative* evaluation of ESARA, assessing the quality of the system and related support, should be undertaken with experimental users. The evaluation criteria should involve the following:

- Structuredness
- Self-descriptiveness
- Knowledge of archaeology
- Familiarity with archaeometric techniques
- Help screen
- Consistency
- Conciseness
- Communicativeness
- Portability
- Modifiability (augmentability)

As of November, 1990, ESARA is being tested and evaluated by resident archaeologists at the U.S. Army Construction Engineering Research Laboratory (USA-CERL).

Conclusions

Archaeology, in a planning context, is a prospective application for expert systems. ESARA is only one example of how the expert systems approach can be used in a multi-objective decision making process for urban and regional planning. Even though ESARA may be used to help select archaeometric techniques for artifact analysis, it doesn't encompass the interpretation of materials analyses.

Development of ESARA began with the identification stage and continued through all the expert system building phases to the testing stage. Some common problems in implementing the expert system approach (Baker, 1988) were encountered during ESARA's development:

- *Acceptability*: ESARA is not meant to replace professionals, but rather to help them solve problems that would otherwise require the advice of an acknowledged expert.
- *Uncertainty*: By supplying full ranges of user defined answers and providing multiple options. Although Personal Consultant Plus provides the capacity of handling uncertainty, it has not been used for ESARA.
- *Conflict among experts*: This is sometimes a problem with human experts who may disagree among themselves. In ESARA development, conflict among experts was observed in the interview stage.
- *Maturity of the technology*: ESARA development and implementation requires large memory capacities in personal computers. The system should be divided into a smaller sub-domain system instead of a full domain system when it is to be run on IBM XT level computers that must fit into the 640 K-byte available on DOS operating systems.

Future research opportunities apparently lie in the methods of knowledge acquisition. Even though the induction method was useful to expedite the knowledge acquisition process, rules derived from induction are not flexible. Efficiency of the derived rules will decrease if a new parameter is added, or if even a small portion of knowledge is changed. Therefore, alternative methods which enhance efficiency and flexibility deserve attention in future knowledge-base construction.

References

Baker, K.G., 1988. "Towards an archaeological methodology for expert systems," in C.L.N. Ruggles and S.P.Q. Rahtz, eds., *Computer and Quantitative Methods in Archaeology 1987*, 229-236, Bar Int'l. Series 393.

Christopher, F.C., 1989. *Artificial Intelligence and Turbo C*, Dow Jones-Irwin Publishing Co., Hornewood, IL.

Forsyth, R. and R. Rada, 1986. *Machine Learning Applications in Expert Systems and Information Retrieval*, Ellis Horwood Limited, West Sussex, England.

Hunt, E.B., J. Marin, and P. Stone, 1966. *Experiments in Induction*, Academic Press, New York, NY.

Jackson, A.H., 1988. "Machine learning," *Expert System* 5(2):132-150.

Kim, T.J., L.L. Wiggins, and J.R. Wright, 1990. *Expert Systems: Applications to Urban Planning*, Springer-Verlag, New York, NY.

Liang, T.-P., 1989. *A Composite Approach to Inducing Knowledge for Expert Systems Design*, faculty working paper, Bureau of Economics and Business Research, University of Illinois at Urbana-Champaign, No. 89-1534.

Michalski, R.S., J.H. Davis, V.S. Bisht, and J.B. Sinclair, 1985. "Plant/ds: An expert consulting system for the diagnosis of soybean diseases," in L. Steels and J.A. Campbell, eds., *Progress in Artificial Intelligence*, Ellis Horwood Limited, New York, NY.

Michalsky, R.S. and R.L. Chilausky, 1980. "Learning by being told and learning from examples: An experimental comparison of the two methods of knowledge acquisition in the context of developing an expert system for soybean disease diagnosis," *International Journal of Policy Analysis and Information Systems* 4(2).

Paterson, A. and T. Niblett, 1982. *ACLS User's Manual,* Technical Report, Oxford, UK.

Quinlan, J.R., 1979. "Discovering rules by induction from large collection of examples," in D. Michie, ed., *Expert System in Microelectronic Age*, 168-201, Edinburgh University Press, Edinburgh, UK.

Quinlan, J.R., 1983, *Learning Efficient Classification Procedures*, Chap. 15, 463-482, Vol. 1, Kluwer Academic Publishers, Inc., Palo Alto, CA.

Shapiro, A.D., 1987. *Structured Induction in Expert Systems*, Addison-Wesley Publishing, Wokingham, England.

Shaw, M.J. and J.A. Gentry, 1988. "Using an expert system with inductive learning to evaluate business loans," *Financial Management*, 45-46, Autumn.

Waterman, D., 1986. *A Guide to Expert Systems*, 339-341, Addison-Wesley Publishing, Massachusetts.

Wisseman, S., J. Fittipaldi, T. Good, Z. Budic, and T.J. Kim, 1989. "Expert systems in archaeology: Choosing an archaeometric technique for ceramic analysis," paper presented at the *Annual Meetings of the Archaeological Institute of America,* Boston, MA, December 28.

Appendix A

Sample decision tree generated by ACLS for ESARA

```
manu-ut
      <2 : destruc
           <2 : budget
                <4 : number
                     <3 : budget
                          <2 : time
                               <3 : number
                                    <2 : 23
                                   >=2 : time
                                         <2 : 23
                                        >=2 : 10
                              >=3 : 23
                         >=2 : budget
                               <3 : time
                                    <4 : number
                                         <2 : 21
                                        >=2 : time
                                                    <2 : 21
                                             >=2 : time
                                                     <3:6
                                                   >=3 : 21
                                   >=4 : number
                                         <2 : 21
                                        >=2 : 23
                              >=3 : time
                                    <2 : 23
                                   >=2 : time
                                         <3 : 10
                                        >=3 : 23
                    >=3 : budget
                          <2 : time
                               <2 : 23
                              >=2 : time
                                    <4 : 10
                                   >=4 : 25
                          .
                          .
                          .
                          .
```

manu-ut, destruc, budget, ... are ACLS attributes for ESARA. Figures after colon (:23, :10, :21 ...) are outcomes.

Appendix B

Sample attribute file of ACLS for ESARA

```
# of parameters:     7
name of parameter  type
   artifact       integer
    manu-ut       integer
     zero         integer
    destruc       integer
     number       integer
     budget       integer
      time        integer
# of outcomes:      22
outcomes: 1 2 3 4 5 6 7 8 9 10 11 12 13 14 15 16 21 22 23 24
25 26
```

Appendix C

Sample matrix input of ACLS for ESARA

no.	artifact	manu-ut	zero	destruc	number	budget	time	class
1.	2	1	0	1	1	1	2	23
2.	2	2	0	1	3	4	3	15
3.	2	1	0	1	1	2	1	21
4.	2	1	0	1	2	4	3	23
	⋮					⋮		
91.	2	1	0	2	2	4	2	13
92.	2	1	0	2	1	2	4	21
93.	2	1	0	2	2	4	3	13
94.	2	1	0	1	1	4	1	23
95.	2	1	0	1	1	4	4	23
96.	2	1	0	1	2	4	1	23
97.	2	1	0	1	2	4	4	23
	⋮					⋮		
171.	2	2	0	1	3	4	2	15
172.	2	1	0	1	1	1	3	23
173.	2	1	0	1	1	1	4	2

Appendix D

Typical ACLS run session. User inputs are bold faced.

c:\pcplus> **acls**

attribute description file ? **stone.att**

command ? read

read from what file ? **stone.dat**

command ? **induce**

start with ? **1**

increment by ? **1**

```
Start with        1 and increment by up to      1 exceptions
-----------------------------------------------------------
Iteration  |Number in  | Wrong      |Unsure      |Rulesize
   No.     |Window     |            | Clash      |
-----------------------------------------------------------
     1           1         136           0            1
     2           2         121           0            3
     3           3         109           0            5
           .                                    .
           .                                    .
           .                                    .
    92          92           1           0          119
    93          93           0           0          119
           .                                    .
           .                                    .
           .                                    .
```

command ? **rule**

The output of this command is as shown in Appendix A.

command ?

14

Knowledge Acquisition from Multiple Experts: A Case of Transport Planning in Poland

Marian Tracz and Bronislaw Wawrzynkiewicz

Introduction

Knowledge acquisition is often regarded as a major bottleneck in expert systems development because the expert system can provide advice only as good as the knowledge it has in its knowledge base. The knowledge acquisition process is difficult and time consuming using technology that is currently available. Research on effective knowledge acquisition from experts, and/or knowledge aggregation among experts, may be more urgent in public sector planning, including environmental planning. The domain of problems in environmental planning is relatively large, difficult to confine to boundaries, and generates frequent conflicts among experts.

The purpose of this chapter is to present a method of acquiring knowledge to use in building an expert system. A common problem in environmental planning is to evaluate and choose between alternatives. This process is often done by: (1) selecting criteria on which to evaluate alternatives, (2) weighting criteria according to their importance, (3) rating each alternative according to the criteria, and (4) using some combination rule (e.g., additive or multiplicative) to combine the ratings and weighted criteria into composite scores.

This exercise is relatively straightforward if only one expert's opinion is to be used. However, in many environmental planning problems it is often desirable to consider the opinion of several experts as well as members of the public. The alternative rankings developed from such a procedure are often used as input to expert systems (Suh, Kim, and Kim, 1988).

At the same time, traditional technical and economic criteria alone are no longer adequate for a decision maker to select the best alternative, and non-technical and non-economic criteria must also be taken into consideration. Such criteria, however, are very often subjective and experts frequently do not agree on their relative weights of importance. In addition, these variables are usually difficult to quantify.

We apply Saaty's (1977) analytic hierarchical process (AHP) to derive ranking for criteria based on multiple expert responses. Specifically, we attempt to evaluate the effectiveness of alternative transport choices by acquiring knowledge from multiple experts and aggregating their opinions on relevant criteria in terms of their relative importance. The method was applied to a case study in the city of Zakopane, Poland. Aggregated results based on experts' opinions are compared to the results based on opinions from members of the public.

Knowledge acquisition in expert systems: A brief review

One of the most challenging ways of facilitating knowledge acquisition is to make computers learn by themselves. While knowledge acquisition in a completely automated manner seems to be an ambitious goal, there are several types of learning strategies studied in artificial intelligence (Waterman, 1986) and applied to expert systems building (Han, Kim, and Adiguzel 1992). Nevertheless, the most frequently used method to construct knowledge bases in expert systems is the traditional interview method and its derivatives.

The traditional interview-style method requires asking the expert to describe a prototypical problem for each category of answer in the domain. During an interview, basic factors which play an important role in selecting an appropriate technique for each sub-domain are usually chosen. The number of these basic factors (attributes or parameters) depends on individual sub-domains. The process of obtaining knowledge from experts for building an expert system can be described as follows:

1. Obtain the initial sub-domain knowledge.
2. Encode the knowledge data into a knowledge base using a language such as the abbreviated rule language (ARL) of an expert system shell such as Personal Consultant Plus.
3. Try a sample session with the prototype knowledge base.
4. Let the experts comment on the prototype knowledge base.
5. If the knowledge base is reasonably complete and covers the diverse consultation combinations, then exit.
6. Otherwise, let the experts comment on what is incorrect or missing from the current knowledge base.
7. Acquire the missing knowledge and add it to the knowledge base.
8. Go to step 3.

While the interview technique is most often used to acquire experts' knowledge, the method suffers from the following constraints:

1. It is difficult and costly to gather experts in the same place at the same time, or to visit each individual expert.
2. No clear means exist to overcome conflicting expertise, and conflicts occur frequently among experts in the public planning domain.
3. It is very difficult to aggregate experts' opinions, particularly when they differ, and yet it is very important for a decision maker to have aggregated opinions from experts.

One way to overcome this cumbersome and difficult procedure, yet facilitate knowledge acquisition, is to develop a knowledge acquisition aid that effectively captures the knowledge of an expert through interactive interviews, distills the knowledge, then automatically generates a knowledge base. Some available software (e.g., Auto-Intelligence) supposedly helps experts capture their own knowledge by: (1) identifying the structure of the knowledge, (2) discovering the relative importance of decision making criteria, and (3) classifying information and inducing knowledge (Intelligence Ware, Inc., 1987). This kind of tool can be useful in building expert systems where human experts who supply knowledge to the expert system perceive complex relationships or come to a conclusion without clearly knowing how they did it and/or cannot articulate the methods they used. To effectively reveal the knowledge structure of human experts, Olson (1987) suggests using several cognitive methods, including "indirect methods" that do not rely on the expert's ability to articulate the information that is used. Without employing knowledge acquisition aids such as Auto-Intelligence, in this chapter we attempt to acquire knowledge from multiple experts through interviews using AHP (Saaty, 1988).

Hierarchical structure of the transportation planning process

The hierarchical structure for evaluating alternative transport choices is assumed to be represented by Saaty's (1977) model of the evaluation process. The hierarchical process consisting of five levels is shown in Figure 14.1. The first level represents the overall welfare of an urban environment, while the second level represents the experts who evaluate the transport alternatives. The third level comprises groups of objective criteria (denoted as the kth level in this chapter), the fourth lists the precise measurement of evaluation criteria in each group (j^{th} level), and the final level contains transport alternatives (i^{th} level).

The weight of an element in each level depends on the weight in each of the subsets to which it belongs on the preceding levels. For example, the final weight of a particular criterion on the fourth level is derived from its relative weight within its own group, the weight of the group at the third level, and the level of competence of the experts making the estimation at the second level.

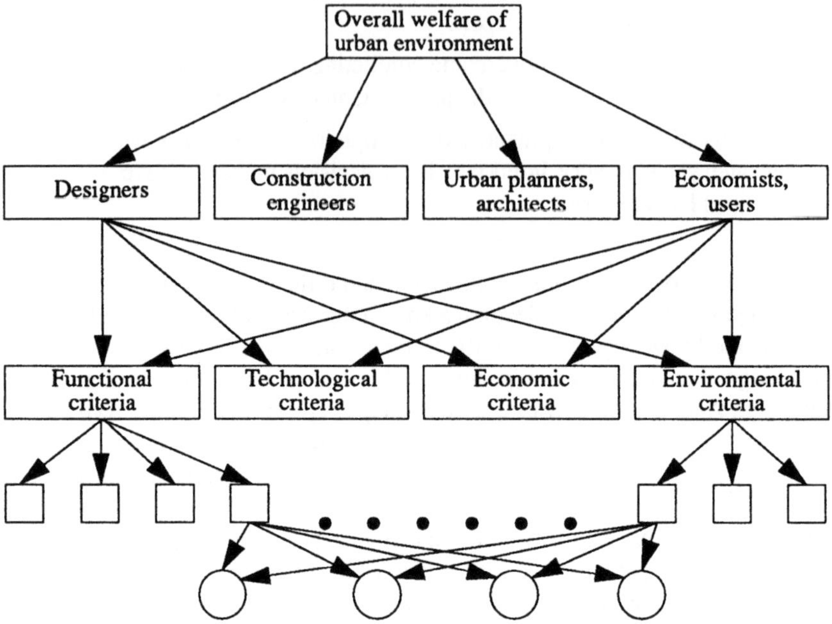

Figure 14.1. Hierarchical structure of the evaluation process.

Combinatorial rules

To obtain an overall ranking of transport alternatives, three combinatorial rules based on a hierachical structure were used. The first is the standard additive rule expressed as follows:

$$\mu_i = \sum_{k=1}^{k} \mu_k \left(\sum_{j=1}^{k_j} \mu_{k_j} \mu_{ik_j} \right) \tag{14.1}$$

where

μ_i = represents the overall rating for alternative i,

μ_k = represents the weight on the kth group of criteria (in this chapter, the third level),

μ_{k_j} = represents the weight on the jth criterion in the kth group (the fourth level), and

μ_{ik_j} = represents the rating of the ith alternative on the jth criterion in the kth-group.

This rule means that the criteria are combined in a linear function.

The second possible rule is a multiplicative rule expressed as follows:

$$\mu_i = \prod_{k=1}^{k} \mu_k \left(\prod_{j=1}^{k_j} \mu_{k_j} \mu_{ik_j} \right) \tag{14.2}$$

where the notation is the same as in equation (14.1).

This rule means that when a rating for an alternative on any criteria is equal to zero, the overall score of that alternative is equal to zero. In other words, we assume that each criteria has some minimum threshold and if this is not met by an alternative, that alternative is scored as zero overall.

The third rule is a proposed new supplementary rule. The underlying hypothesis is that the weights of criteria importance, as well as the ratings, are stochastic variables described by the normal distribution N(m,s). Let's introduce the following notation for this stochastic variable:

$$P(|x - m| > s) \approx 0.3174. \tag{14.3}$$

Equation 14.3 states the probability of a weight or rating falling outside one standard deviation. Then we assert that one criterion is much more important than others when

$$\mu_{k_j} > \bar{\mu}_{k_j} + s_{k_j} \tag{14.4}$$

and an alternative's degree of single criterion rating is lower than the rating of other criteria if

$$\mu_{ik_j} < \bar{\mu}_i - s_i \tag{14.5}$$

where

μ_{k_j} = weight of criterion k_j,

$\bar{\mu}_{k_j}$ = average of criteria weights,

s_{k_j} = standard deviation of criteria weights,

μ_{ik_j} = rating of alternative i on criteria k_j,

$\bar{\mu}_i$ = average rating of alternative i, and

s_i = standard deviation of ratings of alternative i.

As a result of this supplementary rule, a criterion k_j may be identified for the alternative i such that it is more important than any other criteria but is fulfilled to a lower degree by alternative i.

Evaluation of weights

In many circumstances, when criteria cannot be objectively measured nor alternatives objectively evaluated, experts are required to serve society as 'measuring instruments'. In this study, an interdisciplinary group of experts was asked to compare each pair of criteria by answering the question: "How many times one criterion is more important than another?" and to compare each pair of alternatives by answering the question: "How many times one alternative is better than another?" with respect to each criterion, using a scale ranging from 1 to 9. Matrices of pairwise comparisons were obtained from these judgements.

The relative importance of the criteria and the ratings of alternatives are obtained by calculating eigenvectors of the matrices. Judgements of experts are often not transitive because of the great complexity of the problem and the fuzziness of experts' perceptions. In this case, the problem of matrix inconsistency arises. Inconsistency of matrix data is measured according to Saaty (1978) by the expression:

$$i = \frac{\lambda_{max}}{n} \tag{14.6}$$

where λ_{max} is the maximum eigenvalue of the matrix and n is the row of the matrix.

It appeared from the experimental tests that the error levels of 80 percent of the experts did not exceed 0.1. In the case that the measure exceeds 0.1, a supertransitive approximation procedure to the binary comparison matrix can be used according to Narasimhan (1982) as follows:

1. Let $A = \{a_{ij}\}$ be the initial matrix of pairwise comparisons.
2. Construct supplementary matrices $A^1, ...A^k, ...A^n$, such that the kth row of matrix A^k is the same as the kth row of initial matrix A, and the rest of the rows of matrix A^k are computed as follows:

$$a^k_{ij} = a^k_{kj} \cdot (a^k_{ki})^{-1} \tag{14.7}$$

3. Construct the supertransitive approximation $A^S = \{a^S_{ij}\}$ by taking the geometric average of corresponding elements from the supplementary matrices A^k such that:

$$a^s_{ij} = (a^1_{ij} \cdot ...a^k_{ij} \cdot ...a^n_{ij})^{1/n} \tag{14.8}$$

If the degree of inconsistency of the supertransitive approximation matrix still exceeds the value of 0.1, the expert's opinion is omitted.

Evaluation of expert opinions

The main problems in group decision making are how to select the best experts, how to evaluate the worth of a single expert's opinion, and how to obtain a composite group opinion from individual orderings. An approach to the first problem is to have all experts list participants in an evaluation process, and those experts who are most frequently listed are chosen.

An expert's ability to evaluate criteria and alternatives with respect to a single criterion is checked by calculating an indicator of fuzziness defined with respect to relative Euclidean distance (Kaufman, 1975):

$$W_A = \frac{2}{n}\sqrt{\sum_{i=1}^{n}\left[Min(\mu_A(x_i), \mu_{\bar{A}}(x_i))\right]^2} \tag{14.9}$$

where:

W_A = indicator of fuzziness defined with respect to relative Euclidean distance,

$\mu_A(x_i)$ = membership function for the element x_i with respect to fuzzy subset A (weight of importance of criterion k_j, or rating the alternative i), and

$\mu_{\bar{A}}(x_i)$ = membership function for the element x_i with respect to the complement of fuzzy subset A.

This ability may also be checked by the calculation of another indicator of fuzziness using an entropy formulation (Kaufman, 1975) as follows:

$$H_A = -\frac{1}{ln\ n}\sum_{i=1}^{n} P_A(x_i)\cdot ln\ P_A(x_i) \tag{14.10}$$

where

$$P_A(x_i) = \frac{\mu_A(x_i)}{\sum_{i=1}^{n}\mu_A(x_i)} \tag{14.11}$$

with notations the same as above.

The fuzziness indicator W_A assumes a value from the set [0,1]. It reaches a maximum value of 1 when the membership characteristic function $\mu_A(x_i)$ has a value near 0.5, and decreases to 0 when the value of $\mu_A(x_i)$ approaches 0 or 1.

On the other hand, the indicator of fuzziness using an entropy measure is not dependent on the values of individual $\mu_A(x_i)$ but rather on their interaction. It reaches a maximum value of 1 when all values of $\mu_A(x_i)$ are the same, and decreases when the values of $\mu_A(x_i)$ differed from each other. In the work laid out, when the indicator of fuzziness W_A exceeds 0.7, or the indicator of fuzziness H_A exceeds 0.9, the criterion or expert opinion are omitted. It is also possible to cause the experts' opinions to accord more closely with their intentions by calculating a concentration of opinion:

$$CON\ A = [\mu_A(x_i)]^2 \tag{14.12}$$

where the notation is the same as in equation (14.9).

In the computation of concentration, one obtains new values of opinion where reductions of the magnitude of the weighted criteria, or the rating of an alternative on a criteria, is relatively small for those scoring high values and relatively large for those scoring low values.

In a similar way one may improve the consistency of an expert's opinion by applying contrast intensification as follows:

$$INT_A = \begin{cases} 2\mu_A(x_i)^2 & \text{for } (0 \le \mu_A(x_i \le 0.5)) \\ 1 - 2(1 - \mu_A x_i) & \text{for } (0.5 \le \mu_A(x_i \le 1)) \end{cases}$$

where the notation is the same as in equation (14.9).

The operation of contrast intensification differs from that of concentration in that it increases the values of the weight of a criterion, or the rating of an alternative on a criterion above 0.5, and diminishes those which are below this threshold.

Finally, the aggregation of experts' opinions are computed by calculating the geometric average.

A case study for Zakopane, Poland

Zakopane is a town situated in the southern part of Poland in the Tatra Mountains. It is well known as a tourist and ski center, and therefore environmental attributes play an important role in the planning process of the town. The purpose of the this case study is to select the best public transport system from three alternatives: street-car, trolley bus, and high standard bus.

The study also considered four alternatives regarding the accessibility to the central business district (CBD) by tourists in private cars: (1) no private cars allowed—only public transportation or taxis, (2) partial accessibility—one entrance where cars might park for the duration of the visit and one exit, (3) payable access—tourists pay a toll when crossing through the gate to the border of the town and special parking zones are provided, and (4) no restrictions to access.

It is impossible to include in this chapter all the factors which might affect the selection of alternatives. Therefore, only those criteria that could express the main differences between the alternatives are presented here.

1. *Functional criteria:* overall travel time (door to door), accessibility, number of train changes, comfort of travel, damageability.
2. *Technological criteria:* range of redevelopment and renovation, access to equipment, difficulty resulting from introducing a new means of transportation.
3. *Economic and financial criteria*: land value, overall cost of construction, cost of equipment, operating costs.
4. *Environmental criteria*: noise and air pollution, landscaping opportunities, energy type, land development opportunities.

Thirty experts were chosen for the evaluation process, including transportation engineers, economists, urban planners, local government officials, and ten citizens from the Zakopane community. Experts were asked to answer the questions and complete pairwise matrices in a questionnaire. The evaluation process for the four transportation alternatives was programmed on a personal computer to simplify the procedure and demonstrate results of calculations both numerically and graphically. The process of acquiring knowledge from the experts for the evaluation of transportation alternatives and aggregating it into a single score, as illustrated in this chapter, can be briefly described as follows.

1. Select criteria that help distinguish between the alternatives, make a list of groups of criteria, and list criteria in each group.
2. Construct the questionnaire.
3. Choose a knowledgeable, interdisciplinary group of experts for the evaluation and determine, if possible, the level of competence of each expert.
4. Distribute the questionnaire to the experts so that they may answer the questions and complete the matrices.
5. Calculate the eigenvalue of matrices. Omit in cases of inconsistency or use the supertransitive approximation procedure.
6. Calculate eigenvectors of matrices which represent priorities of group, hierarchy of criteria, and ratings on single criterion by alternatives.
7. Calculate indicators of fuzziness through entropy for each expert and omit the experts' opinions in case the indicators exceed a prespecified threshold. Eventually, to improve the opinions, use operations of concentration or contrast intensification.
8. Calculate in the same manner as above the indicators of fuzziness for each criterion and leave out the criterion in case the indicators exceed a prespecified threshold. Again use operations of concentration or contrast intensification.
9. Aggregate experts' opinions by calculating the geometric average.

10. Calculate the single score of alternatives according to the standard additive or multiplicative rules shown in equations (14.1) and (14.2). The final single score of an alternative is the result of the level of competence of the experts, priority of the group of criteria, weight of particular criterion in the group, and how the alternatives' degrees of all criteria are fulfilled.

First, the priorities for groups of criteria were obtained. The results are shown in Table 14.1. Notice the differences in expert and citizen opinions.

Table 14.1. Priority for Criteria: Expert vs. Citizen Opinions

	Expert opinions		Citizen opinions	
Criteria	Rank	Composite score	Rank	Composite score
Environmental	1	0.362	2	0.320
Functional	2	0.277	1	0.324
Economic/financial	3	0.194	4	0.148
Technological	4	0.166	3	0.207

Evaluation results for public transportation alternatives in Zakopane

Results for a hierarchy of twenty main criteria are shown in Table 14.2. The evaluation results of alternative transport options in Zakopane, with respect to the hierarchy of criteria importance, is shown in Table 14.3. The experts concluded that the "high standard bus" option is the best alternative for public transportation in Zakopane. However, the score given by the supplementary rule indicates the fulfillment of criterion 1 (noise and air pollution) is not sufficiently met for this alternative.

Evaluation of alternative choices for tourist access to the CBD

An evaluation of alternative choices for tourist access to the CBD was also completed. The alternatives were: (1) no access, (2) partial access, (3) access with payment, and (4) no access restrictions. A hierarchy of fifteen criteria was calculated based on the procedure described above, and composite scores were obtained by geometric average from the questionnaires of the experts as shown in Table 14.4. The result of the evaluation of alternatives for accessibility in Zakopane, with respect to the hierarchy of criteria importance, is also shown in Table 14.4.

Table 14.2. Priority for Main Criteria: Expert vs. Citizen Opinions

	Expert opinions		Citizen opinions	
Main criteria	Rank	Composite score	Rank	Composite score
Noise and air pollution	1	0.101	3	0.075
Landscaping opportunities	2	0.088	6	0.075
Accessibility (avg. distance to bus stops)	3	0.075	1	0.094
Overall travel time (door-to-door)	4	0.061	5	0.058
Difficulty with pulling down and redevelopment	5	0.060	4	0.065
Energy type	6	0.059	7	0.056
Land development opportunities	7	0.054	14	0.040
Damageability	8	0.052	2	0.076
Overall cost of construction and equipment	9	0.050	15	0.031
Number of train changes	10	0.047	11	0.048
Self-financing opportunity of undertaking	11	0.046	9	0.049
Range of infrastructure reconstruction	12	0.045	8	0.055
Comfort of travel	13	0.044	12	0.047
Overall cost of using and repairing equipment	14	0.041	18	0.027
Access to equipment	15	0.035	10	0.049
Range of pulling down and redevelopment	16	0.033	13	0.046
Possibility of constructing undertaking in stages	17	0.030	17	0.028
Land value	18	0.030	20	0.018
Cost of equipment	19	0.027	19	0.022
Difficulties resulting from introduction of transportation	20	0.024	16	0.030

Table 14.3. Alternative Transport Choices: Composite Scores from Expert and Citizen Opinions

	Additive rule		Multiplicative rule		Supple-mentary rule
Alternative	Expert opinions	Citizen opinions	Expert opinions	Citizen opinions	
Street-car	0.254	0.233	0.01	0.0	
Trolley bus	0.278	0.220	0.01	0.0	
High standard bus	0.467	0.547	0.98	1.0	Criterion not ful-filled

As shown in Table 14.5, the conclusion is that the favored alternative for tourist accessibility in Zakopane is through partial accessibility to the CBD (i.e., one entrance, parking for the duration of the visit, and one exit). The result of using the supplementary rule shows that alternative number 3 (payable access) does not adequately fulfill criterion (2) "protection from noise and air pollution." Alternative number 4 (no access restriction) does not sufficiently fulfill either criterion (1) "agreement with environmental goals" or criterion (2).

Conclusions

The application of a procedure for obtaining knowledge from multiple experts is demonstrated to be useful for evaluating transportation alternatives. Such applications provide an easily understood numerical or graphical representation of the overall effectiveness of each transportation alternative. They also allow an analysis of the importance of evaluation criteria and the influence of those differences on final results. The analysis can include technological, economic, and environmental criterion.

Table 14.4. Alternative Choices for Tourist Access to CBD: Criteria Ranking

	Expert opinions		Citizen opinions	
Criteria	Rank	Composite score	Rank	Composite score
Agreement with environmental goals	1	0.149	3	0.129
Protection from noise and air pollution	2	0.113	2	0.129
Level of service	3	0.087	1	0.163
Overall travel time to the town center	4	0.081	5	0.076
Cost of using a new accessibility organization	5	0.072	6	0.074
Comfort of travel	6	0.068	7	0.058
Public opinion acceptance	7	0.064	13	0.024
Influence on citizens' environmental behavior	8	0.055	11	0.040
Cost of introducing new accessibility organization	9	0.052	10	0.041
Legal/technological concerns	10	0.050	9	0.046
Possibility of using up the existing state	11	0.049	4	0.090
Difficulties with changing form of land use	12	0.046	12	0.034
Indispensable time for introducing alternative	13	0.042	8	0.055
Good influence of alternative on profits	14	0.040	14	0.021
Good influence of alternative on social bounds	15	0.034	15	0.019

Table 14.5. Alternative Choices for Tourist Access to CBD: Final Results

	Additive rule		Multiplicative rule		Supple-mentary rule
Alternative	Expert opinions	Citizen opinions	Expert opinions	Citizen opinions	
No access	0.278	0.265	0.015	0.01	
Partial access	0.289	0.328	0.900	0.98	
Access with pay-ment	0.238	0.213	0.085	0.01	Criterion not ful-filled
No access restrictions	0.196	0.194	0.0	0.0	Criterion not ful-filled

References

Han, S.Y., T.J. Kim, and I. Adiguzel, 1992. "Machine Learning, Expert Systems, and an Integer Programming Model: Application to Facility Management and Planning," in Wright, Wiggins, Jain, and Kim, eds., *Expert Systems in Environmental Planning*, Springer-Verlag, New York, NY.

Kaufman, A., 1975. *Introduction to the Theory of Fuzzy Subsets*, Academic Press, Inc.

Olson, J.R., 1987. "Extracting expertise from experts: Methods for knowledge acquisition," *Expert Systems 4*(3).

Narasimhan, R., 1982."A geometric averaging procedure for constructing supertransitive approximations to binary comparison matrices," *Fuzzy Sets and Systems 8.*

Saaty, T.L., 1977. "A scaling method for priorities in hierarchical structures," *Journal of Mathematical Psychology*, June.

Saaty, T.L., 1978. "Exploring the interface between hierarchies, multiple objectives, and fuzzy sets," *Fuzzy Sets and Systems 1.*

Saaty, T.L., 1988. *Multicriteria Decision Making: The Analytic Hierarchy Process*, RWS Publications, Pittsburgh, PA.

Suh, S., M. Kim, and T.J. Kim, 1988. "ESMAN: An expert system for manufacturing site selection," *Computers, Environment, and Urban Systems 12*:239-252.

15
Testing a Prototype Expert System for Diagnosing Wastewater Treatment Plant Operating Problems

Catherine D. Perman and Leonard Ortolano

Introduction

SLUDGECADET, an expert system prototype, diagnoses operating problems at wastewater treatment facilities. The system implements a model of the diagnostic techniques of a group of expert treatment plant operators. As a step in developing the prototype, two experiments were implemented to evaluate the system's ability to represent the participating experts' diagnostic techniques, to appraise the system's performance using the standards of an expert who was not involved with system development, and to judge the system on the basis of an empirically established standard of performance. The tests were implemented on a limited, experimental basis, and thus they do not constitute a full scale evaluation effort. Despite their modest scope, the tests yielded information useful in refining the prototype and in designing future expert systems' validation studies.

Description of the SLUDGECADET prototype expert system

SLUDGECADET diagnoses treatment plant operating problems by combining a treatment plant model, heuristics used by expert plant operators, and theoretical knowledge of wastewater treatment processes. The intended user of the system, a novice operator, supplies the system with information concerning specific operating problems. General information about the plant is stored permanently by the system. At the end of an interactive session, the expert system displays its diagnosis, a recommended remedial action, associated explanations, and detailed information about the facility.

The system maintains a clear separation between information concerning a particular treatment plant and information that applies to all plants of a given class. The plants considered by SLUDGECADET are those using trickling filters for secondary wastewater treatment and anaerobic digesters for sludge treatment. Information applying to all plants in this class can be instantiated for an arbitrary number of individual wastewater treatment facilities.

There were several reasons for restricting the scope of SLUDGECADET to treatment plants using trickling filters and anaerobic digesters. First, there are many such facilities that are frequently unable to meet their national effluent discharge permit requirements (U.S. Government Accounting Office, 1983), and a system that improved a facility's ability to meet permit requirements would provide significant payoff. Second, the task of diagnosing operating problems at such plants involves an appropriate degree of difficulty for building an expert system. Third, because the U.S. Army has many plants of this type, its Construction Engineering Research Laboratory (CERL) was interested in supporting the development of an expert system. CERL provided financial resources, access to experts, and a wastewater treatment plant that could be used for test purposes. The U.S. Army wastewater treatment facility at Fort Lewis, Washington, is used as a test site for SLUDGECADET. The participating experts are from ES2 Environmental Services of Berkeley, California (referred to herein as ES2) and the engineering and operations staff at the Fort Lewis wastewater treatment plant.

The participating experts were selected using a variety of criteria. Principal among these were the experts' practical experience and acknowledged expertise in operating treatment plants. In addition, experts had to have a thorough understanding of the design and theoretical aspects of wastewater treatment.

The ES2 experts played a key role in defining the scope of the prototype system. Their main concern was that the prototype remain small, while being representative of a more general set of problems. They recommended that the scope of the prototype be limited to certain portions of a wastewater treatment facility and to the diagnosis of a small number of operating problems. The experts emphasized that the selection of appropriate operating problems should be tied to particular portions of a wastewater treatment plant. For selecting portions of the plant to be included in the prototype system, a typical plant was defined by the following unit processes: headworks, primary clarification, trickling filters, secondary clarification, chlorination, and anaerobic digestion (see Figure 15.1). One criterion for selecting appropriate unit processes for the prototype expert system was that, as a set, the processes should encompass both physical-chemical and biological processes as well as both wastewater and sludge treatment. It was also considered important to include unit processes at different stages of treatment. For example, primary clarifiers are close to the beginning of wastewater treatment and chlorination is at the end.

Other criteria used in selecting problems for the prototype concerned representing the complete range of issues involved in diagnosing operating problems.

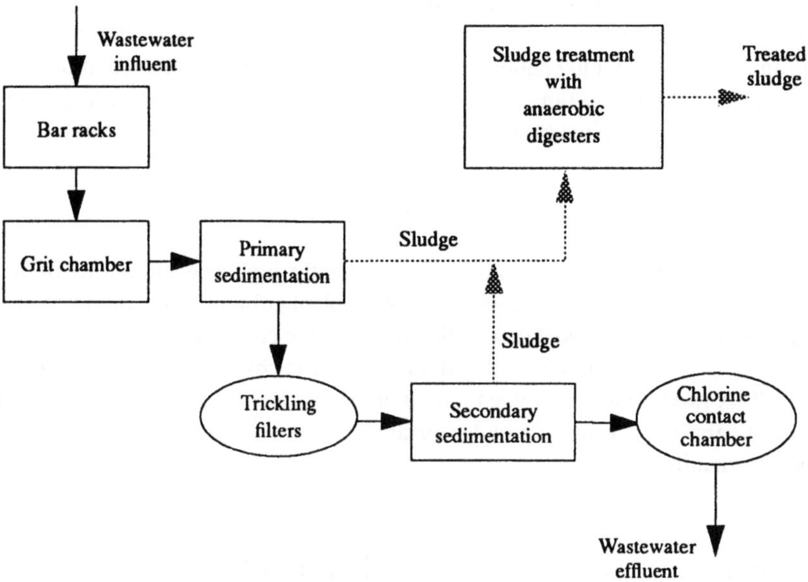

Figure 15.1. Typical schematic for the type of wastewater treatment plant modeled in t SLUDGECADET system.

Experts participating in the research identified these issues in terms of physical-chemical processes, biological processes, mechanical problems, interactions among unit processes, data analysis and interpretation, and external or environmental factors (e.g., changes in the community served by the plant). Based on discussions with the experts, it was decided that the following three "problem modules" would provide a basis for developing a prototype that would demonstrate the feasibility of building a complete system.

Primary clarifier suspended solids module

This module is concerned with abnormally high suspended solids in the primary clarifier effluent. It represents problems with physical-chemical processes that are inherent to the operation of primary clarifiers, and mechanical failures, for example, breakdowns in the primary clarifier sludge pumping system.

Anaerobic digester supernatant module

This module addresses poor quality supernatant, i.e., the effluent from sludge digesters that is returning to the wastewater stream for additional treatment. The model considers: biological processes in an anaerobic digester, the interaction of failures in one unit process with others in the plant (e.g., if the supernatant from the anaerobic digester is of poor quality and is returned to the plant headworks, then the supernatant can adversely impact the rest of the plant), the reaction of the anaerobic digester to changes in the local climate (an environmental factor) or the service community (an external factor), and the analysis of data (e.g., use of volatile suspended solids concentration in diagnosing digester problems).

Chlorination fecal coliforms module

This module examines the causes of abnormally high concentrations of fecal coliform bacteria in the wastewater treatment plant effluent that result from physical-chemical process breakdowns that occur in the chlorination unit process and mechanical problems associated with failures in the chlorination feed control system. This module also involves data analysis to identify trends in wastewater quality parameters (e.g., pH or temperature) used in diagnosing problems with the chlorination process.

Objectives and scope of the validation experiments

In devising a plan to test SLUDGECADET, three test objectives, set at the start of system development, were established. The first objective concerned testing the correctness of the system in representing the problem solving techniques of the experts who participated in developing it; these individuals are referred to herein as "internal" experts. The second objective dealt with appraising the acceptability of the system's results to experts that were not involved with system development (i.e., "external" experts). The third objective concerned how well the prototype performed against an empirically defined standard, one that minimized interpretation and bias from individuals.

Testing related to the first objective, i.e., the performance of SLUDGECADET with respect to internal experts, was conducted as a routine part of knowledge acquisition for system development. Knowledge acquisition followed the pattern typically used in the "rapid prototyping" approach to system development. Using this approach, knowledge engineering proceeds in cycles, each of which aims to augment and improve a partially developed expert system (i.e., a "rapid prototype"). This approach has been thoroughly described by McGraw and Harbison-Briggs (1989), among others.

Knowledge was elicited from participating experts using cycles, each consisting of the following steps: interview the experts, transform interview results into programmable form, implement the new knowledge by making it a part of SLUDGECADET, demonstrate the resulting problem to the experts to obtain comments, and use review comments to refine the system. The experts' reactions to the demonstration were useful in testing the correctness of the work done by the knowledge engineer in transferring knowledge from the experts to the program. In this context, the internal experts could verify that newly implemented portions of the prototype emulated their problem solving techniques. This does not constitute a complete verification because the demonstrations were limited in scope. Nevertheless, this style of knowledge acquisition attempts to make knowledge base validation an integral part of system development. Within this cycle, the experts and knowledge engineer worked together to minimize discrepancies before any implementation occurred. The program demonstrations allowed the experts to examine the program results and explanations. Sometimes the internal experts examined detailed reports of the results and reasoning of the expert system. These reports allowed the results of the program to be critiqued without the distraction of running using the system. For details on how knowledge acquisition was implemented, see Perman (1989).

Although the approach followed for system development has a built-in degree of knowledge base testing by internal experts, it does not provide a thorough test of the correctness and acceptability of the knowledge base. Formal validation studies are required for this.

The research reported below took the form of experiments using different approaches to validate SLUDGECADET's knowledge base. The experiments are modest in scope and are intended mainly as a vehicle to provide insights into expert system validation; they do not constitute a full test of SLUDGECADET. As is made evident by reviews of the literature on knowledge base validation (e.g., Ortolano, Le Coeur and MacGilchrist, 1990), experiments of the sort conducted can contribute to development of a validation method that is both simple enough to be practicable and rigorous enough to gain acceptance. As argued by Green and Keyes (1987), among others, a widely accepted methodology for knowledge base validation does not yet exist.

Fort Lewis test cases

Each of the validation experiments reported on herein relied on a set of case study operating problems that were based on problems that occurred at the Fort Lewis wastewater treatment facility during the summer of 1988. During that period, the staff at the Fort Lewis plant experienced a total of seven operating problems relating to primary clarifiers, anaerobic digesters and chlorination units, and they collected data on each. Documentation for these seven problems included the initial symptoms that led to the staff's identification that a problem existed, the staff's diagnosis of what caused the symptoms, and the approach followed in eliminating the problem.

Of these seven problems, only four were applicable to the kind of operating problems covered by the prototype problem modules. Two of the four concerned the primary clarifier suspended solids problem, and two concerned the anaerobic digester supernatant problem. One primary clarifier problem and one anaerobic digester problem (in each case the simpler of the two problems) were used for calibrating the SLUDGECADET system (i.e., refining the program's input features to assure that real test case problems could be entered without difficulty).

The remaining two operating problems, one for the primary clarifier module and one for the anaerobic digester module, were used for the validation experiments. Documentation for each of these two problems was put in a standard format and made part of a test case dossier. Each dossier contained pertinent design and operating data for the Fort Lewis wastewater treatment facility at the time that the problem occurred. Design information came from the Fort Lewis Operating and Maintenance Manual (U.S. Army, 1983) and the operating information came from the plant records.

Based on a review of the dossiers, it appeared that the primary clarifier test case might be too easy to diagnose because the symptoms practically made the problem cause evident to an experienced operator. Consequently, this case was divided into two slightly different cases. One primary clarifier test case, herein called "scenario number one," used only one of the two symptoms listed by the Fort Lewis staff in their documentation of the problem; a second test case called "scenario number two," retained both of the symptoms reported by the Fort Lewis staff. The anaerobic digester test case is called "scenario number three."

Two validation experiments

Testing was set up in two experiments. In each case, one internal expert (from ES2) and one external expert (from a university) participated. The external expert, a professor of environmental engineering, had no prior knowledge of SLUDGECADET. In comparison to the internal expert, the external expert had more experience in wastewater treatment research and less experience in operating treatment plants. Aside from gathering information related to SLUDGECADET, the experiments were designed to gain insights into the process of validating a knowledge base. Each experiment used different techniques for interacting with the experts, with the intention of resolving some difficulties that arise in testing.

The first experiment involved all three test cases and included a comparative analysis of the problem diagnoses and recommended solutions offered independently by four "agents": the internal expert, the external expert, SLUDGECADET, and the staff at Fort Lewis. The diagnosis and recommendations for the Fort Lewis staff were taken directly from documentation the staff provided regarding what actually took place at the plant. The second experiment consisted of a structured evaluation of SLUDGECADET's outputs for each of the three test cases; these evaluations were conducted using standard forms completed independently by both the internal expert and the external expert.

The experiments were designed in the context of a more general research goal, which is to characterize the knowledge of experts in diagnosing operating problems at wastewater treatment plants. Viewed in this broader context, the experiments represent still another form of knowledge acquisition. They provide insights into how the experts participating in the experiments go about solving operating problems. In some sense, the experiments provide partial answers to two fundamental questions: What do experts know? What makes them experts?

Experiment one: Comparison of results from four agents

This experiment rests on comparisons of four sets of results for each test case. One set of results, which represents an empirical benchmark for comparative analysis, consists of the documentation of how the Fort Lewis treatment plant staff diagnosed and solved the operating problems used in the validation exercises. The other three sets of results were generated independently by the internal expert, the external expert and SLUDGECADET. In each instance, the information used was restricted to that contained in the case study dossiers.

This exercise treats the three objectives established for testing the knowledge base. Comparing the internal expert's results with that of SLUDGECADET reveals how well SLUDGECADET reproduced the expert's problem solving strategies. Comparing the external expert's results with SLUDGECADET's outputs provides insights into whether the external expert would find SLUDGECADET's outputs acceptable. Finally, comparing SLUDGECADET's recommendations with the actions taken at the Fort Lewis plant provides a test of SLUDGECADET based on empirical data.

Pre-testing activities

Originally, the procedure for implementing experiment one was to have been as follows. Interviews were to be conducted with each expert to obtain their diagnoses and recommended actions based on information in the dossiers prepared for each of the three scenarios. In addition, SLUDGECADET was to be run using the same information. Finally, a comparative analyses was to be conducted.

Before implementing this approach, a series of pre-tests were conducted to make sure that the dossiers were clear and complete and that the participating experts would understand what was expected of them. Because the knowledge engineer had prepared the dossiers and knew the results yielded by both the Fort Lewis staff and SLUDGECADET, she was not a suitable person to conduct either the pre-test or the experiment itself. The fear was that she would unintentionally bias the experts because her knowledge of possible responses would make SLUDGECADET's outputs appear favorable.

All interviews conducted in connection with this experiment (as well as experiment two) were conducted by a Stanford University graduate student specializing in infrastructure planning and management who had no prior knowledge of SLUDGECADET's knowledge base. This person, referred to herein as "the interviewer," was informally trained in interview techniques before and during the pre-tests. All pre-test interviews were taped and participants were asked to write down their results. Two experts from the Environmental and Water Studies program at Stanford University served as pre-testers.

The interviewer presented the three scenarios to the first pre-tester with a minimum of information, just a few sentences describing each operating problem and some design information concerning Fort Lewis. The knowledge engineer was present at the pre-test interview, her role was limited to answering technical questions raised by the pre-tester.

The pre-testing process seemed to be distracting and confusing for the first pre-tester. He had trouble organizing his thoughts and remembering the answers to all his questions. When he got down to diagnosing operating problems in the dossiers, he exercised great care and was very detailed (as if it were a formal exam) and provided as much theory and explanation as he could. The process appeared to resemble an oral exam in a way that was counter-productive to the validation exercise. Afterwards, the pre-tester commented that the interview "felt really strange." In addition, he found the scenario two test case trivial to solve; this had been anticipated by the knowledge engineer and is why scenario one was created.

A second session was held to allow the first pre-tester to analyze scenario three. This time the knowledge engineer had prepared a detailed listing of all data concerning the operating problem. However, a summary of the initial symptoms was not provided, in order to see how the pre-tester performed using only the data set. Since the only information available to the pre-tester consisted of tables and lists of data, he examined it very carefully and discovered unintentional errors and inconsistencies.

Generally, the format used in the second session was a success. However, the first pre-tester indicated that the interview was confusing in terms of whether he was to imagine himself at the plant trying to solve the problem in "real time" or whether he was supposed to be "sitting in an office" trying to solve the problem at a later date. He also was confused by not knowing how much technical knowledge the interviewer had. The pre-tester wanted to know this in order to judge how much technical jargon he could use in providing his explanations.

In response to the first pre-tester's comments, a section was added to the validation exercise instructions describing that the interviewee was to imagine being at the plant solving the problem in real time and the interviewer was a novice wastewater treatment plant operator. The formal clarification of roles eliminated a major source of confusion.

The three scenarios were presented to the second pre-tester. Each dossier now contained a front page that explicitly "set the stage" and provided roles for the interviewee (the second pre-tester) and the interviewer, a summary of the problem's initial symptoms and other technical information. The second pre-tester went through scenarios one and two smoothly, and this indicated that the testing process was more clearly defined than the one used in the first pre-test. The second pre-tester found the role playing easy and clear in intent. During the session, the pre-tester discovered inconsistencies with the data in scenario three. The meeting was halted, and re-started at a later time after the inconsistencies in data had been clarified. The follow-up pre-test session proceeded smoothly.

The pre-testing provided an opportunity to clarify the expert's role in the validation study and eliminate ambiguities and inconsistencies in dossiers describing the scenarios. The importance of pre-testing should not be underestimated. As shown by Ortolano, Le Coeur, and MacGilchrist (1990), inadequate pre-testing can lead to problems for experts carrying out knowledge base validation.

Conduct of experiment one

The original plan called for individuals participating in the test to both read the dossiers and present their results in a interview setting. Instead, based on reactions of the pre-testers and requests made by the two experts participating in the experiments, the dossiers and instructions for the first experiment were mailed to both experts. The procedure called for scheduling interviews after the experts had a few weeks to complete their work on the experiment.

The external expert completed his part of experiment one without any rescheduling or change in the above noted procedures. However, the internal expert rescheduled interview dates several times; finally, he said he preferred conducting the entire test in the context of an interview. Accordingly, the knowledge engineer and the interviewer met with the internal expert and completed experiment one by following the same interview procedure used in the pre-test. Results of experiment one were reviewed with each expert to make sure that the knowledge engineer understood their diagnoses and recommended actions. Results from experiment one are summarized below in tables that compare the Fort Lewis staff's results with the diagnoses, problem causes and solutions of the internal expert, the external expert and SLUDGECADET.

Comparative analysis of results

Scenario one. The first two cases were based on an operating problem that the Fort Lewis staff had with their second primary clarifier on July 20, 1988. In scenario one, this problem was described to the experts as having one initial symptom: gas bubbles in primary clarifier two (not in any of the other clarifiers). The Fort Lewis staff described their analysis in a problem report and that report is summarized at the top of Table 15.1. The staff diagnosed the problem as "broken flights" and they described the cause as "worn out rail that the flights travel on." (Flights are paddles attached to a moving chain. They travel along rails and their function is to push sludge to a collection area at the end of a rectangular clarifier.) The staff's solution, which was effective, was to repair the rail.

Table 15.1 also contains results offered by the internal expert, the external expert and SLUDGECADET. Each of these agents produced the same diagnosis, problem cause and solution. In each instance, the focus was on a mechanical problem in the second clarifier at the Fort Lewis plant. The main difference among the three agents is that the internal expert did not restrict his attention to the mechanical problem as a possible cause. He felt that two additional causes (a jump in influent total suspended solids and a hydraulic imbalance) should be looked into, even though there was no data in the dossier to suggest these possible causes. Another difference was that the internal expert and the external expert both considered interactions among treatment processes as a part of their diagnostic reasoning. For example, the external expert considered the possibility that the symptoms in the clarifier were due to either a failure in the headworks or impacts from returned supernatant. SLUDGECADET did not consider problem

Table 15.1. Results for Scenario One: Gas Bubbles in Primary Clarifier 2

	Diagnosis	Problem cause	Solution
Report by Fort Lewis Staff	Broken flights in clarifier two	Railings that flights travel on are worn out	Repair railings and flights
Internal expert	Too much sludge in clarifier two	1. Jump in influent total suspended solids* 2. Hydraulic imbalance among clarifiers* 3. Mechanical problem with sludge collection	1. Check for short term increase to clarifier two 2. Adjust loading to distribute evenly 3. Repair mechanical problem
External expert	Sludge accumulated and went anaerobic in clarifier two	Mechanical problem with sludge collectors or pumps	Repair mechanical problem
SLUDGECADET	Sludge blanket became anaerobic in clarifier two	Mechanical problem with sludge collection system (which includes pumps, valves, and pipes)	Repair mechanical problem

*Expert recommended this even though data does not provide evidence.

causes due to interactions because interactions among units were not within the scope of the primary clarifier two problem module. As indicated in the first column of Table 15.1, the external and internal experts diagnosed scenario one in a manner similar to SLUDGECADET.

Table 15.1 also shows that the wording used by the Fort Lewis staff in their problem report was different from that used by the three other agents. This difference in wording can be explained by differences between how the Fort Lewis staff discovered the problem and how the problem was presented in the dossiers. At Fort Lewis, the operating problem was discovered on a routine inspection that occurred during a change in work shifts. An operator noticed that the flights were broken and both the cause and solution were evident on further inspection. The operator also noted gas bubbles in the primary clarifier, but this data was of secondary importance since the operator knew that the broken flights would cause sludge to accumulate become anaerobic, and produce gas bubbles. Scenario one excluded the information on the broken flights. This was done because the case study might have appeared too simple to the experts if information on the broken flights had been included.

Considering that the information in the dossier for scenario one includes only gas bubbles in primary clarifier as a symptom, the differences in results in Table 15.1 are understandable. The experts and SLUDGECADET attempted to discover a cause of bubbles in the clarifier. All three agents concluded correctly that sludge was accumulating in the clarifier and that the sludge collection system needed repair.

Scenario two. The second case study involved the same operating problem as the first, except this time the full set of initial symptoms noted by the Fort Lewis staff was provided in the dossier. The experts and SLUDGECADET knew that primary clarifier two contained gas bubbles and its flights were broken. (The dossiers presented scenarios without indicating that they concerned the same incident; however, both scenarios included the same date of occurrence.)

Having the additional symptom (i.e., broken flights) caused both experts to emphasize mechanical problems as they reasoned about the likely source of difficulty. As indicated in Table 15.2, the resulting diagnoses and recommended actions for scenario two are precisely the same as for scenario one. The expert systems' outputs are fully consistent with the results provided by both the experts.

As in scenario one, results for scenario two show differences between the advice of the Fort Lewis staff and that of the experts and SLUDGECADET. The Fort Lewis staff diagnosed broken flights, while all other diagnoses concern sludge accumulation in primary clarifier two. In contrast to the scenario one results, the differences in diagnoses for scenario two are not easily explained because the dossier contained all the information available to the Fort Lewis staff.

A possible explanation for the difference in diagnostic reasoning in Table 15.2 has to do with the difference between being at the plant and seeing the broken flights and getting the information on a sheet of paper in an office. At the plant, the physical reality of failed machinery might seem more important than the accumulating sludge. From the internal and external experts points of view, the appearance of gas bubbles triggers a line of reasoning that includes problem causes in addition to mechanical failures. Perhaps they were better able to consider these alternatives because they were not at the plant and were not physically confronted with the broken flights. The different diagnoses may also be explained by differences in training and experience between the Fort Lewis staff and the external and internal experts.

Scenario three. The third test case concerned a situation at the Fort Lewis plant where 48,000 gallons (gals) of sludge were pumped in one day (July 22, 1989) from the sludge thickener to the primary anaerobic digesters. Under normal con-

Table 15.2. Results for Scenario Two: Gas Bubbles and Broken Flights in Primary Clarifier Two

	Diagnosis	Problem cause	Solution
Report by Fort Lewis Staff	Broken flights in clarifier two	Railings that flights travel on are worn out	Repair railings and flights
Internal expert	Too much sludge in clarifier two	Obvious mechanical problem	Repair mechanical problem; look for cause of problem; increase monitoring until cause is discovered
External expert	Sludge accumulated and went anaerobic in clarifier two	Mechanical problem with sludge collectors (not including the pumps)	Repair mechanical problem
SLUDGECADET	Sludge blanket became anaerobic in clarifier two	Mechanical problem with sludge collection system (which includes pumps, valves, and pipes)	Repair mechanical problem

ditions, the sludge pumping rate is about 20,000 gals per day. As noted in Table 15.3, the Fort Lewis staff reported this situation as caused by operator error. The sludge pump was mistakenly left at the manual setting. The solution to the problem was to reset the sludge pump to an automatic setting at the normal pumping rate. No other problems were reported i.e., the excessive pumping did not cause problems with the digestion process.

Scenario three is complex. On the one hand, more than twice the normal quantity of sludge was pumped to the digesters in a single day; on the other hand, the digesters were functioning normally.

The internal expert used data on organic and hydraulic loading rates to conclude that the digesters were working properly despite the excessive pumping.

He concluded that an operator had made a mistake: either the sludge pumps were on automatic at an improper setting or they were incorrectly left on manual

The expert system's approach was similar to that of the internal expert in that hydraulic and organic loading data was analyzed to conclude the digesters were working properly. However, SLUDGECADET offered a diagnosis and recommendations that were far less specific than those of the internal expert. As noted in Table 15.3, SLUDGECADET called for reducing the sludge feed rate to a normal level and attributed the difficulty to a "mechanical or process control problem in digester."

The external expert followed a completely different course. Notwithstanding that the dossiers contained no information on relationships among the three scenarios, the external expert assumed that the excess pumping in scenario 3 was a deliberate decision by the Fort Lewis staff to drain excess sludge from primary clarifier two (the problem noted in the first two scenarios). Having made this assumption, the external expert then checked the digesters' detention time, found it "conservative," and concluded that the digestion process was normal. As detailed in Table 15.3 the external expert recommended either taking primary clarifier two out of service or increasing the percent solids in the raw sludge to reduce the sludge pumping rate while maintaining the same quantity of solids removed per unit time.

The analysis of scenario three by the external expert was perfectly reasonable, but it was unplanned for in the experimental design. In developing the experiment, no thought was given to the possibility that an expert might reason that because the excess sludge pumping occurred two days after the sludge accumulation problem in primary clarifier two, the scenarios might be linked. The dossiers gave no information indicating the scenarios represented independent events and that was a mistake. Because of the external experts reasonable (but unanticipated) assumption, his analyses of scenario three was difficult to utilize fully.

Discussion of results and procedures

A comparison of results summarized in Tables 15.1, 15.2, and 15.3 yields information along several different lines. In general, there was greater consistency in responses among the two experts and SLUDGECADET than there was between those three agents and the Fort Lewis staff. This is understandable in that the operating problems were quite evident to the staff. For the first two scenarios, the flights on primary clarifier two were clearly broken, and for the third scenario the Fort Lewis operator quickly saw that he had incorrectly set the sludge pump. In all cases, little analysis was used by the staff to discover problem causes or solutions.

Although the two experts and SLUDGECADET often agreed with each other, there were notable differences which provide insights into how wastewater treatment plant operating problems are diagnosed and how SLUDGECADET might be improved. In general, the experts used a broader frame of reference than

Table 15.3. Results for Scenario Three: Excessive Quantity of Sludge Pumped from Sludge Thickener to Primary Sludge Digesters

	Diagnosis	Problem cause	Solution
Report by Fort Lewis Staff	Sludge pump was set to manual	Error by operator	Reset sludge pump to automatic at normal pumping rate
Internal expert	Operator error: sludge pump set improperly or pumps turned on manually	Not causing a digester problem because no indication of organic or hydraulic overload	Since pumping did not cause a problem, take no action on digester Check gas production in a month to be sure digesters are stable Find cause of operator error and correct
External expert	Operator increased pumping rate to drain clarifier two (Assumed relationship among scenarios)	Not causing a digester problem; detention time is conservative	1. Take clarifier two out of use. Manually pump about 30,000 gal/day of sludge from clarifier two; or 2. Increase % solids of raw sludge to 6% and manually pump 30,000 gal/day
SLUDGECADET	Mechanical or process control problem in digester	Data does not indicate a digester problem; organic loading is higher than normal for the Fort Lewis plant	Reduce sludge feed rate to a normal level

SLUDGECADET in seeking problem causes. In particular, the experts put greater emphasis on analyzing interactions among unit processes. In particular, the experts put greater emphasis on analyzing interactions among unit processes. By design, SLUDGECADET only considers interactions among units in the anaerobic digester module. The expert system could clearly benefit by incorporating interactions among units in all problem modules.

Another difference is that the experts often provided recommendations that were more detailed than those of SLUDGECADET. This is especially evident in scenario three, in which SLUDGECADET's recommendations, although reasonable, are very general.

Tables 15.1, 15.2, and 15.3 also demonstrate the commonplace result that experts can disagree among themselves. The disagreements that surfaced in this study were quite minor, except for scenario three. In that instance, the external expert's assumption that the high pumping rate was intentional, and not an operator error, led him on a line of reasoning that diverged substantially from that of each of the other agents. The external expert's reasoning was correct, given his assumption, and his assumption was plausible even though it did not correspond to what took place. The domain treated by SLUDGECADET is complex enough that differences among experts will be unavoidable. Whether or not such differences will be frequent or significant is an open question. In this instance, the differences point to a shortcoming of SLUDGECADET: the program is incapable of reasoning about relationships among operating problems over time.

The results in Tables 15.1, 15.2, and 15.3 can also be discussed in terms of the three objectives of the validation exercise. In terms of the first objective, the internal expert's diagnosis for the three problem scenarios agreed with those made by SLUDGECADET, except for the diagnosis for scenario three. Although the prototype was able to reason about the problem cause and the solution in the same way as the internal expert, its diagnosis was not as specific as that of the expert. This deficiency can be corrected in future versions of the expert system. A consideration of the second test goal involves comparing the external expert's diagnoses with those made by the SLUDGECADET. As with the internal expert, the external expert's diagnoses agreed with those made by the prototype except for scenario three. By assuming that all scenarios involved the same problem, the external expert reasoned that the excessive sludge pumping in scenario three was intentional. Under the circumstances, it is difficult to use his scenario three results to comment on the validity of SLUDGECADET with what actually occurred at the Fort Lewis facility. For scenarios one and two, SLUDGECADET emphasized the anaerobic conditions in primary clarifier two and recommended repair of mechanical problems in the sludge collection system. For reasons noted above, the staff at Fort Lewis observed the broken flights and called for their repair, without concentrating on the anaerobic conditions that caused gas bubbles in primary clarifier two. For scenario three, SLUDGECADET identified the problem and proposed a solution consistent with what occurred at Fort Lewis, but the expert system's findings were more general than those of the Fort Lewis staff.

Experiment two: Structured evaluation of the expert system's outputs

In the second validation experiment, the external and internal experts formally evaluated the results of the expert system for each of the three scenarios. SLUDGECADET's results for each scenario were presented in the form of a detailed report that described the data that the system obtained form the user, the problem diagnosis, a recommended solution and an explanation of the diagnostic reasoning. Experiment two used an evaluation form that asked questions about SLUDGECADET's diagnosis, solution and explanation. (A sample form is given in Appendix A.) For each of the three scenarios, each expert was given an evaluation form, the SLUDGECADET detailed report, the dossier used in experiment one and the expert's own results from experiment one. The outputs from experiment two consist of the answers provided by the external and internal expert on the evaluation form for each scenario.

A pre-test phase of experiment two was conducted to eliminate errors and ambiguities from the evaluation form. The same two pre-testers from experiment one reviewed the SLUDGECADET detailed reports, the evaluation forms, and the dossiers in the presence of the interviewer and the knowledge engineer. The latter's role was limited to answering questions. The pre-test indicated that the evaluation form designed for the experiment was not clear, but the procedures proposed for carrying out the experiment were otherwise workable. The main problem with the evaluation form concerned questions to elicit the experts' views on whether the sets of diagnoses and solutions reviewed by SLUDGECADET were complete. These questions were re-worked based on suggestions made by the pre-testers.

Experiment two was implemented as follows. The external expert requested that he be allowed to review the documents and complete the evaluation form at his leisure. Accordingly, the knowledge engineer prepared a notebook with all the necessary information for the experiment and gave it to the external expert. The procedure for the internal expert was slightly different. Because the internal expert chose to complete experiment two immediately after completing experiment one, the interviewer and the knowledge engineer waited in a nearby room while the internal expert filled out the evaluation forms. This procedure was not ideal since it meant the evaluation forms were completed following a long interview session. What made matters worse was that both experiments were conducted in the evening, after the internal expert had completed a long day of work. Given the demanding work schedule that the internal expert was following during this period of the study, there was no alternative circumstance for conducting the validation experiments.

The internal expert's formal evaluation of SLUDGECADET's output was quite positive. He agreed with the diagnoses and solutions offered by SLUDGECADET for each of the three scenarios. He was especially enthusiastic about SLUDGECADET's performance on scenario three. In his view, the "answer and analysis" for scenario three was "particularly good" and it "covered all the bases."

As can be seen from Appendix A, the evaluation forms used in experiment two included questions asking the experts if there are diagnoses or solutions that they would have considered that are different from those considered by SLUDGECADET. Despite the changes made based on pre-testing of the form, this question was misinterpreted by the internal expert so that his responses for these items are unusable. It was surprising that the internal expert did not mention additional diagnoses for the primary clarifier problem he examined in his own analysis of scenario one: interactions among treatment units and imbalanced hydraulic loading among units. Failure to mention these in experiment two can be explained by both misinterpretation of the question and by having completed the evaluation forms late in the evening after a long day of work and a long interview session (experiment one).

The external expert's evaluation of SLUDGECADET's outputs were generally positive for scenarios one and two. For these cases, he felt that the diagnoses and solutions offered by SLUDGECADET were valid. In response to questions on the form concerning diagnoses and solutions that should have been considered by the expert system, the external expert raised two points. One concerned SLUDGECADET's failure to treat interactions among treatment units. The external expert felt that SLUDGECADET should have made an investigation of upstream treatment units (e.g. the headworks) to determine the cause of failure in the sludge collection system of primary clarifier two. The external expert's second point concerned the need to consider purchase of a larger sludge pump as a possible solution. In his view, the accumulation of sludge in primary clarifier two may have been caused by an inadequately sized pump. Both these points involve issues that are outside the scope of the current version of the expert system. SLUDGECADET does not treat interactions among treatment units (except for the anaerobic digester module) and it does not treat inadequacies in plant design as a cause of operating problems.

As might be expected from the results of experiment one, the external expert had a negative view of SLUDGECADET's performance on scenario three. When the external expert made his evaluation, he was unaware of the actual cause of excess pumping (i.e., an error by the Fort Lewis staff) and consequently he critiqued SLUDGECADET for failing to discover a relationship between the three scenarios. Since the external expert based his analysis of the scenario on the assumption that the excess sludge pumped to the digester was intended to reduce the sludge in primary clarifier two, there is no way to use his evaluation of SLUDGECADET's reasoning on scenario three. SLUDGECADET is unable to link problems over time and it analyzed scenario three as a case that was independent of other scenarios.

Although the external expert's critique of SLUDGECADET's performance on scenario three is difficult to utilize, it is informative. His position points out that some experts rely on analysis of problems across both time and unit processes. This is an issue that should be considered in future work on SLUDGECADET.

Conclusions

Results from the experiments can be interpreted using the test objectives. The expert system matched the reasoning of the internal expert closely for all scenarios. This is an expected result, since the program is partially based on the problem solving strategies of the internal expert. Notwithstanding the similarities in outputs, the expert system's responses were often less specific than those of the internal expert. Comparisons between the expert system outputs and the results provided by the external expert were more discordant. However, much of the disagreement concerned scenario three and the outputs from that scenario are not comparable because of different assumptions regarding the independence of the scenarios. Finally, comparisons between the outputs of the expert system and the empirical benchmark, the responses of the Fort Lewis staff, contained many overlaps. However, the expert system, like the two experts, tended to use more complex reasoning than the Fort Lewis staff. For the staff, both the diagnoses and the solutions were practically self-evident.

A comparison of experiments one and two indicates that experiment two provided relatively little new information in evaluating SLUDGECADET's performance. This is understandable, since the experts were free to use their results from experiment one. Although it is possible to infer how the experts would appraise SLUDGECADET using only experiment one, this inference rests on interpretations of interview data. The contribution of experiment two is that it provides a direct way of learning whether the experts approve of SLUDGECADET's performance. For the three case studies reported herein, the results inferred from experiment one are confirmed by experiment two.

Validation experiments do more than test the acceptability of expert systems. They also provide insights into how those systems can be improved and thus they constitute a form of knowledge acquisition. This is clearly demonstrated by the numerous insights the validation exercise provided into how SLUDGECADET's knowledge base can be improved. An example is the great reliance both experts placed on analyzing interactions among treatment units. This is an area where SLUDGECADET needs additional work.

The two experiments also provided insights into procedure used to validate an expert system's knowledge base. The improvements in procedures that resulted from pre-testing highlight the importance of pre-test activities. The experiments also reveal some of the practical difficulties in implementing a validation study. Experts are typically busy and scheduling was a continuing problem. Moreover, the experts had different styles of work and that made it difficult to implement a uniform procedure. The internal expert preferred an interview process that was interactive and had a fixed time interval. The external expert preferred having the flexibility to complete work at his own pace. Even the pre-testers had different styles of work and views of their roles in the study. The original research design had to be molded to accommodate the work styles and schedules of the participating experts.

Although this work provided numerous insights into the strengths and weaknesses of SLUDGECADET, those insights are only part of the yield. The research also provided insights into procedures that can be useful in attempting to validate a knowledge base.

Acknowledgments

This research would not have been possible without the help of the participating experts, the Fort Lewis wastewater treatment facility staff and ES2 Environmental Services of Berkeley, California. Financial support was provided by the U.S. Army Corps of Engineers Construction Engineering Research Laboratory (Grant number DACA-88-86-D0008) and the UPS Foundation. We also thank IntelliCorp of Mountain View, California for providing all software and hardware used in this research.

References

Green, C. and M. Keyes, 1987. "Verification and Validation of Expert Systems," *Proceedings of WESTEX'87*, June 2-4, Anaheim, CA, pgs. 38-43.

McGraw, K.L. and K. Harbison-Briggs, 1989. *Knowledge Acquisition: Principles and Guidelines*, Prentice Hall, Englewood Cliffs, NJ.

Ortolano, L., G. LeCoeur, and R. MacGilchrist, 1990. "Expert System for Sewer Network Maintenance: Validation Issues," *Journal of Computing in Civil Engineering* 4(1):37-54.

Perman, C.D., 1989. "Improving the Performance of Wastewater Treatment Plants: an Expert Systems Approach," Ph.D. Dissertation, Department of Civil Engineering, Stanford University, Stanford, CA.

U.S. Army Corps of Engineers Facilities and Support Agency, 1983. "Operation and Maintenance Manual: Fort Lewis Wastewater Treatment Plant," Fort Lewis, Wash., E-83093, Fort Belvoir, VA.

U.S. Government Accounting Office, 1983. "Wastewater Treatment Dischargers are not Complying with EPA Pollution Control Permits," GAO/RCED-84-53, U.S. Government Printing Office, Washington, D.C.

Appendix A: Sample evaluation form used in experiment two

Introduction

This form will help us to evaluate the results of the SLUDGECADET system, which are included as a detailed problem report produced by the SLUDGECADET. The detailed problem report, scenario data, and a copy of your previous diagnosis and explanation for this scenario are included in this section of the notebook behind the evaluation form.

The SLUDGECADET detailed problem report has the following sections:

- Problem summary
- Dialogue with the plant operator
- Summary of the diagnosis
- Diagnostic reasoning

You are asked to answer questions on the section concerning summary of the diagnosis and diagnostic reasoning.

Please follow the evaluation form during your analysis of SLUDGECADET's results for each scenario. If you have any questions, please call Don Duke (415/328-5323) and he will get back to you as soon as he can.

How SLUDGECADET found out about problem in scenario three

A novice operator notified the SLUDGECADET program that she had seen an operating problem. The operator reported that 48,000 gals sludge was pumped in one day to the anaerobic digesters. With this information, SLUDGECADET attempts to diagnose the problem with information listed in the section of the report called "Dialogue with the plant operator." SLUDGECADET already has design and general information about the Fort Lewis facility. When it is done asking questions, SLUDGECADET proposes a diagnosis and a solution as stated in the "Summary of the diagnosis" section. In the "Diagnostic reasoning" section, SLUDGECADET prints a step by step description of the reasoning it used in making the diagnosis.

Please feel free to refer to the detailed problem report, scenario data, and a copy of your previous diagnosis and explanation as much as you want while filling out the evaluation form.

A. *Questions about the diagnosis for scenario three*

1. Given the data available to SLUDGECADET and reasoning presented in the report:

   Does the diagnosis (see summary of the diagnosis in report) provide a valid cause of the operating problem?

   Yes

   No

   If no, explain why not?

   ____

   ____

   ____

2. Would you change the diagnosis made by SLUDGECADET in any way?

   Yes

   No

   If yes, how?

   ____

   ____

   ____

3. Here is a list of some possible diagnoses for this problem:

   a. Excessive mixing rate

   b. Mechanical problems with heater or mixing systems

   c. In plant toxic spill

   d. Toxics in wastewater (plant influent)

   e. Overpumping to the primary digesters

   f. Withdrawing too much sludge

   g. Increase in plant loading

   h. Increased organic loading of digesters

   i. Increased solids in raw sludge

Are there any diagnoses missing from this list?

Yes

No

If yes, what are they?

____

____

____

4. Using the list of possible problem causes (and including any others that you added on the previous page), please indicate how appropriate they are as diagnoses to the primary digester problem by distributing 100 points among them. (More points means more appropriate, e.g., 50 points each for two diagnoses means that both are equally likely to be a correct diagnosis for this problem.)

a. Excessive mixing rate ________

b. Mechanical problems with heater or mixing systems ________

c. In plant toxic spill ________

d. Toxics in wastewater (plant influent) ________

e. Accidental sludge overpumping to the primary digesters ________

f. Withdrawing too much sludge ________

g. Increase in plant loading ________

h. Increased organic loading of digesters ________

i. Increased solids in raw sludge ________

j. ____________________ ________

k. ____________________ ________

l. ____________________ ________

m. ____________________ ________

TOTAL=100

B. Questions about the problem solution

1. Given the data and the diagnosis as stated in the report, is the proposed problem solution (see summary of the diagnosis in report) appropriate for the diagnosis?

   Yes

   No

   If no, please explain why not?

   ____

   ____

   ____

2. See Table 1 of problem diagnoses below (same as questions A2). Please match each diagnosis in Table 1 with the code of the appropriate solution from Table 2 below. Please add any solutions that you think are missing.

<u>*TABLE 1*</u>

Diagnosis	*Solution code*
a. Excessive mixing rate	________
b. Mechanical problems with heater or mixing systems	________
c. In plant toxic spill	________
d. Toxics in wastewater (plant influent)	________
e. Accidental sludge overpumping to the primary digesters	________
f. Withdrawing too much sludge	________
g. Increase in plant loading	________
h. Increased organic loading of digesters	________
i. Increased solids in raw sludge	________

<u>*TABLE 2*</u>

Solution code	*Problem solutions*
A	Adjust sludge pumping rate
B	Repair or replace failed component
C	Adjust sludge withdrawal rate

D	Add seed sludge
E	Extend mixing time
F	Adjust temperatures
G	______________
H	______________
I	______________

C. Questions about the explanation

1. Do you understand the explanation of the diagnosis given in the diagnostic reasoning section?
   Yes
   No
   If not, why?
   ____
   ____
   ____

2. Would a novice operator understand the explanation of the diagnosis given in the diagnostic reasoning section?
   Yes
   No
   If not, why?
   ____
   ____
   ____

3. In order to further assist the novice operator, would you add or change anything in the explanation?
   Yes
   No
   If yes, what would you add or change?
   ____
   ____
   ____

D. Additional comments

Please feel free to add any comments you have about the SLUDGECADET report or this evaluation form.

16

Evaluating an Expert System in the Field: Experience with the CORA Expert System

Lyna L. Wiggins, T. John Kim, and Ravinder Jain

Introduction

Most expert systems are built through a process of iteration: from "toy" or prototype systems to more complete and polished systems. At each iteration step, a validation may be conducted. The validation may be as simple as sitting down with the source expert for a run-through of the system to a complex external validation by outside experts (Castore, 1987). Less frequent in the expert systems literature are discussions about how a variety of end users, with varied computer skills and diverse information needs, actually use an expert system in practice. Two recent papers addressing this issue are Berry and Hart (1990) and Irgon, et al. (1990).

In this chapter, an evaluation of an expert system in practice is discussed. For this evaluation, field interviews with users of the system were used to collect information about user learning patterns, perceptions, and patterns of use.

Expert systems have some intrinsic problems with testing. Unlike many traditional computer systems, it is not easy to define the paths through a program in an expert system, making it difficult to test for completeness and correctness (Hart 1986). And in most algorithmic programming, validity questions may arise regarding only the assumptions on which models are based. Once the end user accepts the assumptions, it is not difficult to derive valid answers, i.e., correct optimal values. Expert systems, however, are designed to provide acceptable solutions using knowledge obtained from experts, which raises the question of who determines the level of acceptability and how.

This problem may not be so serious in other fields. In the medical field, for example, the validity of an expert system which diagnoses a particular disease may be relatively easily determined by several doctors through field tests. In environmental planning and engineering, however, it is difficult for planners and engineers to agree on the solution provided by expert systems, mainly because their subjective judgments can differ greatly from each other and the expertise used to develop the knowledge base tends to be extracted from various sources. For example, the expert system ESSAS (Expert System for Site Analysis and Selection), which is designed to determine the suitability of a particular site for the construction of military facilities, was built by extracting expertise from various sources, including a master planner in a military installation, numerous field manuals, and safety and environmental regulations (Han and Kim, 1989). Clearly, it is difficult to determine the validity of this system because human experts who possess the knowledge used in this expert system are rarely available.

A distinction between the terms "evaluation" and "validation" may be helpful in understanding the necessary tasks required for the effective testing of expert systems. According to Liebowitz (1986), evaluation measures the system's accuracy and usefulness, while validation determines whether the correct problem was solved. Thus, the evaluation process involves users who determine the utility of the system and domain experts who determine the accuracy of any advice and conclusion.

In some cases, the evaluation or validation methods developed in other fields may not be applicable to the expert systems in environmental planning and engineering, because the subject areas covered by the expert systems can be very different in nature, and the decisions made using expert systems can affect a relatively broad range of people. Several evaluation techniques have been developed for other fields by Liebowitz (1986). We shall examine them for their applicabilities to environmental planning and engineering. If not applicable, then we need to devise our own evaluation/validation methods to produce acceptable expert systems in environmental planning and engineering.

Expert systems are being developed for a range of activities related to planning and engineering (Kim, Wiggins, and Wright, 1990; Maher, 1987). In an evaluation of the use of expert systems for this chapter, we focused on operational expert systems in planning and engineering, with particular emphasis on systems in the area of environmental analysis.

The goal of our research was to evaluate how expert systems are being used in practice in environmental planning and engineering. The research was conducted using a six step research design. The research design was to: (1) identify expert systems with extensive use in environmental practice; (2) become familiar with the systems to be evaluated; (3)identify field users of the system and develop a sampling plan; (4) design an interview instrument; (5) conduct field interviews; and (6) summarize results and develop conclusions.

Identification of expert systems with extensive use in environmental practice

The first research step was to identify expert systems in the environmental planning and engineering area with extensive use in practice. This step was conducted through a literature review and phone calls to government environmental agencies, academic institutions, and private industry. Four criteria were used to find systems that: (1) had been used in the field for at least one year, in order to have an adequate evaluation period, (2) had been used by at least ten or so users, to enable us to conduct multiple user interviews, (3) had public access, so that we could review both the operational system and its documentation, and (4) had a system goal that was of high potential interest to environmental planners and engineers, in order to ensure the pertinence of this evaluation. In general, it proved more difficult to locate systems than we had expected. Many calls and follow ups were made to universities, labs, environmental agencies, and private sector firms. We often found systems that met one or several of the criteria, but not all four. Although we found quite a number of systems in academic institutions and laboratories, these were often prototype systems that had only limited field testing. In the private sector, we found quite a number of systems in use, but they were generally proprietary. Several of the private vendors also expressed a distinct lack of interest in an academic evaluation of their systems. In one case, we discovered an interesting expert system for a hydro dam operator to aid in dam operations, but the system had only been used by one operator.

After many hours of searching, we did find one system from the Environmental Protection Agency that met all four of our criteria. This system, Cost of Remedial Action (CORA), was developed for the Environmental Protection Agency, Office of Solid Waste and Emergency Response, by CH2M-Hill, an environmental consulting firm. CORA is intended to provide "order of magnitude" cost estimates for remedial actions to be taken at EPA Superfund sites. The system had been in the field for over one year, had been used by several hundred users to evaluate over 1000 sites, was in the public domain and available to us for review, and was designed to provide expert assistance in the important environmental area of cleaning up hazardous waste sites.

Becoming familiar with the CORA system

Copies of the CORA expert system and its documentation were obtained from CH2M-Hill. The system was installed at MIT and the University of Illinois, and graduate students were encouraged to use and explore the system and report back their evaluation and any problems with the system. The installation procedures were straightforward and well-documented.

The specific purpose of CORA, as stated by the EPA, is to help project managers develop order of magnitude cost estimates for remedial actions at Superfund sites. A broader use of these cost estimates is to assist the EPA Regional Offices in developing remedial action budget estimates for sites which do not yet have cost estimates from detailed feasibility studies. Ultimately, the goal is to provide consistent and timely estimates for annual budgeting purposes for the Office of Management and Budget (OMB).

CORA was designed with two independent subsystems: an expert system component and a cost component. The overall system design is illustrated in Figure 16.1. The expert system component recommends cleanup technologies for separate waste matrices. The final output from the expert system is a range of recommendations for alternative remedial actions, with a separate set of recommendations for each waste matrix. The set of recommended actions is not all inclusive, but is limited to the set of actions covered in the cost component of the system. The cost component develops the order of magnitude cleanup costs based on 42 technology-based cost modules (summarized in Table 16.1). Estimated costs are given for capital costs and first-year operation and maintenance.

The procedures that CH2M-Hill used to cost out each technology are given in some detail in a set of fact sheets in the system documentation. The CORA developers state that the system is designed to handle 90 percent of the typical Superfund site conditions, but is not as site-specific as feasibility studies.

The CORA project was initiated by the EPA in 1985. The Phase I model (Version 1.0) was developed and validated by independent engineering consultants in 1986 and continued through 1987. The Phase II model (Version 2.1) was developed, refined, and released starting in late 1987 and continuing through 1988.

The CORA system cost approximately $1.5M dollars (as of 1989) and took approximately eight person-years to develop. The Phase I model underwent an independent validation study by a separate engineering contractor. The system has extensive documentation, and in addition, the CH2M-Hill developers may be contacted by users via a telephone hotline. In addition, the CH2M-Hill staff provided extensive on-site support upon the request of the regional EPA offices. This on-site support consisted of everything from system installation and one-on-one tutorials with users to actually running the system directly for the Superfund project managers (e.g., the consultants actually entered the data at the keyboard with the project managers providing the input on paper). Because it was important to implement the system using hardware and software that would be available to the final users (Superfund project managers at EPA Regional Offices) the choice was made to use a readily available microcomputer platform for the system. The platform selected was the IBM PC or compatible (graphics card optional) with a full 640K RAM available to the system and at least five megabytes of hard disk space available for the programs and cost database. Because of this hardware choice, the range of software choices was limited. Rather than writing original code, the choice was made to use a commercially available "shell," in particular, Level 5 from Information Builders, Inc. The costing system was developed using dBASE III Plus and compiled with Nantucket Clipper.

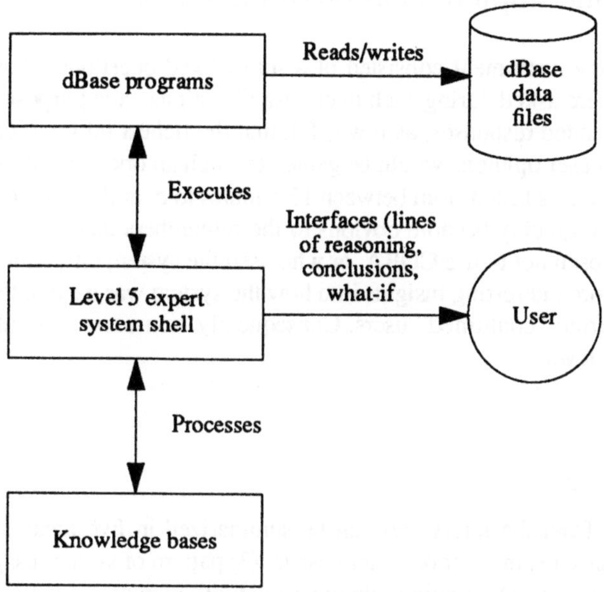

Figure 16.1. The overall system design of CORA (Source: CH2M-Hill, 1989. *Cost of Remedial Action Model Users Manual*, Version 2.1, p. 2-6, prepared for the Environmental Protection Agency under contract no. 68-01-7090).

Identification of users

In order to evaluate the CORA system in the field, we then identified actual users. We discovered that the actual implementation and use of CORA varied widely among the EPA regional offices. Some had used the system very little while others had made quite extensive use of the system. After consultation with the CH2M-Hill system developers, we chose to conduct field and phone interviews with users at the Chicago, Boston, and San Francisco regional offices. Interviews at the EPA regional offices were predominately with Superfund project managers and related staff.

We also learned that the Navy at Port Hueneme had used the CORA system to estimate the costs of cleanup of a large number of military waste sites. Therefore field interviews at Port Hueneme were also conducted. The CH2M-Hill developers also sold copies of CORA to approximately 15 and 20 state agencies, but this group of users was not interviewed.

Design and implementation of the interviews

The interview instrument consisted of a set of fixed questions (shown in Table 16.2) that were asked during each interview. Questions were purposely framed to elicit open-ended responses, as it was felt that the richest information about system use and user opinions would be gained through an open-ended exchange.

The interviews lasted from between 15 minutes to over three hours. In the field interviews, it quickly became obvious to the researchers that interviewing users who had chosen not to use CORA, or who used the system only minimally, often provided more interesting insights into how the system was used in the field, than comments from "confirmed" users. Consequently, non-users were also included in the interviews.

Results

The results from the interviews can be summarized in five areas: (1) adoption characteristics, (2) manner of system usage, (3) pattern of system usage, (4) overall acceptance, and (5) opinions about productivity increases due to system use.

Adoption

The most striking feature of the adoption pattern of CORA was the role of a "champion." The importance of a strong, vocal advocate in technology adoption has been noted by many researchers (Feigenbaum, McCorduck, and Nii, 1988). The role of the champion was important to CORA at two levels: first, at EPA in its funding and development, and second, in the regional offices in its implementation and use patterns.

At the EPA level, the role of the champion was played by Mr. Kirby Biggs, the Task Manager. A long-term advocate of technological innovation and computer applications, Biggs was a vocal advocate for CORA at several crucial points in its development. In its initial phases, Biggs supported its development funding, and in later phases, he advertised its use through seminars and personal contacts at EPA headquarters. Feigenbaum, et al. (1988), describe such advocates as "remarkable people, tirelessly championing their ideas throughout their companies. There seemed to be no end to the energy they brought to the task of championing. Talk, talk, talk, dozens of talks, even hundreds. Fight the battle. Or find a workaround. Never give up." This description fits Mr. Biggs well.

Mr. Biggs was crucial in the system development and adoption. Integral to the widespread dissemination of CORA was Mr. Biggs' idea to require its use in regional offices for budgeting and to fund the on-site consultant support by CH2M-Hill at the EPA regional offices.

At the regional office level, the role of local champion may be essential to actual final implementation and widespread use. When a local champion was

Table 16.1. Technologies available in the CORA cost modules

Special modules	
	Site preparation
	Health and safety
	Site administration
Containment technologies	
101	Soil cap
102	Asphalt cap
103	Multilayered RCRA cap
105	Soil/bentonite slurry wall
106	Surface water diversion/collection
Removal technologies	
201	Soil excavation
202	Sediment excavation/dredging
203	Pumping contained wastes
204	Drum removal
205	Active landfill gas collection
206	Groundwater extraction
Treatment technologies	
301	On-site incineration
302	Off-site incineration
303	Soil flushing
304	In-situ biodegradation
305	Soil vapor extraction
306	Flaring
307	Air stripping
308	Vapor phase carbon
309	Granular activated carbon
310	Activated sludge
311	Metals precipitation
312	Ion exchange
313	Pressure filtration
314	Residential activated carbon units
315	Off-site RCRA treatment and recycling
316	Solidification
Disposal technologies	
401	Off-site RCRA landfill
402	On-site RCRA landfill (above grade)
403	On-site RCRA landfill (below grade)
404	Off-site solid waste landfill
405	Discharge to POTW
406	Discharge to surface water
407	Water reinjection
408	Water infiltration
Miscellaneous modules	
501	Transportation
502	Municipal water supply
503	Groundwater monitoring
504	Site access restrictions
505	User-supplied costs

Source: CH2M-Hill, 1989. *Cost of Remedial Action Model Users Manual*, Version 2.1, p. 3-2, prepared for the EPA under contract no. 68-01-7090.

Table 16.2. List of Open-Ended Interview Questions used in Evaluation of CORA

1. *Familiarity.* Are you familiar with the CORA system for estimating the costs of remedial actions at Superfund sites? Would you say that you have used the system frequently or infrequently? For about how many sites have you used CORA to help develop costs?
2. *Learning.* How did you learn about CORA? (Detailed probes about any seminars, training sessions, information from consultants, co-workers, etc.) How did you learn how to use CORA? (Detailed probes about whether users entered the information themselves or whether others helped them.)
3. *Expert system component.* Did you have any difficulties going through the questions asked by the expert system? Did you make use of the explanation functions provided by CORA? Did you have questions about the logic behind the recommendations that the expert system was making? If so, who did you ask these questions? About how long did it take you to run through a session using the expert system for one of your sites? Did some sites take longer than others? Why? Without CORA, how would have completed the analysis to recommend particular actions? How long would your former procedures have taken you? Did the expert system component provide recommendations that you felt comfortable with? If yes, why? If not, why not? After using CORA several times, could you go through the answers more quickly? How did you feel about the interface with the expert system? Easy to use? Did it keep your attention? After repeated uses? Any other comments about the expert system component that you would like to make?
4. *Cost component.* Did you make use of the default costing options provided by the CORA cost component? Did you make use of the ranges of costs provided by the system? Did you have any difficulties providing the input required for the cost component? If so, what were they? Who did you ask about potential solutions to these difficulties? About how long did it take you to run through a session using the expert system for one of your sites? Did some sites take longer than others? Why? How long do you estimate such a cost estimate would have taken to generate using manual procedures? What procedures would you previously have used to generate such cost estimates? Did the cost component provide recommendations that you felt comfortable with? If yes, why? If not, why not? After using the CORA cost component several times, could you go through the answers more quickly? How did you feel about the interface with the cost component? Easy to use? Did it keep your attention? After repeated uses? Any other comments you would like to make about the cost component?
5. *Usage (hardware).* Where did you use CORA? At your desk? On another computer? (If so, where was the computer you used? Who else uses that computer? Is easy or difficult to access the computer resources you needed to run CORA?) What kind of computer did you run CORA on (e.g., XT, AT, etc.)?
6. *Usage (frequency).* How frequently have you used CORA? If only for one-time budgeting, why? If for daily use, what have you used CORA for? Have you been monitoring the actual costs at any of your site? Have you compared your actual actions taken and costs incurred to the recommendations and cost estimates provided by CORA? How have they compared?
7. *Other comments.* Any other comments you would like to make about CORA?

present in an office there was greatly increased use. Again, the major role of the local champion is verbal support and "spreading the word." In each regional office, one staff person was assigned to be a point of contact with CH2M-Hill and was provided with training. This contact person was not, however, in each case the identifiable local champion. For example, in the Chicago regional EPA office the champion was self-appointed, and took on the role of training and running the system for other project managers.

Manner of system usage

The two components of CORA, the expert system and the database, had very different usage patterns. The expert system was used less frequently than the database, and after a user had considerable experience with CORA, the expert system was often bypassed altogether and the required information was simply input into the database component. In the expert system component, the help function was used frequently. The explanation facilities were only partially implemented in CORA, Version 2.1. Most commercial expert system shells allow explanatory texts to be assigned to particular terms or questions. These explanations are generally accessed by the user using an assigned function key.

In Version 2.1 of CORA, some terms and rules had quite complete explanations, while others were less incomplete or missing. Even though these explanations were available to them, they were rarely used by the project managers. None of the interviewees were aware that most expert systems are provided with facilities which allow a user to trace through the logic used by the system in reaching a conclusion. In fact, the CORA expert system as developed by CH2M-Hill did not provide this feature.

Pattern of system usage

The striking finding regarding the pattern of use of CORA was that it was used predominantly for one-time only budget estimates for OMB and not for day-to-day consultation by the project managers. When questioned at some length as to the reasons for this pattern, the interviewees expressed a range of views. The most common reason given was the lack of access to computer resources. The second most common reason given was time constraints and day-to-day pressures of managing many projects. Most of the project managers did not have PCs on their own desktops.

While all responded that they could access PCs in their office, there was competition for their use. Sometimes this competition was with clerical staff who made heavy time demands on the available machines. Other interviewees expressed a lack of knowledge about just which machines had the system up and running on them. Since CORA requires significant disk space, it was not generally available on all of the PCs in the office. Another group of interviewees suggested that they had never really run CORA on their own. They said that they had indeed completed the budgeting exercise for their sites using CORA, but that the CH2M-Hill on-site consultants had in fact entered all of the data using the project managers' information. This meant that many (and perhaps those with computer-phobic tendencies) had never run the system totally on their own.

Acceptance

Many of the less enthusiastic interviewees expressed that their day-to-day time constraints mitigated against running large number of cost scenarios using CORA. Some suggested "let the consultants do it." On three occasions, respondents noted perceived inaccurate recommendations from the expert system component. It appeared that even one inaccurate recommendation was enough to discourage these users and make them negative.

When questioned further about these inaccurate recommendations, it was clear that none of these interviewees had read the documentation in detail, where it is clearly stated that the CORA recommendations are not inclusive. Many of these discouraged users began skipping the expert system component altogether. Others noted that the repetitive quality and lack of screen interest in the expert system discouraged them from using it. Other users who began to skip the expert system component suggested that, even before a detailed feasibility study had been done for a site, they had already used their "best engineering judgement" about a possible set of actions.

When asked about the quality of the database component of CORA, the responses were much more positive. Several respondents made strong positive statements that the minimum "bottom line" estimates from the database had come very close to actual bids from their contractors. They expressed overall confidence in the cost database and praised its ability to provide default values and a range of values for individual cost components. Only one interviewee expressed a dissenting view of the cost component. This particular project manager had an extremely large and complex site. He suggested that everything was "too site-specific" for confidence in the default cost estimates.

Another interviewee suggested that for one site, there was "only one contractor possible in the area." Since this one contractor was considered truly expert about the specific waste contamination problem, the project manager felt that the contractor's knowledge of costs would be much more accurate than CORA's estimates. None of the respondents had gone back to check on CORA's estimates using actual cost data. Generally, this was because most of these site clean-ups go on over extended periods of time, and real cost information was not available for such a user validation.

Productivity increases

The respondents gave extremely consistent estimates of the time required to run CORA. For typical sites, the expert system component took about one hour to run. After receiving recommendations from the expert system component, the cost database took one to two hours to run. In many cases, additional detailed information required for the cost database had to be obtained before the run could be completed. Many interviewees were unwilling to hazard a guess about how long it would have taken them to complete the analysis manually. Some said they would have gotten such estimates from their contractors, others were new project mangers who had never completed such a preliminary estimate.

Several more experienced project managers suggested that it would have taken them from four to five times as long to do manually. This group of users tended to be pro-CORA in general and unanimously praised the use of the system to check on the bids coming in from consultants.

Implications for other potential users and systems developers

It seems clear from our interviews that CORA is both used and useful to environmental practitioners in the field. A particular characteristic of CORA is that the expert system component seems to have been of less importance than the database component. We may surmise that this is because of the clear need among Superfund project managers for good disaggregate data about costs to use in budget estimates. Since the overall goal of CORA is to provide order of magnitude cost estimates for sites without detailed feasibility studies, the core of real information is found in the cost database. Many of the users mentioned how useful it was to simply have this cost information automated in a database.

In addition, the CORA database design is flexible and allows project managers the option of entering known costs directly, using default values provided by the system, or looking at low, medium, and high cost ranges. This flexibility replicates the often-cited advantage of spreadsheet analysis to allow "what-if" questions to be addressed. On the other hand, the expert system component of CORA is much more limited in scope. Using simple backward chaining, the system recommends a fixed set of possible remedial actions.

Suggestions for future improvements

Since expert systems are still relatively new, many research issues need to be resolved for them to become practical computer tools. This section examines two research issues for possible future improvements of CORA: reasoning with uncertainty and conflict resolution.

Reasoning with uncertainty

Human experts possess knowledge, a repository of human expertise that, for the most part, is imprecise and uncertain. It is therefore useful for any expert system to have the capability of representing and processing uncertain information in a correct way. Although the role of probability theory has been important in expert system development, the way in which expert systems handle uncertainty tends to be rather ad hoc. This results in criticisms of the poor use of probability theory and other well-established techniques for uncertainty reasoning in expert systems (Hart, 1986).

An ad hoc approach to uncertain reasoning in other disciplines, such as medicine and law, may perhaps be acceptable because their outcome domains are relatively well-defined with less likelihood of new situations. In these fields, uncertain reasoning methods that successfully replicate the behavior of human experts (doctors or lawyers) and that pass empirical tests may well be acceptable. In CORA, however, the outcome domain of problems is difficult to define completely, as is the case in most environmental planning problems; there is significant likelihood of new situations arising. A need exists to devise a consistent and justifiable logic that can effectively handle such uncertainty problems in CORA.

Conflict resolution

Most of the expert systems developed for environmental planning have not explicitly provided mechanisms to handle the situation where conflicting knowledge (i.e., the same production rules with different objectives) can occur. In most expert systems, however, it is very typical that several conflicting or different rules apply to the same problem in the reasoning process.

There are two common ways of handling conflict resolution in present expert systems building: one is to apply the first rule encountered in the consultation process and the other is to provide meta-rules. Without conflict resolution strategies explicitly specified, the default strategy embedded in most expert systems is to apply the first rule encountered in the consultation process. This is the case in CORA. But this simple method is likely to ignore other rules that may produce a better solution. One reason that most expert systems developed in environmental planning do not use such "meta-knowledge" is provided by Goodall (1985). Goodall observed that the use of meta-rules inevitably makes the knowledge base harder to read and understand because one meta-rule affects the sequence of other rules and the effect is distributed throughout the rest of the knowledge base. Another important reason may be that meta-knowledge is simply difficult to obtain.

If well designed, however, meta-rules may be useful in systems such as CORA. For instance, when the expert system reasoning is related to the evaluation of a plan from an environmental perspective, a meta-rule which gives first priority to the protection of endangered species over any other knowledge might be useful. And when the reasoning is related to the evaluation of a plan from a legal perspective, a meta-rule which gives first priority to the rules which avoid complex litigation would be useful.

Although this meta-rule approach has the limitation that it requires the developers to express explicitly the order of priority through the whole knowledge base, it may be very similar to the way actual human experts or decision makers deal with conflicting knowledge. Nonetheless, the possible benefits of using meta-knowledge have not yet been rigorously researched in the environmental planning field.

Acknowledgment

Partial support by the Construction Engineering Research Laboratory, U.S. Army Corps of Engineers (DACA 88-88-D-0013-28) is gratefully acknowledged.

References

Berry, D.C. and A.E. Hart, 1990. "Evaluating expert systems," *Expert Systems* 7(4):199-207.

Castore, G. 1987, "Validation and Verification for Knowledge-Based Control Systems," presented at SOAR Conference, NASA/JSC, Houston, TX.

CH2M-Hill, 1989. *Cost of Remedial Action Model, Users Manual*, Version 2.1, prepared for the U.S. Environmental Protection Agency under contract no. 68-01-7090.

Feigenbaum, E., P. McCorduck, and H.P. Nii, 1988. *The Rise of the Expert Company: How Visionary Companies are using Artificial Intelligence to Achieve Higher Productivity and Profits*, Times Books, New York.

Fenves, S.J. 1986. *What is An Expert System. Expert Systems in Civil Engineering*, edited by Celal N. Kostem., American Society of Civil Engineers, New York.

Goodall, A. 1985. *The Guide to Expert Systems*, Learned Information Ltd., Oxford, England.

Han, S. and Kim, T.J. 1987. "An application of expert system in urban planning: Site selection and analysis," in T.J. Kim, L.L. Wiggins and J.R. Wright, eds., *Expert Systems: Applications to Urban Planning*, 1989, Springer-Verlag, NY.

Hart, A., 1986, *Knowledge Acquisitions for Expert Systems*, McGraw-Hill Book Co., New York.

Irgon, A., J. Zolnowski, K.J. Murray, and M.G. Bellocore, 1990. "Expert System Development: A Retrospective View of Five Systems," *IEEE Expert*, June, pp. 25-39.

Kim, T.J., L.L. Wiggins and J.R. Wright, eds., 1989. *Expert Systems: Applications to Urban Planning*, Springer-Verlag, New York.

Leibowitz, J., 1986. "Useful approach for evaluating expert systems," *Expert Systems*. *3*(2).

Maher, M.L. (ed.), 1987. *Expert Systems for Civil Engineers: Technology and Application*, American Society of Civil Engineers, New York.

MIX
Papier aus verantwortungsvollen Quellen
Paper from responsible sources
FSC® C105338

If you have any concerns about our products,
you can contact us on
ProductSafety@springernature.com

In case Publisher is established outside the EU,
the EU authorized representative is:
Springer Nature Customer Service Center GmbH
Europaplatz 3, 69115 Heidelberg, Germany

Printed by Libri Plureos GmbH
in Hamburg, Germany